STUDIA TRADITIONIS THEOLOGIAE

Explorations in Early and Medieval Theology

Theology continually engages with its past: the people, experience, Scriptures, liturgy, learning and customs of Christians. The past is preserved, rejected, modified; but the legacy steadily evolves as Christians are never indifferent to history. Even when engaging the future, theology looks backwards: the next generation's training includes inheriting a canon of Scripture, doctrine, and controversy; while adapting the past is central in every confrontation with a modernity.

This is the dynamic realm of tradition, and this series' focus. Whether examining people, texts, or periods, its volumes are concerned with how the past evolved in the past, and the interplay of theology, culture, and tradition.

STUDIA TRADITIONIS THEOLOGIAE

Explorations in Early and Medieval Theology

39

Series Editor: Thomas O'Loughlin,
Professor of Historical Theology
in the University of Nottingham

‘The loss of a minute, is just so much loss of life’

Edward Robinson and Eli Smith in the Holy Land

Haim Goren

BREPOLS

This book's title was taken from Rosewell Dwight Hitchcock's address read before the New York Historical Society, March 24, 1863 (Robinson, M. A. [1862-64], 147).

The research for this book was supported by the Israel Science Foundation (Grant no. 239/15). The book has been published with the generous support of the Research Authority, the Dean of Research and the Rector of Tel-Hai College, Israel.

The Special Collections of Burke Library, Hamilton College, in Clinton, New York, are very pleased to have supplied illustrations to Professor Haim Goren's work on Edward Robinson, Hamilton College class of 1816.

D/2020/0095/96
ISBN 978-2-503-58913-8
E-ISBN 978-2-503-58914-5
DOI 10.1484/M.STT-EB.5.119899
ISSN 2294-3617
E-ISSN 2566-0160

Printed in the EU on acid-free paper.

Dr Robinson, I believe it is not too much to say, was the first person who ever saw Palestine with his eyes open as to what he ought to see.

Dean Arthur Penrhyn Stanley in his address 'An American Scholar,' delivered at the UTS, October 29, 1878: Stanley (1879); Prentiss (1889), 63.

TABLE OF CONTENTS

LIST OF ILLUSTRATIONS

ERP Edward Robinson Papers, Hamilton College
Courtesy of Special Collections, Burke Library, Hamilton College

ESP Eli Smith Papers, Harvard University
Courtesy of Houghton Library, Harvard University
Courtesy of the United Church of Christ

PREFACE

In 2010, I embarked on a collaborative research project with Dr Bruno Schelhaas, my colleague from the Leibniz-Institut für Länderkunde (Leibniz Institute for Regional Geography, IfL, Leipzig, Germany), entitled 'Robinson, van de Velde, and German Holy Land Cartography in the mid-19th Century.' The project was generously financed by the German-Israeli Foundation for Scientific Research and Development (GIF). Working on the project, I had the opportunity to pay a short visit to both Edward Robinson's and Eli Smith's archives in the US. I went to Hamilton College in Clinton NY, which, until September 2012, held in its collections only a small amount of Robinson's material. During the visit, Profs. Jay Gomer Williams and Hermine Weigel Williams transferred the seven boxes of archival material that were in their possession for nearly twenty years to the Special Collections of Burke Library, Hamilton College. Consequently, these collections now hold the vast Robinson Archive, together with more than a thousand of his books previously donated, most of which are classified under 'Palestine Literature.'

Once we began investigating documents in the relevant archives for the GIF cartography project, I became convinced that much of the Robinson Holy Land narrative had yet to be unearthed, and answers to important research questions awaited discovery. Moreover, significant parts of the accepted narrative deserved re-evaluation and, consequently, rewriting. Much has been written about Robinson, his personal and social history, religious beliefs, approach to historical geography, and his teaching and writing; yet, several issues remain untouched by scholars. One of the reasons has been, necessarily, the lack of relevant available and accessible sources. The archival sources cited here tell the story behind the scenes of the developments resulting in the Holy Land expedition and the scientific research it yielded.

As I began collecting the documentary material and comprehending its scope and significance, even without considering the huge quantity of printed sources, I made the decision to limit this study in time, from late 1837 to 1841, concentrating on Robinson's first expedition conducted with Eli Smith, and on the subsequent undertaking of the writing and publication of the *Biblical Researches*. This focus enabled highlighting related background issues and outcomes of import. Naturally, some earlier and later events are included when necessary to illuminate the discussion. The vast amount of material convinced me that the period covering Robinson's teaching at the Union Theology Seminary (UTS), his second expedition and its scientific products, as well as the posthumous publication of his *Physical Geography*, would have, I am afraid, to await a future study.

The last two words of the title chosen for the research project leading to this publication, financed so generously by the Israel Scientific Foundation (ISF), 'Robinson and Smith, Generators of Change in Holy Land Scientific Study: New Considerations,' reveal the essence of its objectives. I did not intend to revolutionize what has been written to date about Robinson's Holy Land research and the significance of his scientific achievements. Nonetheless, my working hypothesis is unequivocal: a detailed study and careful analysis of the vast archival material, primarily letters but also other documents, will reveal many new understandings concerning Robinson and Smith's expedition and publications that will differ considerably from those currently accepted. I intended to expand, through this intensive archival research primarily of archival sources considered for the first time, the body of material upon which the Robinson narrative is based, to revise minor as well as more major currently held views, and to answer unresolved questions, some of which are of cardinal importance to evaluating Robinson's work – the work of a scholar who influenced nearly all subsequent Holy Land research. I hope that the issues considered, while in no way pretending to present a full biography of Robinson (various scholars mentioned in this study have already published worthy volumes), will contribute significantly to our knowledge and understanding, not only of Robinson, his world and Holy Land projects, but also to more general research issues concerning the history of Holy Land scientific study.

ACKNOWLEDGMENTS

Yes, it is a pleasure! Upon arriving at the end of a project that has lasted some six years, to acknowledge all people and organizations without whom or which writing and publishing this book would have been impossible. The list is rather long, and I do hope I have not forgotten or neglected anyone.

The project was funded by the Israel Science Foundation. I wish to convey my deepest thanks to the foundation, its director and highly professional and extremely helpful team for understanding the archival nature of the project, particularly the need for travel and deciphering impossible-to-read documents at the core of the study. Their open approach was a continuous source of encouragement during the long and difficult project.

Over the years, Leipzig, located near Halle and Berlin (the three 'Robinson centres' in Germany), and the IfL, became a second home to me, where as a visiting scholar I researched much of this study. At this major research centre, Bruno Schelhaas was, and continues to be, an endless source of information and solutions to repeated queries. I would also like to express my deepest gratitude to Sebastian Lentz, Ute Wardenga, Hans-Peter Brogiato, and Dirk Hänsgen. In spring 2017, I was invited as a guest scholar to SFB 1199, a joint program of the University of Leipzig and the IfL, titled 'German scholars and Robinson's *Biblical Researches.*' My sincerest gratitude goes to both institutes, their management, faculty and staff. Frank Hadler, of the Leibniz Institute for the History and Culture of Eastern Europe (GWZO) in Leipzig, invited me to be a scholar-in-residence for the month of October 2018, during which I wrote the chapter on Talvj; my deepest thanks to him and the institute's team. My gratitude also goes to Susanne Richter in the *Museum für Druckkunst* (Museum of the Printing Arts) in Leipzig.

I made much use of the German archival treasures in Gotha, the Gotha Research Library in Castle Friedenstein and the Perthes Collection, with the generous help of Petra Weigel and Sven Ballentine. In Berlin's State Library (STABI), I was greatly assisted by Wolfgang Chrome and Markus Heinz in the maps collection and Jutta Weber in the archives. In Halle, I visited the archives of the Martin-Luther-University Halle-Wittenberg and of the *Franckesche Stiftungen* ('Francke Foundation'), and wish to thank Britta Klosterberg for her assistance. I also wish to thank my long-time friend and colleague, Bruno Tauché, for his continuous help while working in Halle.

As for the United States, also here the list is rather long. I visited the home of Robinson's archives, the Burke Library of Hamilton College, Clinton, NY numerous times, including a month's residence as Special Collections Research Fellow, for one month during summer 2014. I worked primarily in the special collections and archives, managed by Christian Goodwillie, together with Mike Tilson, special collections coordinator, a continued source of information and help, and Katherine Collett, college archivist. All became good friends and their contribution was highly significant. Randall Ericson and Christian Goodwillie, heading Couper Press of Hamilton College, were extremely generous in permitting the use of their archive material for the illustrations. I would also like to thank David L. Smallen, former vice president for libraries and information technology. I am grateful to Robert L. Paquette, head of the Alexander Hamilton Institute for the Study of Western Civilization in Clinton, NY, for inviting me to stay on the AHI's premises on several occasions. Jay Gomer Williams of Hamilton College, together with his wife Hermine Weigel Williams, was the first to use Robinson's archive. I owe him special thanks for his friendship, tremendous help, and cooperation.

Most of Smith's letters, as well as those of the American Board of Commissioners of Foreign Missions (ABCFM) and many of its missionaries, are safeguarded in the Houghton Library of Harvard University, Cambridge, MA. I could not imagine how much material I would find there and how crucial it would be for my work. My thanks go to all the librarians whose help I received during three separate visits, and to all who ensured publication of the book's illustrations. The United Church of Christ Wider Church Ministries, owner of Smith's archive, also graciously permitted publication of illustrations. At the Burke Library Archives of the Union Theological Seminary in New York, I was helped generously by Betty Molden, reader services for rare books and archives, Matthew Baker, collection services librarian, and Rebecca Nieto, project archivist. At Yale University, New Haven, I am grateful to the librarians of the Sterling Library Manuscripts and Archives, where the papers of missionary Isaac Bird are housed, and of the Yale Divinity Libraries, where I was helped by Kevin Crawford and Martha Lund Smalley.

At the Israel National Library, I would like to thank Ayelet Rubin, manager of the Eran Laor Collection, who continuously helps me locate and acquire the most 'impossible' sources. Here, it is a pleasure to thank my dear teacher and mentor, Yehoshua Ben-Arieh, who at the age of 91, still publishes and closely follows my research. I am grateful, also, to my Israeli friends and colleagues with whom I discussed my research, each contributing important ideas from his field: Rehav 'Buni' Rubin, Israel Bartal, Eran Dolev, Gideon Bigger, Mustafa Abbassi and Ariel Tishbi. Eliyahu Hacohen's unprecedented knowledge and collection of travellers' literature were of tremendous value. Author Haim Be'er supplied me with interesting ideas concerning the literary use of letters and other archival material.

I could not have written this book without the help of Teresa Kilian, formerly of the IfL Leipzig, and now at the *Thüringer Universitäts- und Landesbibliothek* Jena. Her outstanding ability to decipher impossible handwriting unlocked hundreds of the letters and documents that otherwise would have remained inaccessible. My editor, Amy Klein, has been much more than 'only' a language editor. Her meticulous work and her scientific understanding added enormously to this book, and she is responsible for my ability to reach its end.

I received generous financial support from the James Amzalek Fund for Research in Historical Geography, and am deeply indebted to its coordinators, Ruth Kark, Yehoshua Ben-Arieh, Rehav Rubin and Ran Aaronson,. My long-time friend, archaeologist Shimon Gibson, suggested that I present my book to Rosie Bonté, a publishing manager at Brepols Publishers. Through her mediation, the manuscript reached Thomas O'Loughlin, Professor of Historical Theology at the University of Nottingham, who so generously accepted it for publication in his series, 'Studia Traditionis Theologiae (STT): Explorations in Early and Medieval Theology'. O'Loughlin and Publishing Manager Bart Janssens led me throughout the difficult process from manuscript to finished product, a published book.

Last but not least, I wish to express my gratitude to Tel-Hai College in Israel's Upper Galilee, my academic home for over three decades. The

Research Authority helped with financing, and other authorities enabled me to cross this technical and administratively complicated river. I would like to thank here, personally, Shalom Termechi and Udi Grosman, my research assistants, for their ongoing help.

True to endeavours of this sort, that take up endless hours, I owe the biggest thank you to my wife, children, and grand-children, who suffered silently, here and there even showing interest and, together with me, awaited the crossing of the finish line.

EDWARD ROBINSON 1794–1863: A SHORT BIOGRAPHY

Ill. 1. Portraits of Edward and Therese Robinson. Date, country where painted and name of artist are unknown. The two portraits, designed to be hung as a pair, may have been painted as late as 1839 by Caroline Bardua.

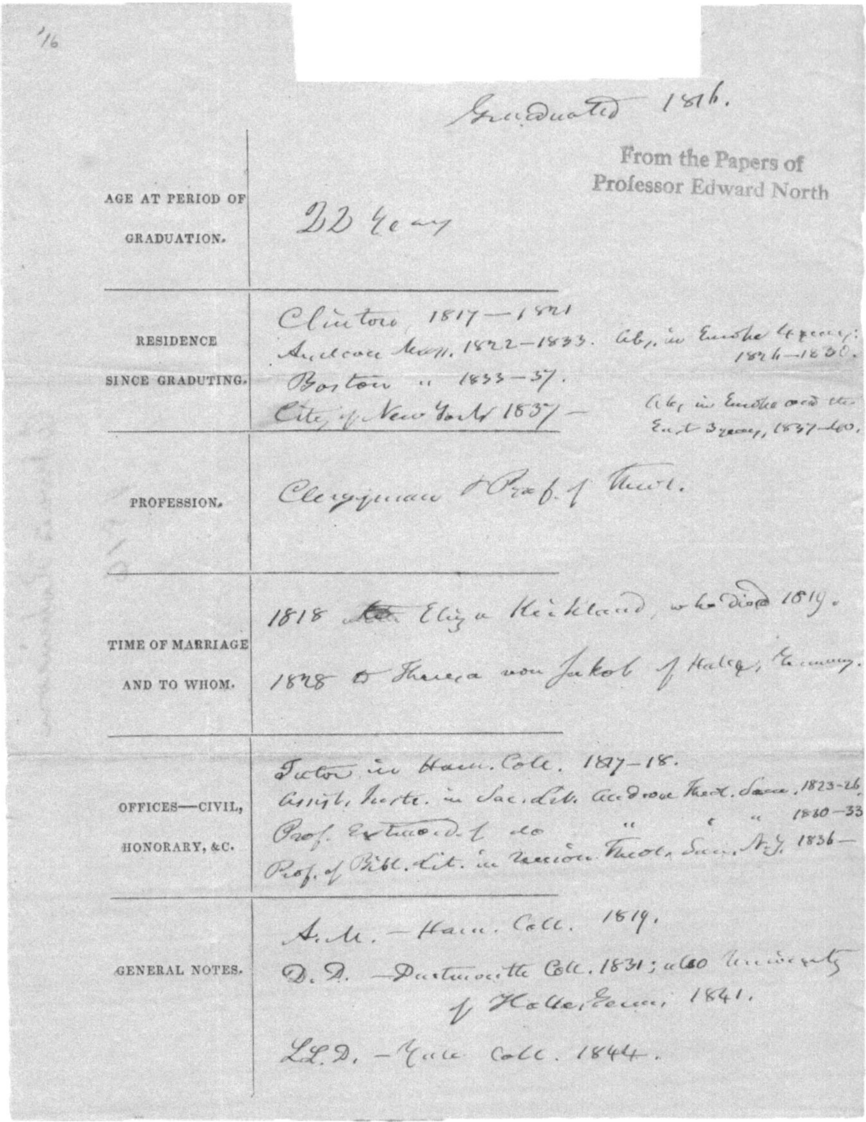

1/16

Graduated 1816.

From the Papers of Professor Edward North

AGE AT PERIOD OF GRADUATION.	22 years
RESIDENCE SINCE GRADUTING.	Clinton, 1817–1821 Andover Mass. 1822–1833. Abr. in Europe 4 years: 1826–1830. Boston " 1833–37. City of New York 1837— Abr. in Europe and the East 3 years, 1837–40.
PROFESSION.	Clergyman & Prof. of Theol.
TIME OF MARRIAGE AND TO WHOM.	1818 [struck] Eliza Kirkland, who died 1819. 1828 to Theresa von Jakob of Halle, Germany.
OFFICES—CIVIL, HONORARY, &C.	Tutor in Ham. Coll. 1817–18. Assist. Instr. in Sac. Lit. Andover Theol. Sem. 1823–26. Prof. Extraord. of do " " 1830–33 Prof. of Bibl. Lit. in Union Theol. Sem. N.Y. 1836—
GENERAL NOTES.	A.M. — Ham. Coll. 1819. D.D. — Dartmouth Coll. 1831; also University of Halle, Germ. 1841. LL.D. — Yale Coll. 1844.

Ill. 2. Robinson's Alumni Questionnaire, Hamilton College, 1816.

April 4, 1794 – born in Southington, CT
1812–16 – studies at newly established Hamilton College, Clinton, NY
1817–21 – instructor at Hamilton College
September 1818 – July 1819 – marries Eliza Kirkland, widowed in less than a year
1822–26 – Andover Theological Seminary
1822 – first publication: translation of the *Iliad*
1823 – assistant to Moses Stuart for Biblical Studies and Hebrew
June 1826 – embarks to Europe to begin period of academic studies (Paris, August 1826 to Basel, September 1826 to Göttingen, October 1826 to Halle)
October 1826 – April 1830 – studies in Halle; January – March 1828 studies in Berlin
August 7, 1828 – marries Talvj
August 1828 – May 1829 – honeymoon travels in Europe
June 25, 1829 – birth of first daughter, Mary Augusta
July 1830 – return to the US
November 1830 – September 1833 – Andover Theological Seminary, Prof. Extraordinary for Biblical Literature and Seminary Librarian
November 1831 – son Maximillian dies
October 1833 – January 1837 – Boston, independent scholarly writing and publishing
January 1837 – offer from the UTS, position of Professor of Biblical Literature, accepts conditioned upon a research expedition to the Holy Land, moves to New York
July 1837 – sails with family to Europe, Berlin
November 1837 – embarks on the expedition, arriving in Alexandria December 30, 1837
January 5 – February 28, 1838 – tour of Upper Egypt
March 12 – June 27, 1838 – Sinai and Holy Land expedition, ends in Beirut[1]
August 1838 – August 1840 – Berlin, writing the *Biblical Researches*
September 1, 1840 – manuscript completed
November 1840 – back in New York
January 20, 1841 – Robinson officially inaugurated as Professor of Christian Literature, UTS

[1] For a detailed itinerary of the expedition see chap. 3.b.

1841 – publication of the *Biblical Researches*, in London, Boston, and Halle
December 20, 1851 – embarks on the second expedition
March 2 – July 5, 1852 – second expedition, beginning and ending in Beirut
October 27, 1852 – returns to New York
1856 – publication of the *Later Biblical Researches*, Boston (1857 German edition in Berlin)
November 5, 1862 – in Germany for medical care
January 27, 1863 – death

CHAPTER 1
INTRODUCTION

The nineteenth century in Palestine is often called the 'century of changes,' when the territory underwent a process of tremendous transformation from a medieval-like, sparsely populated, and neglected province of the Ottoman Empire, to a country on its way to modernization. One of the foremost factors of this transformation was the intervention of European powers. This intervention took many forms, including religious, diplomatic, and economic activities, immigration and settlement of Europeans, and intensive scientific research of this relatively unknown country.[1]

Nascent 'modern' scientific investigation of Palestine can be identified as early as the mid-eighteenth century. However, the real beginning of what has been termed 'rediscovery,' 'recovery,' or the 'pioneering exploration' awaited the beginning of the nineteenth century.[2] This early stage continued until 1832, when the Egyptian conquest of Palestine and Syria from the Ottomans triggered significant changes, including the aforementioned European involvement in intensification of scientific research of the country.[3] From the fourth decade of the nineteenth century, this rediscovery stage was followed, primarily, by sporadic and accidental studies of a general nature, increasing in number and scientific significance. Naturally limited in scale and scope, those works were initiated and conducted by individuals or small groups, such as religious

[1] e.g. Marriott (1940); Anderson, M. S. (1966); Ben-Arieh (1976); Bartal (1983); (2018); Shamir (1984); Carmel (1985); Kushner (1986); Makdisi (2008).

[2] Bliss (1907), 168–83; Ben-Arieh (1972), 81–84; (1979), 12; Goren (2003), 13–27.

[3] Shamir (1984); Fahmy (1997); Kutluoğlu (1998); Farah (2000). Contemporary descriptions: Rochfort Scott (1837); Prokesch-Osten (1877).

and social organizations, and only occasionally also supported, but not initiated, by governments.[4] The main actors involved were scholars from Europe (UK, France, Russia, and Germany) and the US. As the nineteenth century advanced, discovery and exploration gave way to programmed scientific research, focusing on special topics or regions. The picture changed dramatically following the establishment of the British Palestine Exploration Fund (PEF) in 1865, of its German counterpart, the *Deutscher Verein zur Erforschung Palästinas* (DPV) in 1877, and five years later of the Russian Imperial Orthodox Palestine Society.[5]

1.a. Robinson in the context of early Palestine explorers

The list of European scholars who visited and researched the Holy Land during the nineteenth century, producing scientific (and also popular) reports, articles, books, maps, and more, is impressive and rather long. It is much shorter when considering only the first phase of modern scientific inquiry, which, as noted, reached its end in the early 1830s.[6] George Basalla's early model for European scholars in countries of the undeveloped world[7] is applicable to this group. Yehoshua Ben-Arieh also published numerous leading studies on nineteenth-century Palestine travellers and explorers and their literature.[8] Prominent early explorers, operating independently or as part of official contingents, are worthy of special mention for their contribution to the development of this modern 'scientific revelation' of the land of the Scriptures. They are featured in various studies dedicated to this process.[9] The two nineteenth-century 'Palestine-literature' bibliographers, the Swiss Titus Tobler (1806–77) who published his *Bibliographia geographica Palaestinae* in 1867, and his successor, the Berlin teacher Reinhold Röhricht (1842–1905) who followed in 1898, even simplified the task.[10] Searching for academicians who studied Palestine from the late eighteenth century, the prominent

[4] Goren (1998); (2001); (2002); (2010).

[5] Conder, C. R. (1880); Bliss (1907), 255–87; Watson (1915); Goren (2001); Silberman (1982), 79–127; Moscrop (2000).

[6] Goren (1998); (1999).

[7] Basalla (1967).

[8] Ben-Arieh (1976); (1989); (1991).

[9] Benzinger (1903); Bliss (1907); Ben-Arieh (1972); (1979); Goren (1998); (2003); Silberman (1982).

[10] Tobler (1867); Röhricht (1890).

names are easy to establish. As reported in detail by Anne Godlewska and others, it was the surveyors of the French army under Napoleon I Bonaparte (1769–1821) who took the first trigonometrically-based measurements in the region, encompassing parts of Palestine. Colonel Pierre Jacotin (1765–1827), the military engineer-geographer who commanded Napoleon's surveyors in Cairo and during the 1799 Palestine campaign, was responsible for the 1810 publication of the maps, including six sheets covering areas in the Holy Land. The measurements that he and his team managed to take in the course of the brief campaign were quite incomplete; hence, they did not produce survey maps worthy of the name, at least concerning Palestine.[11]

Ulrich Jasper Seetzen's (1767–1811) almost legendary story has been told in various studies, and his findings and acquisitions during his excursions from 1802 to 1811 remain an important part of the Oriental collections in Gotha. Gotha was the seat of his scientific guide and mentor, the astronomer Franz Xavier von Zach (1754–1832), as well as his financier, the prince of Sachsen-Gotha and Altenburg. His travels in Syria, Palestine, and Sinai, and his letters and map, were of crucial significance for the scientific rediscovery of the countries.[12]

The third scholar of note is unquestionably the Swiss Johann Ludwig Burckhardt (1784–1817). Travelling in the service of the British Association for Promoting the Discovery of the Interior Parts of Africa during the years 1809–17, like Seetzen, Burckhardt prepared himself for the 'grand plan' of penetrating central-west Africa from the east (Egypt) by adopting Moslem disguise. His early rediscoveries in the region and detailed reports were of a tremendous importance for the study of the Near East.[13]

Naturally, many others travelled the Holy Land during the first four decades of the nineteenth century, a number contributing significantly to the accumulating body of scientific data. Most, including the British traveller James Silk Buckingham (1786–1855), the German theologian Johann Martin Augustin Scholz (1794–1852), and the French no-

[11] Godlewska (1988); (1999), 77–78. Cf. Jacotin (1810); (1818); Alderson (1843), 21–38; Lord (1901), 146–62; Laurens (1989); Gichon (2003); Strathern (2009), 314–81, p. 315 for the savants accompanying the Syrian campaign. For the French surveyor measurements in Palestine, cf. Berghaus (1835a), 1–4; Karmon (1960); Godlewska (1988); (1994).

[12] Seetzen (1805–6); Lindenau (1810); Kruse (1854). Cf. Ritter (1850), 15, 21–24; Ben-Arieh (1972); Goren (2003), 36–48; (2011), 139–43; Schäbler (1995); Schienerl (2000), 59–63; Pfullmann (2001), 402–08.

[13] Burckhardt (1822); Gesenius (1823);. Cf. Burckhardt-Sarasin (1956); Hallett (1965), 366–78; Sim (1969); Ben-Arieh (1972); Goren (2003), 48–55. Henze (2011), 1, 399–409.

ble Léon Emmanuel Simon Joseph, comte de Laborde (1807–69), are discussed at greater length in the coming chapters, as Robinson relied measurably on their descriptions, data, and maps when writing the *Biblical Researches.*

Titus Tobler, the Swiss physician, earned the title 'father of the German Palestine-research' for his four periods of research in the Holy Land and over sixty publications, including monographs, maps, and a scientific edition of early pilgrims' translations.[14] His first visit, in 1835, produced a book titled *Lustreise in Orient*, and ignited a lifelong passion for the study of the country. In 1845, he returned in order to concentrate on Jerusalem.[15] As noted, Tobler was the first to publish a bibliography of what he called Palestine Literature. Perhaps difficult to believe, but he did read every book, paper, and pamphlet and checked every map he could find. And, he was highly critical, exceptionally critical, bestowing 'titles of honour' like *unbedeutend* [un-important] or *werthlos* [worthless]. This fact lends stronger impact to his complementary words describing Edward Robinson (1794–1863): *Genauigkeit* [accuracy], *beharrlichkeit* [steadfastness], *gelehrsamkeit* [scholarship], *scharfsinn* [astuteness] *und umbefangenhait im urtheile* [and impartiality to opinions], and *sind eigenschaften, welche den wahrhaftig hochwürdigen hochschullehrer auszeichnen* [are qualities, which honour the real dignitary college teacher].[16]

Certainly, Tobler is far from unique in his praising. In subsequent chapters, I will present views of Robinson's contemporaries and of twentieth and twenty-first century scholars, including those of critics and academic rivals.[17] It seems that no other Palestine explorer has received as much attention as Robinson. American minister, philologist, theologian, scriptural- and historical-geographer, he was, and remains, one of the most well-known nineteenth-century Holy Land explorers, responsible for laying the foundation for its modern historic-geographical study. Not in vain, Robinson was declared by his contemporaries and

[14] Furrer (1878); Heim (1879); Stern (1988); (1988a); Goren (2003), 222–43.

[15] Tobler (1853–4), I, iii–iv; Furrer (1878), 54.

[16] Tobler (1867), 163. Writing in German, Tobler had the habit of avoiding capital letters for nouns. Tobler's book (Tobler (1839)) is absent from Robinson's bibliographical list (Robinson (1841), III, 1st Appendix). See chap. 4.c. All 'Robinson' bibliographical references are to Edward R., the other Robinson's are referenced with their initials.

[17] For an early study listing Robinson's supporters and critics, his strengths and achievements (Conder: 'It is in his steps that we have trod') as well as mistakes ('his fixed ideas (...) controlled his investigations'), enumerating misidentifications and other weaknesses, see Bliss's very early brilliant paper (Bliss (1907), 203–14).

later by recent scholars the 'father of Eretz Israel's scientific study.'[18] This leading Biblical scholar and scriptural geographer conducted two expeditions (1838, 1852) and published three epoch-making books and numerous papers.[19] His expeditions to the Holy Land, his approach to choosing travel routes, and method of documentation, are included in almost every publication dedicated to the history of the scientific study of the country or of special regions within it.[20]

In his research and books, Robinson adopted many of the approaches and theories of the leading Berlin geographer Carl Ritter (1779–1859). Ritter influenced him more than any other researcher and hence is significant in this book. Robinson did not set for himself the goal of discovering unknown places; he sought to investigate a known, historical land connected to the Scriptures that were the constant focus of his interest. Accordingly, his geographic space lay in the area west of the Jordan, between Dan and Beersheba. Ritter, a determinist and teleologist, maintained that the physical conditions are the most important factor in the historical development of any land, all the more so of the development of the Holy Land, and Robinson adopted this view. He believed that research of this country required understanding the reciprocal relations between the physical backdrop and the historical development. Even though the primary motive for his touring was biblical research, he dedicated special attention to the morphological lines of the Holy Land. His interest in this latter topic, in its own right, increasingly developed. Thus, if before his first journey (1838) he concentrated primarily on Biblical geography, his second expedition (1852) strengthened his determination to publish on topography and physical and historical geography. The first volume of a book on the physical geography of the Holy Land, encompassing the land of Israel, Lebanon and the Sinai, was the only volume published and that only after Robinson's death. In the second volume, Robinson had intended to broaden the format to include neighbouring countries.[21]

[18] See, for example, Bliss (1907), 184–223; Ben-Arieh (1979), 85–91; Goren (2003), 83–91.

[19] Robinson (1841); (1841a); (1841b); (1856); (1857); (1865). Cf. Tobler (1867), 162–63, 238–39; Röhricht (1890), 391–93, 636, 638–39. For the best bibliography of his numerous publications cf. Williams (1999), pp. 350–59. All 'Williams' bibliographical references are to Jay Gomer W., other Williams are referenced with their initials.

[20] Additional examples include: Schubert, G. H. (1841); Ritter (1866), II, 69–74; Schaff (1891); Benzinger (1903); Bliss (1907), 184–223; Brown, J. W. (1969), 113; Soler (2014).

[21] Robinson (1865).

And thus, Robinson's Biblical-toponymy is quite remarkable, with a list of about 170 discoveries to his credit, principally re-location and -identification of Biblical sites.[22] To grasp the breadth, examples include: Eleutheropolis at Beit Jibrîn and, subsequently, other settlements in the area; portions of Nabatean cities during the expedition from 'Akabah to Jerusalem; and renewed identification of the southern Mt. Hebron settlements and various settlements north of Jerusalem, in Judea, Samaria, and the Galilee. Of course, prominent are the arch that carries his name with the supposed bridge extending close to the southwestern corner of the Temple Mount, and the first report of the twisting, turning route of the Siloam Tunnel-Aqueduct. He also searched for remains of the third wall surrounding ancient Jerusalem. Robinson was the first to turn his attention to the large city built around Petra, beyond the sites carved in the rock and cliffs; he re-identified a number of structures in the Galilee as Jewish synagogues; and he explored the temples of the Hermon, in the Damascus area, and opposite the mountains of Anti-Lebanon.

A significant contribution springing, of course, from his achievements above, was the great improvement Robinson introduced in the map of the Holy Land when he gave young Berlin cartographer Heinrich Kiepert (1818–99) access to his expedition journals and data. It is sufficient to compare his map to the map of Syria compiled by Heinrich Berghaus (1797–1884) in 1835 to see that Kiepert's map had become the new standard, summarizing all that was known up until that point about mapping the Holy Land.[23]

1.b. Archival material for the study of Robinson's expeditions and publications

For a scholar so influential and highly regarded, there is a great deal of data missing from the many studies devoted to Robinson. After his death, two Union Theological Seminary (UTS) colleagues published his biography, and his daughter followed with her own edition.[24] In the

[22] Abel (1939); Goren (2014). For a prominent example of a later, extensive use of Robinson's works and toponymy see Smith, G. A. (1974). Robinson is mentioned so many times, that in the Index of Authorities (701) Smith gives up counting, and writes 'etc.'

[23] Berghaus (1835a); (1840); Kiepert (1841a); Heller (1836); Engelmann (1960). Cf. in detail Partsch (1901); Zögner (1999); Goren (1999a); (2015); (2017).

[24] Smith, H. B. (1863); Robinson, M. A. (1862–64).

ensuing years, colleagues eulogized Robinson on different occasions, evidence of his stature and career.[25] One outstanding document is an 1878 address to UTS students by Arthur Penrhyn Stanley (1815–81), Dean of Westminster, himself an important British Palestine scholar, co-founder of the PEF, and an old acquaintance of Robinson.[26] Correspondence from 1849 reveals Stanley's esteem for Robinson. With the mediation of George Barncroft (1800–91), secretary of the Navy 1845–46 and then American envoy in London, 'Stanley, son of the Bishop of Norwich and one of the worthiest men in England' sought Robinson's advice before setting out for the Holy Land.[27] Stanley opens his long query of Robinson this way: 'I trust that the common interest that we feel in the subject on which I venture to address you will concern the liberty I take in putting to you a few questions [in] relation to the geography of Arabia & Palestine.'[28]

Entries on Robinson are included, of course, in various biographical collections and encyclopaedias, such as the one by leading Holy Land archaeologist William Foxwell Albright (1891–1971) in the *Dictionary of American Biography*, and that by Philip Schaff (1819–93), from the 1870 faculty of the UTS.[29] Since the early twentieth century, scholars have published research on various aspects of Robinson's life and scholarship,[30] including a book (in French) dedicated to Robinson's lead role in the emergence of the field of Biblical archaeology.[31] His name appears frequently in geographic-historical, theological, and archaeological Holy Land studies, as well as in those dealing specifically with American involvement and activity in the Near East.[32]

25 His archive holds a collection of these obituaries, probably collected by his widow, under ERP, MM; e.g., Robinson (1863).

26 Stanley (1879); Robinson to Smith, New York, 20.8.1853, ESP, ABC 60 (63).

27 Barncroft to Robinson, London, 24.8.1849, ERP, L.1 London correspondence 1840–58.

28 Stanley to Robinson, London, 22.8.1849, ERP, L.1 London correspondence 1840–58.

29 DAB Robinson; Schaff (1891).

30 Abel (1939); Bewer (1939); Stinespring (1939); Dearman (1991); Ibbotson (1993); Williams (1996); (1997); Goren (2014); (2017).

31 Soler (2014).

32 e.g. Benzinger (1903), 585–91; Bliss (1907), 184–223; Macalister (1925), 21–25; Brown, J. W. (1969), 111–24; Ben-Arieh (1972a), esp. 84–85; (1979), 85–91; Handy (1981), 3–36; Silberman (1982), 37–47; Shepherd (1987), 78–89; Lipman (1989), 29–40; Vogel (1993), 190–94; Davis, J. (1996), 32–37; Bartlett, J. R. (2008), 109–18, 132–46;

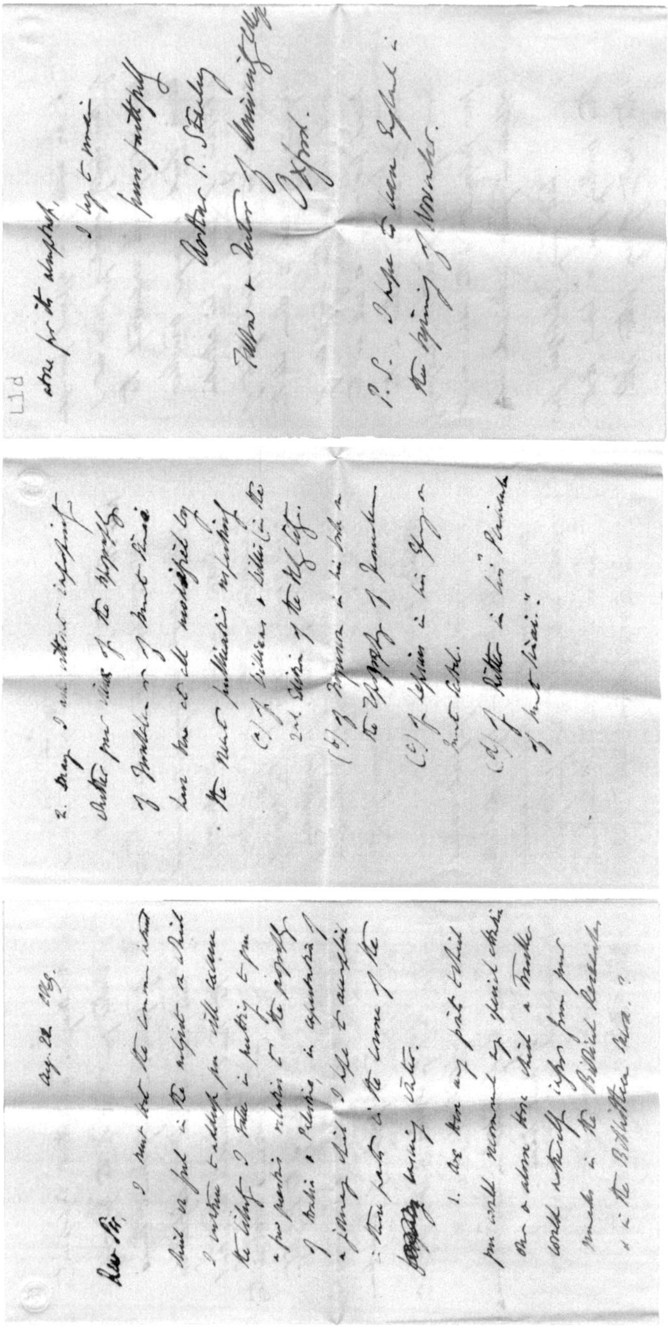

Ill. 3. Arthur P. Stanley, London, to Robinson, NY, August 22, 1849.

Throughout this period, it appears that Robinson's archive remained untouched, the personal property of his family. No one examined the archive for useful material on Robinson's beliefs, expeditions, or scholarship. This changed significantly only in 1994, when Robinson's descendants agreed to transfer his vast archive to Hermine Weigel Williams and Jay Gomer Williams, both professors at Hamilton College, Clinton NY. The story of Robinson's archive and its transfer to Hamilton College is recounted in studies the Williams wrote on Robinson and Talvj and in documents of Hamilton's archive and special collections.[33] During the twenty years that the archive was in their possession, the Williams achieved the enormous task of organizing and cataloguing the vast collection. Hermine Weigel Williams also published collections of many of the letters, with comments explaining Robinson's writing style, his underlying for emphasis, use of the dash, 'a trademark of his letter writing (...) to separate ideas within a multi-phrase sentence' or between two sentences, his commonly used abbreviations, and more.[34] Between 1997 and 2009, both Williams published books based on an extensive use of Robinson's vast archive.[35] These publications focus on Robinson's early life and education up until the early 1830s and Robinson's and Talvj's life and scholarship after settling in the US. They provide considerably less information on Robinson's Holy Land expeditions and publications, his process of writing, publishing, and of compiling his maps or the acceptance of his scholarship in the academic world.

Without recourse to the archival material, it might be difficult to understand Robinson's obsession with the writing and publication of the *Biblical Researches*; he worked painstakingly in Berlin and Halle for twenty-two months, from November 1838 to September 1, 1840. The story of his extended stay in Germany to write the manuscript, while continuously consulting Smith, expert scholars, collections, and libraries has been relatively neglected in modern research. This omission is probably due to the fact that except for Jay and Hermine Williams, who had greater interest in Robinson's early years and in his wife's history, no other scholars have consulted the vast archival material. No attempt has been made to answer

Goren (2003), 83–91; Hallote (2006), 9–13; Oren (2007), 134–41; Oliver (2014), 235–41. Near East and East are used interchangeably to refer to the Near (Middle) East.

33 e.g. Williams (1999), 1–3; ERP, 1816 Robinson- 2, Edward Robinson, His Life and Times, Captions and Legends. Print, program of the exhibition; ERP, 1816 Robinson- 3, Remarks made by Hermine Weigel Williams on the occasion of the 200th Anniversary Celebration October 9, 1994, Hamilton College Chapel.

34 Weigel Williams (2003), introduction; (1994); (2009).

35 Williams (1996); (1999); Weigel Williams (2009); (2014).

basic questions relating to Robinson's development of ideas and production of his groundbreaking nineteenth-century historical and Scriptural geography study of the Holy Land. A painstaking reading and deciphering of the mostly handwritten archival material, numbering hundreds of letters and other documents from Robinson's as well as other archives, revealed the narrative behind the publications. On many points, this account is not only more detailed and complex than what has been written to date, it also differs from the accepted version in the existing literature.

No less important than Robinson's archive is that of American Board of Commissioners of Foreign Missions (ABCFM) Beirut missionary Eli Smith (1801–57). Fluent in Arabic and highly familiar with the country and its inhabitants, Smith was Robinson's singularly well-suited travel and research companion.[36] Smith's well-organized main archive is housed in the Houghton Library at Harvard University. Despite their accessibility, to date I have not uncovered any attempt to conduct a detailed study of the archives of either scholar, separately – and as complementing each other – for studying and reconstructing their expeditions and publications. Even the last publication concerning Robinson, *et l'émergence de l'archéologie biblique*, was written, as the author himself admits, without the use of any archives.[37] Margaret Leavy did use Smith's papers intensively for her studies, but they do not deal with his Holy Land connection.[38]

The chapters that follow are intended to contribute considerably to the history of the scientific study of the Holy Land within its Scriptural borders 'from Dan to Beer-Sheba' as well as to that of neighbouring regions. Robinson was always a leading actor in this narrative; the story of his research in the modern era and his expeditions and publications had a lasting influence on nearly all who came after him. This study aims at adding important new insights to this narrative, as well as to issues concerning the reality behind early nineteenth-century exploration of Sinai and Palestine, relations between European travellers and explorers and the local population, technical and financial issues of early nineteenth-century travel in the East, and the development of Scriptural Geography over nearly two centuries.[39]

[36] Leavy (1993). Cf. Kawerau (1958); Abu-Gazaleh (1990); Farah (2000), esp. 299–303; Stoddart (2009). See chap. 5.a. for detail.

[37] Soler (2014), 291: 'Des sources manuscrites existent, mais leur consultation nous a été impossible.'

[38] Leavy (1992); (1993).

[39] Important studies include: Smith, G. A. (1974); Bliss (1907); Noth (1962); Brown, J. W. (1969); Goren (1998); Robey (2009); Aiken (2010); Soler (2014).

Ill. 4. Eli Smith and his (third) wife, Hetty Butler.

Given the centrality of archival research to this study, a few words on the origin and use of the sources are in order. The two obvious starting points were Robinson's archive in the Special Collections of the Burke Library, Hamilton College, Clinton, NY, and Smith's archive in Houghton Library, Harvard University, Cambridge, MA. At Houghton Library, I also found the ABCFM and other personal missionaries' archives quite helpful. In northeast US, I located additional sources at the Burke Library Archives of the Union Theological Seminary in New York, at Yale University Library, New Haven, CT, and at the Sterling Library. I researched smaller archives in London and Dublin. In Germany, I located much relevant material in the Gotha Research Library in Castle Friedenstein and in the Perthes Collection, in Berlin's State Library (STABI, Prussian Cultural Heritage Foundation) Maps Collection and Manuscript Department, and in Halle, at the archives of Martin-Luther-University Halle-Wittenberg and the *Frankesche Stiftungen*.

After a long period of intensive archival work in these public and private collections, I found myself with an overwhelming and totally unexpected amount of material. Deciphering the handwritten documents was a long process, hampered by their condition, being partially torn or covered with dirt, and other impediments such as illegible handwriting and a crisscross pattern of writing to save space. These led to lacunas and surely to some mistakes, as well, in the given citations. In many cases, in order to decipher a particular text, it was necessary to read and re-read it with the help of colleagues. In this, I was fortunate to have the expertise of Teresa Kilian, who performed many a miracle in reading the most impossible documents. Where the words were missing, torn or impossible to decipher, we used brackets with a question mark, to signify that we were unsure about the correct reading. Brackets with an ellipsis between them are used primarily in cases of missing text deemed unnecessary for the argument and\or in order to shorten long citations. Square brackets are used only for additions and comments of the author. Surely, mistakes remain in the given citations, and one can only hope that they are minimal and do not affect the correct reading and argumentation.

Finally, a word about the large number of characters and citations in this narrative. Robinson had a huge array of connections consisting of impressive circles of colleagues, collaborators and acquaintances, a point noted previously by Williams. To this group one must add the hundreds of Holy Land historians and travellers whose works were used and cited by Robinson. At times, names may overwhelm the text, to a point where it may feel impossible to follow. As I believe that these names are an important part of Robinson's world and deeds, I decided to leave them in and help the reader with an index. When first noted, each person is listed by full name and, if possible, year of birth and death.

The exceptional number of citations included is an aspect that differs significantly from many studies. The central argument, that this new material reveals much that was previously unknown concerning Robinson and his Holy Land expeditions and publications, prompted the decision to let the participants speak in their own words, as much as possible. It seemed preferable to bring Robinson's and Smith's and other important figures' own words, idioms, and thoughts.[40] Following Robinson's statement that 'after all the American edition will be the Standard,' I relied on and reference the American edition of the *Biblical Researches* as my text-book for the entire work.[41]

[40] The citations are brought in two forms: The shorter ones, up to four lines in length, are interwoven in the text, notated by single quotes ('') and the longer citations are set off by indented paragraphs.

[41] Robinson to Ritter, New York, 15.3.1840, SBB-PK, Nachl. 133 (Carl Ritter), K 5, Q – SCH, Bl. 30. This letter is an enigma! Clearly, it is signed 'New York, March 15. 1840,' but Robinson returned to NY only in November! My estimation, also due to the letter's contents, is that it should have been 1841!

CHAPTER 2
THE 1838 EXPEDITION TO THE HOLY LAND: ORIGINS AND PREPARATIONS

I can most truly say (...) that my journey to the east
was undertaken in weakness & fear & much trembling...[1]

Edward Robinson's education, perceptions, and beliefs impelled him to embark on a historical geographic study of the Scriptures. His contemporary biographers Henry Boynton Smith (1815–77) and Rosewell Dwight Hitchcock (1815–87), as well as later historians, devote significant attention to these influences in their writings.[2] 'Of athletic stature and imposing shape, but a bit bent with the age,' he has been described, 'of strong, healthy personal understanding, sober and dry, but in educated society very conversing and not without humour; full of good-hearted and plain sympathy, – a simple, serious, solid, through and through respected character.'[3] Robinson, himself, professes in the introduction to the *Biblical Researches* and elsewhere that it was 'a journey, which has been the object of my ardent wishes, and has entered into all my plans of life, for more than fifteen years.'[4] Yes, it was certainly an abiding dream. As early as September 1827, at the beginning of his 'German period' following a long overland journey in Germany and Austria, Robinson writes to his sister that Vienna was 'probably the most easterly point of

[1] Edward Robinson to Secretary of the Royal Geographical Society William Hamilton, upon receiving the society's gold medal, New York, ?.?.1842, ERP, I, 4.

[2] Robinson's biographies: Smith, H. B. (1863); Robinson, M. A. (1862–64); Prentiss (1889), 243–51; Williams (1999); Soler (2014), 13–27 ('La formation d'un savant Américain et Chrétien'), 179–81 ('Le pèlerin et missionaire').

[3] Cited e.g., by Kaarsberg Wallach (2001), 248.

[4] Robinson (1841), I, 1; almost in the same words, published two years earlier: Robinson (1839), 400; Robinson to Hamilton, New York, ?.?.1842, ERP, I, 4.

all my journeyings, unless God should so order it, that I may yet visit Palestine.'[5]

Robinson began fulfilling his wish in 1832. Eli Smith, 'former pupil and friend,' had just returned from a 16-month pioneering missionary research expedition in Armenia with missionary Harrison Gray Otis Dwight (1803–62) of the ABCFM.[6] Before returning to Lebanon, where he was becoming a leading figure in ABCFM Middle Eastern activity,[7] Smith went home for a visit. He married Sarah Lamman Huntington (1802–36) on July 21, 1833, and they sailed together for Malta exactly three months later.[8] Naturally, while home, Smith, who graduated from the seminary in 1826 and maintained connections with his teachers,[9] visited Andover and met with Robinson. Hermine Weigel Williams writes that Smith's recounting of his Armenia travels convinced Robinson that the former 'was exactly the kind of person he needed as a guide for his proposed travel to Palestine.'[10] Robinson's account of their conversation in the *Biblical Researches*, cited by Weigel Williams and in greater detail by her husband,[11] confirms only that they agreed on a joint expedition to the Holy Land, outlining a general plan starting from Egypt and journeying to Mount Sinai, then 'Akabah, and through Wadi Mûsa (Petra) to Hebron and Jerusalem. In 1838, Robinson notes, in retrospect, that they had not known of others taking this route, though 'it has since become almost a highway for travellers.'[12]

Robinson's biographers have written extensively on his Andover and Boston periods between his return from Germany and his expedition

5 Robinson to [his sister] Elisabeth R., Vienna, Oct. 1, 1827, Weigel Williams (2009), 223.

6 Smith (1833); cf. DAB Dwight; Leavy (1992); (1993); Phillips (1939), 144–46; Gustrau (2016), 214. For Dwight: Struthers (1834), 172–74; Kaplan (1995), 20–25. All 'Smith' bibliographical references are to Eli Smith, other Smiths are referenced with their initials.

7 Smith is central to every study of nineteenth-century American missionaries. See, for example, Tibawi (1966), passim; Makdisi (2008), passim. For detail see chap. 5.a.

8 Hooker (1845), 142, 151.

9 While travelling in Armenia, he sent a detailed letter to Stuart, 'The most interesting questions of Biblical nature connected with Armenia, are the positions of the mountains on which the ark rested…': Smith to Stuart, Tebreez, Persia, 18.2.1831, ERP, F-5 (transcribed version).

10 Weigel Williams (2007), 83.

11 Williams (1999), 194.

12 Robinson (1841), I, 2. Cf. Williams' discussion (1999), 192–94.

to the Near East, citing numerous sources, yet there are no references to the expedition preparations.[13] From this silence it is not possible to determine when Robinson and Smith began making arrangements or whether there is record of any activity. The American archaeologist Frederick Jones Bliss (1857–1939), born in Lebanon to ABCFM missionary Daniel Bliss (1823–1916, founder and president of the Syrian Protestant College, the future American University of Beirut), devotes forty pages to Robinson in his important 1907 published collection of lectures *The Development of Palestine Exploration*. However, with regard to expedition preparations, Bliss' only notation is this quote from Congregationalist clergyman Roswell Dwight Hitchcock, Robinson's colleague and biographer, 'the time has come for a scholar (…) to enter this field [of the Holy Land] with thermometer, telescope, compass and measuring-tape, but, above all, sharp-eyed and sufficiently sceptical.'[14] Hitchcock, a graduate of Andover Theological Seminary, who like Robinson studied for a year in Halle and Berlin (1847, most probably with Robinson's help), became Robinson's colleague as professor of church history at the New York Union Theological Seminary (UTS) in 1855.[15]

Robinson later notes that he considered a research expedition to Palestine during his stay in Europe from 1826–30, but could not go due to the war in Syria.[16] These dates are questionable. The war in Syria began only in November 1831, when Egyptian forces started marching into Sinai and Palestine and the navy sailed from Alexandria, laying siege on Acre and bombarding it from the sea (December 9th).[17] Robinson's 'winds of war' may have been related to the Greek revolt and ensuing European intervention, which reached its climax with the Battle of Navarino on October 20, 1827.[18] The first seeds for the expedition, however, were planted in Germany.

[13] Smith, H. B. (1863), 53–63; Soler (2014), 33–38, 47–52.

[14] Bliss (2007), 184–223, citation 189. On Bliss, father and son: DAB Bliss; Kaplan (1995), 31–38; VanDeMark (2012), 23–81; Hallote (2006), passim.

[15] Hitchcock also travelled to Italy and Greece in 1866, and Egypt, Sinai, and Palestine in 1869. In 1871, he was elected president of the American Palestine Exploration Society (DAB Hitchcock; Hallote (2006), 13, 51–65), and in 1880 president of UTS.

[16] Robinson (1841), I, 1; and see hereafter, chap. 2.b.

[17] Alderson (1843), 38–53; Cvikel (2012).

[18] See the classic works of Marriott (1940), 204–33 and Anderson, M. S. (1966), 53–87, compared to a more recent study, Kutluoğlu (1998), 61–107, and missionaries' letters, for example: Bird to father, Beirut, 18.4.1825, IBP-YSM, Box 1.

2.a. The German period: seeds of the Holy Land expedition

Robinson's letters confirm Jay Williams' conclusion, that Robinson began formulating the idea of a Holy Land expedition during his German period, from 1826 to 1830. His interaction with the German scholars and exposure to their ideas and research methods most likely catalyzed Robinson's unique thinking. Robinson travelled, almost always and nearly everywhere. Touring mostly by coach, he crossed Europe repeatedly, describing in detail his experiences and impressions both in his diary and in his letters. Robinson reveals his excitement and even envy of primary education in Germany, 'but this much is fact, as to learning, no country is at least half a century behind Germany; the slow patient & thorough training of the youthful mind which is here common, we have not, & cannot have.'[19] European, mainly German, higher education, universities and university cities, and methods of teaching and learning were primary topics.[20] Well-known is his long letter sent from Halle on April 2, 1827, in which Robinson details his impressions of academia from his short periods in Paris, Basel, and Göttingen. This letter was analyzed by Williams:

> Göttingen is on the whole the most renouned university of Germany, has the largest library, 320,000 volumes & 1,500 students (...). The great theological school however is in Halle, and of its 1100 students, 800 are theologians, & here especially is the Hebrew & the general philosophy of the Bible cultivated. Gesenius is without comparison the first biblical oriental philosopher of the age;- he has devoted his life solely to the illustration of the O. Testament;- & there he is at home.[21]

Robinson's educational encounters at this early stage of his travels engendered an interesting conclusion that certainly influenced his future thinking and studies:

> In summary, the education of the learned however, is far from being practical, - the mind is not cultivated in reference to its connection with influence upon the external world, it is rather abstracted further from it. & hence of the from 10,000 to 15,000 young men who are found in

19 Robinson to Norton, Halle, 2.4.1827, ERP, ER 1.19.

20 See e.g. Howard (2005).

21 Robinson to Norton, Halle, 2.4.1827, ERP, ER 1.19; Williams (1999), 116–36.

> the different universities, the very great majority are taught to think & reason & build theories, without any firm practical foundation on which to build. This view, I think, accounts in some measure for the wild speculations to wh.[ich] the Germans as a nation are home, & wh. form a distinct feature of their religion & philosophy.[22]

In Basel, Robinson, who had been trained at Andover, cradle of the ABCFM, also paid a visit to 'the interesting Missionary Seminary,' established in 1816. Robinson had a letter of introduction to Württemberg Pietist Christian Gottlieb Blumhardt (1779–1838), the first inspector of the Basel Mission Seminary, a three-year training school for missionaries deeply embedded in the Pietistic philosophy and belief.[23] Robinson writes, '[I] found myself almost at home (…) under the care of Mr Blumhardt; – it resembles much of the theol. Seminaries of our own country, & is exerting a most beneficial influence in the interest of religion in Switzerland & the S. of Germany.'[24] Blumhardt introduced Robinson to his first 'major Biblical scholar' and one of the early founders of biblical criticism, Wilhelm Martin Leberecht de Wette (1780–1849), at the time in 'academic exile' from Berlin for being too liberal and radical.[25] In his book, Williams devotes much attention to Robinson's lack of interest 'in questions of critical scholarship which bothered Germans as De Wette.' Moses B. Stuart (1780–1852), Robinson's teacher and tutor in Andover, had read De Wette's writings but did not adopt his approach and neither did Robinson.[26]

Stuart's influence upon Robinson is well-documented.[27] When the young Edward Robinson came to Andover in 1821 planning to finish his translation of and commentary to the *Iliad*, Stuart brought Robinson into his field, the study of the Bible, in conjunction with an in-depth study of the Hebrew language. Robinson became Stuart's first 'student assistant to hear the Hebrew and Greek grammar recitations of the juniors.' He was also 'asked to correct the proofs' for the second edition of

[22] Robinson to Norton, Halle, 2.4.1827, ERP, ER 1.19.

[23] Schlatter (1916), I, 11–22; ADB Blumhardt; Akel (1978), 15–23; Blaser (1981), 131–36; Lindt (1981), 115–20.

[24] Robinson to Norton, Halle, 2.4.1827, ERP, ER 1.19.

[25] ADB De Wette.

[26] Williams (1999), 125–28, 336; for De Wette as an Old Testament scholar, cf. Cheyne (1971), 31–53.

[27] For Stuart's place in American scholarship see DAB Stuart; Kitto (1853); Dearman (1979–80); Giltner (1988).

his mentor's *A Hebrew Grammar*, published in 1823. Stuart also hosted Robinson at his large home. His skills as a linguist quickly established, Robinson began assisting Stuart with his scientific works,[28] launching his career as a scholar and educator. Stuart, who studied German theological works, eventually introduced Robinson into that world, and was responsible for the latter's studies in Germany.[29]

To his impressions from Paris, Basel, and Göttingen, Robinson adds those from the universities of Halle and Berlin where he studied.[30] He repeats much of this information in 400 pages of letters written between 1826 and 1830 to his sister Elisabeth, which were transcribed, edited, and published by Hermine Weigel Williams.[31] In an effort to spread his ideas in the US, Robinson began publishing articles on the topic in 1831. The first volume of the *American Biblical Repository* (*ABR*) contained a long, four-part essay on 'Theological Education in Germany.' This volume was followed, among others, by the 1834 publication of 'Outlines of a Course of Theological Study for the Use of Students prepared by the Theological Faculty of the University of Leipsic' and translated by the editor – Robinson.[32]

Within a year in Germany, Robinson had acclimated to his new surroundings. He had come as a total stranger, yet writes to his sister in November 1827, 'now I have here warm friends (...) I know the manners & customs (...) & as to the language, I mingle in the society (...).'[33] This change occurred not only in his personality and private life, but also in his academic perceptions and methods. His most influential teachers in Halle were Gesenius and Tholuck. Heinrich Phillip Wilhelm Gesenius (1786–1842), 'leader in Hebrew philology' and 'one of the most gifted teachers which Hebrew and Oriental studies have ever had,' had been Robinson's host and teacher while in Halle between 1826 and 1829. As early as 1829, Gesenius published a translation of Burckhardt's voyages.[34] Friedrich August Gottreu Tholuck (1799–1877), theologian and

[28] Stuart (1823). For Stuart's influence on Robinson: Dearman (1991), passim; Williams (1999), 79–111 ('Andover days'), 175–205 ('Andover and Boston'), and passim. Weigel Williams (2003) transcribed and edited Robinson's letters to Stuart written from Germany, 1826–30.

[29] Dearman (1991), 164.

[30] Robinson to Norton, Halle, 4.4.1828, ERP, ER 1.17.

[31] Weigel Williams (2009); Williams (1999), 117.

[32] Robinson (1831); Theological Faculty (1834); cf. Soler (2014), 71–3.

[33] Weigel Williams (2009), 241–42.

[34] Cheyne (1971), 53–65 (citation); Gesenius (1823); NDB Gesenius.

church leader, taught in Berlin and then in Halle from 1826.[35] Robinson attended every lecture possible of Gesenius during his stay in Halle, and in his wonder-arousing habit took notes of everything he heard, now bound in about twelve volumes of 'Notes on lectures in the universities of Halle and Berlin.' The two volumes of notes from Gesenius' lectures are 155 and 267 pages in length each.[36] Naturally, there is also a volume of Tholuk's lectures, containing 381 pages.[37] Robinson maintained correspondence with Tholuk after returning to Andover.[38]

Robinson quickly became a consultant for young American scholars studying in Europe. Charles Hodge (1797–1878), Presbyterian theologian and founder of the periodical *Biblical Repository* in 1825, and later principal of the Princeton Theological Seminary, reached Paris in October 1826.[39] Following Robinson's advice, he chose to continue his studies in Halle, as 'he tells me that for the purposes for which I have come hither there is no comparison between any other university at present and Halle.'[40]

2.b. The role of the ABCFM and other missionaries and officials

The ABCFM played a prominent role in Robinson's decision to embark on a Palestine expedition, both in terms of a shared historic-geographic approach to Scriptures and potential organizational support from the leadership and missionaries in the field.[41] Established in 1810 and with its roots at Andover Theological Seminary, the ABCFM became a leading and influential organization in nineteenth-century Lebanon, Syria, and Palestine. In 1840, it had the following staff and facilities:

35 ADB Tholuck.

36 ERP, Edward Robinson, Notes on lectures in the universities of Halle and Berlin, v. 12: Notes of the lectures of Dr Gesenius on the Psalms, R65N9; vol. 11: Notes of the lectures of Dr Gesenius in Eccles. History, from the birth of Christ, to Gregory VII, 1073, with the crusades, R65N9.

37 ERP, Notes on lectures in the universities of Halle and Berlin, v. 7: Lectures by Dr Tholuck, R65N7. For Robinson's letters to Tholuck: AFSt/N, THOLUCK, B Iii.

38 Robinson to Tholuck, Theol. Sem. Andover, Mass., 28.2.1830, 20.1.1831, 4.7.1831, 16.1.1832, AFSt/N, THOLUCK.

39 DAB Hodge; Hodge (1880), 104–09; Williams (1999), 149–50 and passim.

40 Hodge to wife, Paris, 21.12.1826, Hodge (1880), 110–11.

41 Anderson, R. (1863); Shepherd (1987), 80–83; Vogel (1993), 190–94; Kaplan (1995), 16–19; Soler (2014), 154–63.

Ill. 5. Robinson's Corporate Membership in the ABCFM,
Boston, September 25, 1843.

> It has at pres[en]t. 12 Amer. Miss[iona]r[ie]s. 9 in field & 3 temp[oraril]y abs[en]t. 13 fem[inine] assis[tants] missrs 9 nat. preach[er]s., 16 teach[er]s & 3 others call[e]d helpers. There are six Stations (...). 17 oth[er]. places are call[e]d out-stations, where meet[in]gs for prayer or preach[in]g are, more or less reg[ularl]y held. The schools are numb[ere]d at 30 hav[in]g 1000 pupils...[42]

Early on, Robinson cultivated a relationship with the ABCFM independent of that of Smith, who joined the ABCFM as a missionary after graduating from Andover in 1826. Robinson and his relatives began donating to the organization in 1823. Four years later, he became an 'Honorary Member' and in 1843, a 'Corporate Member.'[43] It is not surprising, of course, that Smith maintained an intensive correspondence with his employers throughout the expedition, primarily with Andover graduate Rufus Anderson (1796–1880). In 1832, Anderson took charge of all overseas work and became the most influential secretary of the ABCFM. He travelled extensively to all stations, visiting the Near East in 1828–29 and 1843–44.[44] On his 1829 voyage to Greece, he took young

[42] Bird's report, IBP-YSM, Box 3, pp. 164–70; Vogel (1993), 99–106.

[43] Henry Hill, Treasurer, Boston, 1.11.1823, ERP, ER; Documents of membership, 28.2.1827, 25.9.1843, ERP, ER, GG.

[44] ABCFMA. Harris (1999); Phillips (1939) (comprehensive history of the first 50 years of the ABCFM); Makdisi (2008), passim, esp. 190–200.

Eli Smith as his travel companion, looking for potential missionary station sites.[45] Anderson's important two-volume narrative of the organization and its works is an outstanding source, not only for the history and whereabouts of the Mission in the Near East, but also for its ideology, struggles, and continuous defeats.[46] The relatively poor achievements after tremendous efforts, financial difficulties, and personal sacrifice led Ussama Makdisi, in 2008, to write about the *American missionaries and the failed conquest of the Middle East*.[47]

In addition to Smith, other ABCFM missionaries had decisive roles in Robinson's travels and publications, and are accordingly mentioned throughout this work. George Backus Whiting (1810–55) and his wife Matilda S., born Ward (1805–73), arrived in Beirut in 1830 and were sent to Jerusalem in 1834 following the end of the 'Peasants' Revolt' against the Egyptians, who ruled the country from 1832 to late 1840. They established a mission school, but in 1838 'were forced to return to America because of Mrs Whiting's poor health.' George Whiting returned to Syria sometime before 1844 and died there eleven years later.[48] Isaac Bird (1793–1876), a Connecticut native and graduate of Yale College, married in 1822, and arrived in the East the following year, serving first in Malta, then Beirut from November 1823, and then in Smyrna until 1836.[49] In 1837, Bird was back in the US and awaiting Smith's arrival. A letter of his dated Andover, December 8, 1837, reveals the fact that he had no idea about Robinson's travel plans and that Robinson had already left New York.[50]

Another important member of the missionary community of much relevance to the current study is William McClure Thomson (1806–94). Thomson entered the Princeton Theological Seminary in 1829, but left in 1831 for Syria as an ABCFM missionary, where he served until 1876(!). A traveller and scholar, known primarily for his *The Land and*

[45] Eli Smith's passport 1829–30, ESP, ABC 60, 134–35; Papers connected with ES and R. Anderson's travels in Greece 1829, ESP, ABC 136; Eli Smith, Notes on Greece taken during a journey in that country in 1829, ESP, ABC 17.3 vol. 1; Phillips (1939), 142–43.

[46] Anderson, R. (1862); (1872). Cf. Lindner (2009), 17–19.

[47] Makdisi (2008).

[48] Kawerau (1958), 254–57; Tibawi (1966), 57–62, 74, 111, 130; Abu-Gazaleh (1990), 23–25; Crombie (2006), 121–22. For the American missionary station in Jerusalem: Lipman (1989), 80–86.

[49] Struthers (1834), 130–33; Kawerau (1958), passim; Oren (2007), 122–33; Makdisi (2008), passim. For his archive: IBP-YSM; Box 2, Folder 2: Journal 1827–60.

[50] Bird to Temple, Andover, 8.12.1837, ESP, ABC 60 (9).

the Book, Thomson became a fellow of the London Asiatic Society and Royal Geographical Society.[51]

In addition to ABCFM leadership and missionaries, various government officials were instrumental in assisting Robinson and Smith during the expedition. John Gliddon († 1844) was an English merchant who from 1818 resided in Egypt. In 1832, he was appointed by the US a non-salaried consular agent in Alexandria, and in 1835 promoted to consul. His more famous son, George Robins Gliddon (1809–57), served as American vice-consul in Cairo from 1835 until 1842.[52] Gliddon the son later established his reputation as the first American Egyptologist. George Gliddon had close ties with Smith and managed Smith's account. In a letter dated January 9, 1836, Gliddon thanks Smith upon his appointment as US Consul and refers to his extensive involvement with the American mission and missionaries, 'I am satisfied to find, that my efforts to obtain the order for the Jerusalem burying-ground are appreciated.'[53]

Both Gliddons assisted travellers generously and were famous for the contacts they maintained in Muhammed 'Ali's regime.[54] Unfortunately for Robinson, during his visit to Cairo consul Gliddon (the son) 'was absent in the United States,' but he and his companions were helped by the father, 'we have experienced much kindness and attention from Mr Gliddon & his servants & family.'[55] In addition, they were assisted by British representatives in Cairo, Vice-Consul Mr Walne and resident missionaries Lieder and Kruse of the Church Missionary Society (CMS).[56]

Alfred Septimus Walne (1806–81) came to Egypt in the early 1830s as a physician and was appointed vice-consul to Cairo in 1835 or 1836.[57]

[51] DAB Thomson; Thomson (1880). Cf. Kawerau (1958); Shepherd (1987), 90–91; Kaplan (1995), 20–21; Travis (2012), 142–43.

[52] Oliver (2014), 137–38, 249–58 and passim. Kark (1994), 46–55, deals with the beginning of the American consular service in Alexandria, not mentioning the Gliddons.

[53] Gliddon to Smith, Alexandria, 9.1.1836.

[54] Vivian (2012), 95–112 ('The Gliddons and the beginning of American-Egyptian relations'). Cf. also Paxton (1839) from Palestine, 239; Fagan (1975), 267–69; Dawson, W. R. (1995), 169; Kalfatovic (2011), 240–41. For European physicians in Egypt and in the service of Muhammed 'Ali cf. Jagaillouz (1989); Goren (2004).

[55] Robinson to Smith, Alexandria, 3.1.1838, ESP, ABC 60 (63); Robinson (1841), I, 21, 25.

[56] Robinson (1841), I, 25–26, 52–55.

[57] Walne, Elizabeth, 'Cousin Albert Septimus: Queen Victoria's Consul in Cairo,' https://elizabethwalne.co.uk/blog/2011/4/25/cousin-albert-septimus-queen-victorias-consul-in-cairo.html (red 16.11.2016).

One of many Egypt travellers to require Walne's help was the American lawyer and future Mayan explorer John Lloyd Stephens (1805–52). Stephens reached Alexandria in December 1835, where he had an audience with Muhammed 'Ali, the viceroy. After his travels in Upper Egypt, he journeyed through Sinai to 'Akaba, Petra, Jericho, and Jerusalem.[58] In his book *Incidents of Travel*, first published in 1837, Stephens writes:

> I have seen with great pleasure, in a late English paper, that Dr Walne has been appointed English vice-consul at Cairo. In the close relation now growing up between England and Egypt by means of the Red Sea passage to India, it is a matter of no small consequence to England to have at Cairo as her representative a man of character and talents; and I am sure I but express the opinion of all who know Dr Walne when I say that a more proper appointment could not have been made.[59]

William Kruse and Johann Rudolph Theodore (or Theophilus) Lieder were two of five Basel educated CMS missionaries sent to Egypt in 1825 to convert the Copts. Sources praise them, noting that Lieder 'and his brethren itinerated all over the Delta, into the Fayum, up the Nile into Nubia, selling and distributing scriptures and tracts...,' opened a boy's boarding school, and published linguistic guide books.[60] Surprisingly, missionary Bird's archive contains a short, unsigned, and undated note titled 'P. S. of Mr Jos. Wolff,' which expresses an opposing view:

> For God's sake write to the [CMS] Committee, not to send Muller back to Lieder or Kruse, or what would be worse, to the traitor [missionary Christian] Kugler, for all three are most wretched fellows. I wish you could see Kruse's house, there are divans, & notwithstanding, he complains of being unable to buy more furniture – the belly is his God. Lieder (...) is a lazy fraud, unbridled lubber, who has no other desire but to marry and to carry on a lazy life (...). I must give you only one instance of Lieder's conduct. In the beginning he was exceedingly polite, but soon he displayed his roughness by talking continually

58 Hagen (1948), 28–30, 50; Finnie (1967), 152–60; Oren (2007), 152–60.

59 Stephens (1858), I, 39 note. For England's 'Red Sea passage to India' see Elath (1971), 9–16, 43–49; Goren (2011), 1–59.

60 Stock (1899), 350–51; CMS (1896), 70–71; Carmel (1981), 25, 60-61, 106 (mentions only Kruse and Kugler); Tejirian (2012), 79; Oliver (2014), 106, 281. The other missionaries were Theodor Müller, Christian Kugler, and Samuel Gobat, who later became the Protestant bishop in Jerusalem.

> against the English King & nation to Lady Georgiana [Wolff's wife]. – He throws dissonance between Kruse & his wife, & talks continually about females.[61]

Joseph Wolff (1795–1862), a converted German Jew, first to the Catholic and later to the Anglican Church, was a leading missionary who travelled extensively to different parts of the world. He was also a recognized Oriental scholar and theologian. He travelled in the East from 1821, through the 1820s and 1830s, looking, among other things, for the ten lost tribes of Israel.[62]

Lastly, John Nicolayson (1803–56), a Schleswig-Holstein native, was a missionary of 'The London Society for Promoting Christianity Amongst the Jews' (London Jews' Society, LJS). Nicolayson came to help the resident missionary physician, the Irish George Edward Dalton (also D'Alton), who left Ireland with his wife and infant in June 1824, travelling first to Alexandria and then to Beirut where Dalton studied Arabic, and from Beirut to Tyre. The family reached Jerusalem late in 1825, the first missionaries to settle in the city. Dalton died in January 1827 of fatigue, fever, and mismanagement of his health care, just three weeks after Nicolayson joined him.[63] Staying on a few more months in Jerusalem and marrying Dalton's widow, Nicolayson then moved north and concentrated his activity in and around Beirut. He returned to Jerusalem in summer 1831, and after the Egyptian conquest was able to rent a house bordering the Jewish quarter. Nicolayson became the first permanent resident Protestant missionary in Jerusalem and a key figure in all Protestant activity in the city during the twenty-three years until his death in 1856.[64] Nicolayson was one of the most active and influential Europeans in Jerusalem.[65] He is frequently mentioned in the letters, revealing a strong cooperation between the ABCFM and the LJS missionaries.

[61] IBP-YSM, Box 1.

[62] Wolff (1835). For Wolff, his travels and publications see, Palmer (1935); Kawerau (1958), passim (for his connections with ABCFM missionaries); Henze (2011), 5, 535–36. In 1827, he married Georgiana Mary Walpole.

[63] Gidney (1908), 119–21; Schwake (1983), I, 103, 113–14; Perry (2007), 28–30, 60–61.

[64] Carmel (1981), 17–24; Schwake (1983), I, 117–52; Lieber (1985); Crombie (1991), 20–25, 47–85. Robinson and Smith's archives hold vast correspondence with Nicolayson.

[65] Gidney (1908), 120–22, 174–77, 232–43; Carmel (1981), 17–24; Schwake (1983), I, 117–52; Crombie (1991), 20–22 and passim; (2006), 46–47 and passim; Perry (2003), passim.

2.c. From theory to field study: A historic-geographic study of the Holy Land

The sixth chapter of Jay Gomer Williams' 1999 biography of Robinson is dedicated to the period that he calls 'Andover and Boston,' starting with the Robinson family's return from Germany and continuing with his second Andover period and years as a freelancer in Boston. This chapter is the best and most detailed study of those crucial years for Robinson's development into a Holy Land scholar, which, as noted, undoubtedly began while in Germany.

Initially, it was not clear that a position awaited Robinson in Andover; he received it solely due to Stuart's successful mediation. Only on February 2, 1829, was Stuart able to write Robinson that as a result of the discharge, following a long struggle, of Professor of Sacred Rhetoric and Ecclesiastical History James Murdock (1776–1856), did the way open up for Robinson's appointment.[66] As early as 1832, Robinson demonstrated his newly developed interest in the geography of the scriptures in a rather long and detailed paper considering 'The Exodus of the Israelites,' with special emphasis on an explanation of the miraculous crossing of the Red Sea. It was first published in 'his' periodical, *Biblical Repository*, and reprinted in the American edition of *Calmet's Dictionary of the Holy Bible*.[67] This edition of the leading dictionary came out in Boston, 'Revised, with large additions,' by Edward Robinson. The preface to the American Edition is signed Andover, Oct. 15 1832.[68] It is interesting to note that Robinson used Carsten Niebuhr's 'Plan of the Extremity of the Red Sea Near Suez' for 'illustrating the Passage of the Israelites.' Niebuhr's excursion to Sinai took place in autumn of 1762, and his book was published ten years later, but Robinson could find nothing more

[66] Stuart to Robinson, Andover, 1.12.1828, 9.2.1829, ERP, F-5 (transcribed version); cf. Williams (1999), 62, 97, 157–58.

[67] Taylor, C. (1832). Antoine Augustin Calmet (1672–1757), French Benedictine theologian, historian and biblical scholar; studied Hebrew. His first exegetical work, on which rests his reputation, is the commentary *La Ste. Bible en Latin et en Français avec un Commentaire Littéral et Critique* (1707; 4th ed., 1729). Charles Taylor (1756–1823) was an English engraver, a man of letters, and biblical scholar. He revised Calmet's Dictionary of the Bible, which he began to publish anonymously in 1797, crediting himself only as publisher and engraver. A fourth edition was published in 1824. He was acknowledged as editor only after his death. Cf. ERP, Hamilton 5, R65.

[68] Robinson (1832); (1832a). Cf. Williams (1999), 190.

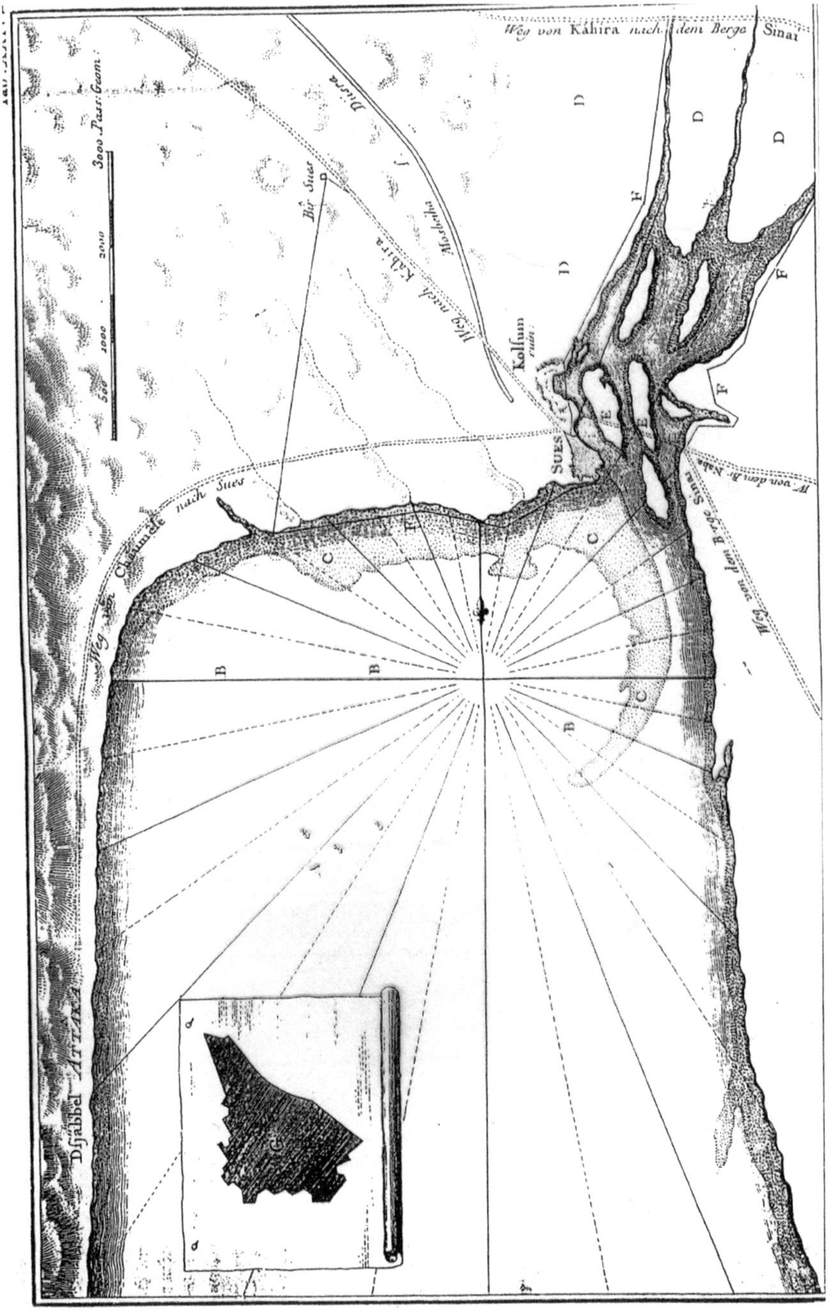

Ill. 6. Carsten Niebuhr, *Grundriss von dem äusersten Ende des Arabischen Meerbusens und der Stadt Sues* [plan of the outer end of the Arabian bay and the city of Suez].

current.[69] The fact that 'the enterprising Publishers put [this edition] into my hands, with the request that I would revise it'[70] is one of the best proofs of Robinson's status in 1832 as America's leading Biblical scholar, a point that will be addressed later in detail.

Three key questions frame the research of this period in Robinson's life: How did Robinson develop his approach to the study of Scriptures as a sacred, historic-geographical document, whose narratives and facts cannot and should not be questioned, rather uncovered, described and explained? How did the idea for the Holy Land expedition originate? When, where, how, and with whom was the project developed until it came into effect?

Robinson's 1832 meeting with Smith was a turning point, according to Williams, for his life, career, and publications, and the most influential event paving the road to Palestine. 'Another seemingly minor event which bore much more valuable fruit,' he calls it.[71] Robinson could not have known when he was Smith's tutor at Andover that Smith would become the most qualified person for his needs. Smith, a devoted missionary with established connections and friendships with local leaders, superb control of the Arabic language and its dialects, and knowledge of local Near Eastern customs and habits was, in Williams' words, 'a godsend' to his tutor. So critical was their encounter that Williams cites the relevant page from Robinson's book[72] and adds, 'the die was now cast. The question was only when.'[73] In time, however, it became clear that the question was also 'how.'

The problem of 'how' generates a second list of questions. A project of such proportions, an expedition to the East in the fourth decade of the nineteenth century, required rigorous planning and preparation. Preparation issues ranged from acquiring scientific proficiency with regard to inquiry, setting objectives, and data collection, including knowledge of field instruments and other equipment, to how and from where to obtain their equipment, to resolution of complicated financial and technical matters to coordinating their timetables given other, primary obligations. Moreover, Robinson and Smith had to pre-

[69] Taylor, C. (1832), facing 409. For the original: Niebuhr, C. (1969), Tab. XXIV, facing 410. Cf. Hansen (1864), 152–89.

[70] Robinson's preface, Taylor, C. (1832), iv.

[71] Williams (1999), 192.

[72] Robinson (1841), I, 2; Williams (1999), 193–94.

[73] Williams (1999), 194.

pare for such a complicated task living on different continents, with Smith moving constantly between missionary stations. Naturally, the only mode of communication was by letter. They had to depend almost solely on a highly unreliable system of the occasional ship sailing between New York or Boston and the eastern Mediterranean. Robinson and Smith address the challenges of communication extensively in their letters.

However, with regard to the issues of scientific inquiry and expedition preparation, one finds only hints, here and there, in Robinson and Smith's publications and later studies. It seems that the only way to uncover the sources for their ideas and details of their planning is by studying their extensive correspondence, deciphering hundreds of letters and extracting from them the relevant information. The most comprehensive study of Robinson and Smith's correspondence to date was done by Jay Williams, who had access to all existing letters housed at Hamilton College Special Collections. Williams did not work with Eli Smith's papers, so he had only limited correspondence concerning the expedition and publication of the *Biblical Researches*. Fortunately for curious scholars, Smith (and his relatives?) was much more diligent than Robinson in building his archive, scrupulously filing nearly each and every document. Smith used carbon paper when writing in order to preserve copies of his letters, a fact which makes them almost impossible to read, but is of tremendous value for scholars. His archive also contains dozens of Robinson's letters to him, sorted chronologically under Robinson's name.[74] Smith's letters to Robinson are spread among his general correspondence, numbering hundreds of letters and organized chronologically.[75]

In his letters to Smith as well as to some of his teachers and relatives, Robinson's primary focus is the scholarly preparation for the expedition. It seemed to be of greater importance to him, certainly more than the technical issues. This focus led Williams to summarize: 'Thus, in a few brief years Robinson accomplished one of his major goals: to provide for American Biblical scholars the tools necessary for well-informed, up-to-date study (…). American Biblical scholarship came of age and gained a recognized preeminence which it has never relinquished.'[76]

74 ESP, ABC 60, Robinson, Edward (63).

75 ESP, ABC 60, II. Letters from Eli Smith, (105).

76 Williams (1999), 197.

2.d. Expedition preparations

When did Robinson conclude that this might be his opportunity to fulfill his longstanding desire to travel to the Holy Land? From the letters, it seems that though the desire long existed, Robinson made a decision only in January 1837, following his visit to New York when the seminary agreed to his conditions and he accepted the position. It is not clear whether he went to New York with a preconceived plan or whether he developed the idea while negotiating with the UTS governors.

On July 17, 1837, nearly five years after Robinson and Smith's pivotal 1832 meeting, Robinson set sail with his family from New York to Europe. In the year prior, Smith, already a leading ABCFM Near East missionary, lost his first wife Sarah, herself 'famous among missionaries for her dedication to the education of Syrian women.'[77] In Beirut, Smith had been working toward his lifetime goal of establishing an Arabic printing house. Due to Sarah's prolonged illness, on June 11, 1836, the Smiths sailed for Smyrna hoping it would improve her health. However, their ship wrecked near the shore of Caramania (southern Turkey) and it took over a month to reach their destination. The adventure-filled journey most certainly aggravated Sarah's condition. Being very ill upon arrival, she was housed in a village residence of the local ABCFM missionaries, John Bailey Adger (1810–99) and his wife Elisabeth Keith (1813–90). Sarah died there on September 30, 1836.[78] Sarah's death was a great loss for Eli Smith as he writes in his letters, this to his fellow missionary George Whiting in Jerusalem:

> You know my heart you know my occupations. I do most sincerely thank you for your very kind expressions of sympathy in that unspeakable loss which I have sustained (...). All I can say, comes so far short of expressing the feelings of my heart, that I have little inducement to speak, & my wounds are yet so fresh that tears generally soon prevent my utterance (...). Often when I am alone, the image of that loved one, in some scene of her benevolent labors, or of her last sufferings comes before me with so much freshness, that my heart seeks immediate relief in sob & tears.[79]

[77] Makdisi (2008), 162, 172–73. See Smith (1845).

[78] 'Death of Mrs Smith,' Isaac Bird's diary, IBP-YSM, Box 2, Folder 25: Journal 1827–60; Hooker (1845), 320–60; Adger (1899), 105–07. For Adger's autobiography: Adger (1899), 70–129. Cf. chap. 5.c.

[79] Smith to Whiting, Beirut, 5.6.1837, ESP, ABC 60 (105/5); Smith to Whiting, Smyrna, ESP, ABC 60 (105/5); Smith to Bird, May 1837, ESP, ABC 60 (105/5). Edward W.

Perhaps Smith sought solace from the death of his beloved and cherished wife in the intensity of the expedition. Reading the biographies of European missionaries in the East during the first half of the nineteenth century, it is a wonder that they were able to overcome the frequent losses they endured, particularly the deaths of their children and wives.

In the *Biblical Researches*, Robinson moves directly from events in 1832 to 'embarking with my family' in 1837, skipping the entire period and planning of the expedition![80] But it is quite clear that during this interim period Robinson never gave up the idea of travelling. In 1834, Talvj shared with an acquaintance that the family was planning to arrive in Germany in the autumn of 1835 or spring of 1836.[81] In 1835, Robinson informed his lawyer that he was planning to go abroad and might be absent for several years.[82]

Robinson formally announces his acceptance of the position of Professor of Biblical Literature at New York Theological Seminary (later UTS) in a January 20, 1837 letter to the board of directors of the seminary.[83] Following a long discussion of his acceptance of the seminary's theological approach, and after adding suggestions for curricular changes, Robinson writes, 'there remains a single point which is personal to myself.' Here, Robinson outlines his two-pronged proposal for a family trip to Europe, during which he would purchase books for the seminary library (see chap. 4.e.), to be followed by his long-awaited expedition to Palestine, 'with reference to the preparation of a Biblical Geography.'[84]

Robinson mentions his plans for the expedition publicly soon after in an early February 1837 letter to Reverend Asahel Strong Norton (1765–1853), minister of the Congregational Church in Clinton, NY. Norton was Robinson's only surviving uncle out of his mother's two sisters and five brothers and had been very close to Robinson ever since his studies at Hamilton College.[85] Robinson writes, 'you indeed, in the

Hooker (1794–1875), an Andover graduate who in 1845 published a book about Sarah Smith, wrote a long letter to Eli Smith on July 9, 1838, discussing the publication, noting that he has in his possession some 700–800 letters that he is sorting according to the planned chapters: Hooker to Smith, Bennington, Vermont, 9.7.1838, ESP, ABC 60 (38).

80 Robinson (1841), I, 1–2.

81 Milović (1941), 94.

82 Weigel Williams (2007), 104.

83 Cited: Prentiss (1889), 246–48.

84 Prentiss (1889), 248.

85 Reverend Asahel Norton of Clinton, NY and Professor Seth Norton, president of Hamilton College from 1816, were the brothers of Elisabeth Norton, Robinson's mother; cf. Robinson, M. A. (1862–64), 146.

providence of God, are the only one left to me, whom I ever viewed as a parent, – all the rest of my family and friends, whom I have known and loved, already sleep in death.'[86] Norton was probably the first to hear from him about the planned expedition, perhaps only after Talvj.

Robinson describes the changes in his life, leaving Andover, his publications, his loss of two sons and the birth of another, and then describes his future plans:

> In December last I received from the directors of the New York Theological Seminary an anonymous invitation to take charge of the Professorship of Biblical Literature in that institution with a salary or $2500 per annum. I went to New York in January to inquire into the state & prospects of the Seminary, before arriving to any decision. Nursling under the circumstances in which I was pleased, this seemed an offering in Providence, with which I might enter with the hope of being useful; **all my previous studies & pursuits having prepared me for just such a station** [emphasis added], if for any. I therefore accepted the appointment, on condition of having leave of absence after July next, in order to go abroad for the purpose of procuring a library and, if possible, of visiting Palestine.[87]

Robinson had declined many former scholarly appointments;[88] this is the first he accepted, seeing in it 'an offering in Providence.' Being cautious, he first went to New York to confirm the details and reassure himself. In his letter, he considers the seminary's theological background and approach, as well as its financial stability and seriousness of its founders:

> The Seminary doubtless is the (?) of the churches there, and rests for its support & direction upon men who are very serious of Christian enterprise & liberality in that city. They are disposed to do every thing which may seem important & expedient, to build up an institution which shall sustain a high character & exert a wide influence (...). If properly conducted, there can be no doubt that the Seminary, by the blessing of God, will prosper & exert a happy influence at the churches & in the city. I go to try; and if God gives success, I see not why that spot should not become my permanent home.[89]

86 Robinson to Norton, Boston, 9.2.1837, ERP, ER 1.13.

87 Robinson to Norton, Boston, 9.2.1837, ERP, ER 1.13.

88 See for example previous offers: ERP, 1816 (2).

89 Robinson to Norton, Boston, 9.2.1837, ERP, ER 1.13.

The tempo of events surrounding his decision was especially quick, even hasty. Robinson, who was residing in Boston, received the initial offer from the UTS Governors in December 1836. He went to New York the following month 'to inquire' and present his conditions, to which his future employers agreed 'unanimously & cordially.' As early as 'the next week,' from the letter to Norton dated February 9, he went to New York 'to enter upon my duties there.'[90]

In the years following the pivotal 1832 meeting, the expedition plans were also always forefront in Smith's mind. During his 1834 and 1835 travels in Palestine (the latter trip together with his wife[91]), he began preparing different lists in Arabic, 'of a large part of the towns, villages & ruins in Palestine, east of the Jordan & north of Damascus as far as Hamah.' In a letter to Robinson written much later, in January 1838, Smith specifies that 'I had yourself in mind in procuring it.'[92] Although some of Smith's travel diaries were lost in the Caramania Shore shipwreck,[93] other handwritten lists survived that became an important part of the *Biblical Researches*, a special appendix in the third volume.[94]

Yet in his letters from Beirut in mid-1837, to brother missionaries Bird (on May 15th) and Whiting (June 5th), Smith does not mention his travel plans, nor the fact that Robinson was already making his way to the East. The letters, exceptionally difficult to read, deal with routine mission issues and with personal matters, as all his brother missionaries sent Smith their condolences upon his wife's death. Robinson is not mentioned![95] In a subsequent letter addressed to Whiting written from Smyrna on September 6, 1837 (two days earlier Smith baptized John Adger's girl[96]), Smith raises an entirely different issue. Smith confiden-

[90] Robinson to Norton, Boston, 9.2.1837, ERP, ER 1.13; Dearman (1991), 168–69.

[91] Sarah Smith's letters, in Hooker (1845), 237–61.

[92] Smith to Robinson, Smyrna, 6.1.1838, ESP, ABC 60 (105); Mehitable Butler Smith, Journal of a journey from Beirut to Jerusalem. IBP-YSM, Box 2, Folder 15.

[93] Hooker (1845), 327.

[94] Smith (1841); Lists of the Arabic Names of Places in Palestine and the Adjacent Regions, ESP, ABC 60 (119) B; Rules of the orthography of Arabic and Turkish names and words, in Roman letters. Adopted by the Syrian Mission of the ABCFM, April 1838, ESP, ABC 60 (116). The list has three sections: 1. Names of Places in the Districts visited or seen during the Journey of E. Robinson and E. Smith in 1838. 2. Names of Places, chiefly in the Pashalik of Damascus, obtained during a journey in the Spring of 1834. 3. Names of Places, chiefly in Mount Lebanon, or Jebel Libnân. Cf. Robinson (1841), II, 106.

[95] Smith to Bird, Beirut, 15.5.1837; Smith to Whiting, Beirut, 5.6.1837, ESP, ABC 60 (105).

[96] Adger (1899), 107.

tially expresses his opposition to John D. Paxton's request to join the mission, which was declined by the 'Commission.'[97] Paxton (1784–1868), a strong abolitionist from Kentucky, went on a grand tour through Europe to the East, arriving in Beirut in mid-June 1836 after visiting Turkey, and married the widow of ABCFM physician Asa Dodge (?-1835). Dodge had arrived with his wife in Beirut in February 1833 and accompanied Smith on his spring 1834 travels in the Hauran. He then went to Jerusalem in late 1834, where he died in January 1835.[98] Paxton resided in Palestine and Syria in '1836, 7 and 8,' travelling extensively in Palestine as well as in Lebanon and Egypt.[99] He was living in Jerusalem during Robinson and Smith's visit, and even joined them on one of their excursions.[100] Naturally, Paxton's rejection sparked much interest and is mentioned in missionary correspondence.[101] As late as spring 1838, missionary printer Homan Hallock (1803–94) asks Smith to give a 'very kind etc. love to Bro. Paxton & his wife.'[102] Only in his letter to Whiting of December 16, 1837, a mere month before leaving for Egypt, did Smith find it necessary to reveal his upcoming plans to his fellow missionaries.

The intricate story of the expedition's evolution does unfold in the intensive letter exchange between the two 'heroes,' although it is not simple to reconstruct in detail, for two primary reasons. First, not all the letters survived, so there are gaps in the available information. Second, there is a significant time gap between letters, given that the postal service was dependent upon sailing ships; six weeks was a relatively short time for a letter to travel between two major cities such as Berlin and Smyrna.[103] In 1835, C. Brown writes to Smith from Alexandria that 'yesterday the American Brig Alexander, Captain Pendleton, arrived here from Boston in fifty-five days;' in 1838, he writes that 'the American Brig Regulator Capt Wm D. Phelps arrived here on the 26th ult, via Malta, in 73 days from Boston.' He calculates the best schedule connections thusly, 'the passage is about sixteen days from England to Malta and four

[97] Smith to Whiting, Smyrna, 6.9.1837, ESP, ABC 60 (105). See Smith's very long letter to Paxton summarizing the issue: Smith to Paxton, Smyrna, 24.1.1837, ESP, ABC 60 (105/5).

[98] Kawerau (1958), 255–56, 285–86, 412.

[99] His summary for the period in the East: Paxton (1839). Cf. Paxton (1870).

[100] Robinson (1841), II, 107.

[101] e.g. Whiting to Smith, Jerusalem, 4.6.1838; Elmira, NY, 22.8.1839, ESP, ABC 60 (91), Whiting, George B.

[102] Hallock to Smith, Smyrna, 28.4.1838, ESP, ABC 60 (32).

[103] Smith to Robinson, Smyrna, 8.12.1837, ESP, ABC 60 (105/7).

from Malta to Alexandria, she remains here six days. By these boats they receive their letters from England in about twenty-one days.'[104] Surprisingly, numerous examples of letter exchanges demonstrate that it always took about two months and one week for delivery. Elisabeth Robinson sent letters from Albany, NY to her brother in the East. A letter sent on December 27, 1837, via Le Havre in France, was received by Robinson in Cairo sixty-six days later. Another letter, dated February 26, 1838, was sent through Gliddon to Robinson in Cairo, who received it after sixty-eight days. Elisabeth sent another letter on April 28, 1838, to Robinson in Beirut, which he received sixty-seven days later.[105]

Hermine Weigel Williams, who prepared the catalogue for Robinson's archive and published his letter exchange with his sister Elisabeth, is an informative source on the challenges of nineteenth-century overseas correspondence and letter deciphering:

> Here perhaps I should add a word or two about overseas mail. Until about the middle of the 19th century, it was common practice for the RECIPIENT in Europe to pay the postage for overseas letters. This placed a considerable burden on the person in America writing the letter. As Edward's brothers explained, if you couldn't think up something of importance to fill up every inch of the paper then you shouldn't bother writing at all – and as a consequence they wrote very little to Edward until the law was changed. Elisabeth, on the other hand, did what others of the period did to maximize space of the standard sized sheet folded to create 4 pages: she filled three pages writing in the usual horizontal fashion; then she turned the paper 90 degrees and on pages two and three wrote over the previous text. So the finished product had page one with just the single horizontal lines, page four was reserved for the envelope informat, and pages two and three had both a horizontal and a vertical text. Now this is not as great a problem to decipher as you might think so long as the handwriting is extremely neat and the ink is neither too light nor too dark. Unfortunately, there are many letters in the collection for which the optimum situation does not hold true and this has caused a great deal of difficulty in deciphering the texts.[106]

[104] Brown to Smith, Alexandria, 29.7.1835, 12.9.1838, ESP, ABC 60 (16): Letters from C. Brown.

[105] Elisabeth to Edward Robinson, Albany, 27.12.1837, 26.2.1838, 28.4.1838, ERP, H-1, scanned p. 2.

[106] Remarks made by Hermine Weigel Williams on the occasion of the 200th Anniversary Celebration, October 9, 1994, Hamilton College Chapel, ERP, 1816 Robinson- 3.

An additional problem was figuring out how to send a package from Cairo to Jerusalem in 1837. The only solution was to find a dependable carrier. Consul Gliddon relates one colourful example:

> A similar disappointment had attended a saddle which at Mr Whitings request was commissioned, and made many months ago, at Cairo; where it had been waiting the frail, but reiterated promise of a nephew of Mr Murad Arutin, to take it with him on an overland journey, which he pretended he was to undertake from day to day. Seeing no end to the delusion, I have at length had it sent down with a determination to forward it by way of Beyroot; and very fortunately an opportunity offers for so doing, through the return of a Polish Gentleman belonging to Ibrahim Pasha's staff, who had kindly promised to place it in the Lazaret, at your disposal. The Gentleman's name is Schultz, and, I may add, that be his character, and talents, he is entitled to every respectable consideration.[107]

Finding a reliable courier was a regular necessity for the consuls. It caused endless difficulties and misunderstandings, even arousing anger, and the consuls refer to it repeatedly in their letters. Consul John Gliddon, posted in Alexandria, writes to Smith in Beirut, 'your letter for the US I sent by the packet via London, my connexion with Gibraltar being at that moment vacillating. Though by such discretionary proceeding the postage be increased [*sic*], your letter will have the advantage of greater certainty, and most probably of greater speed.'[108]

Nearly half a year after his letter detailing the seminary appointment, having solidified his plans as much as possible, could Robinson send them to his uncle Norton. Robinson went to Albany, NY, 'with the intention of proceeding to Utica & Clinton, chiefly in order to have the pleasure of seeing you once more ere my departure for foreign lands. This purpose, however, was frustrated by continuous rain on the only days I could devote to the excursion…'[109] This outcome was typical for the time-conscious Robinson, who calculated every day, as he did on his expedition to the Holy Land.

The primary focus of the letter, naturally, was an account of his itinerary. Robinson and his family embarked for Liverpool on Monday, July

[107] Gliddon to Smith, Alexandria 30.3.1837, ESP, ABC 60 (28): Gliddon, John. Lieutenant Colonel Auguste Schultz, 'Joussef Aga,' served in Muhammed 'Ali's army. See Alderson (1843), 53–54; Goren (2011), 100. Arutin, a wealthy Armenian and resident of Jaffa, served as American consular agent in Jerusalem and Jaffa from 1835–42 (Kark 1994, 100–05, 345–47).

[108] Gliddon to Smith, Alexandria, 30.3.1837, ESP, ABC 60 (28): Gliddon, John.

[109] Robinson to Norton, New York, 1.7.1837, ERP, ER 1.12.

17, 1837, on the *England* with Captain Waite at the helm, who 'informs me that he has never been for (?) more than 19 days in going out.' After travelling through London, Amsterdam, and Frankfurt and leaving the family in Germany, the plan was to continue to 'Egypt, Mount Sinai, and Palestine, - visiting also if possible such points, as Constantinople, Smyrna, Athens, etc.' Robinson also planned his return, 'if God prospers our journey, we hope to return in the autumn of the next year, so that I can enter upon my duties in the Seminary at the opening of the term in October.'[110] Yes, Robinson intended to start teaching in October 1838.

In the meantime, Smith had been tremendously busy casting the types for the Arabic printing press, this in addition to his daily Mission duties (see chap. 5.a.). It seems that a few more months passed before Smith realized that Robinson was already in the Middle East; thus, he hardly had time for even the minimum of preparations. It is quite clear that Robinson's plan was contingent upon Smith's assistance. Moreover, Robinson was almost certain that, based upon their meeting five years earlier, Smith was anticipating the start of their journey together. However, a careful search did not uncover a single letter between them prior to late 1837! Robinson, whose passport was issued on June 21, 1837, sent his first letter to Smith on July 11, six days before departure![111] Unfortunately, this and a second letter from Robinson, dated September 13, 1837, were not found in either of their archives.

On November 10, 1837, as Robinson was leaving Germany ('I set off from Berlin on the 13th Nov. by way of Halle'[112]), Smith, in a very long letter from Smyrna, writes, 'your two letters of July 11th & Sept. 13th were received at this place too late for an answer to reach you at Berlin, & I therefore write now to the care of Dr Robertson of Syra not doubting that the course of the Adriatic Steamer will bring you to that island.'[113] Robinson comments in the *Biblical Researches* that only during his stay in Athens on his way to Egypt, from December 8–25, 1837, 'I was able to have communication with my friend, the Rev. Mr Smith, who was then in Smyrna.'[114]

[110] Robinson to Norton, New York, 1.7.1837, ERP, ER 1.12, all citations.

[111] Robinson's Passport, ERP, J.2; Smith to Robinson, Smyrna, 10.11.1837, ESP, ABC 60 (105).

[112] Robinson (1841), I, 3.

[113] Smith to Robinson, Smyrna, 10.11.1837, ESP, ABC 60 (105).

[114] Robinson (1841), I, 15–16.

Edward Robinson's travel itinerary until meeting Smith in Cairo:

UNITED STATES OF AMERICA
To all to whom these Presents shall come, Greeting
DESCRIPTION
United States of America
DEPARTMENT OF STATE
GRATIS

Vû au Consulat Des Etats Unis D'Amerique en Syrie & Palestine bon pour se rendre a Smirne par voie D'alexandrie.
Beyrout 7 Juillet 1838
Le Consul des E. U.
J. Chasseaud

Ill. 7. Robinson's passport, June 1837, signatures of Chasseaud, American Consul in Beirut and (in Arabic) 'Akkad in Tyre, writing that Robinson reached Tyrus on his way to Beirut.

Dates of consular stamps:[115] Passport issued June 21, 1837. Berlin, November 9, 1837; Leipzig, November 17, 1837, November 19, 1837; Prague, November 20, 1837; Trieste, November 23, 1837, November 30, 1837 (leaving?); Greece, November 30, 1837; Alexandria, January 2, 1838.

1837: July 17 – embarks from New York.
August 18 – disembarks in Liverpool.
September 8 – London.
November 13 – leaves Berlin through Halle.
December 1 – embarks from Trieste.
December 4–5 – Corfu.
December 9–25 – Athens.
December 25–26 – Syra.
December 27–28 – Canéa, Crete.
December 30 – reaches Alexandria.

[115] Robinson's Passport, ERP, J.2.

1838: January 5 – Nile journey from Alexandria to Cairo.
January 10 – arrives in Cairo.
January 19 - leaves Cairo sailing north up the Nile.
February 9–11 – Thebes.
February 16 – Smith arrives in Alexandria.
February 24 – Smith arrives in Cairo.
February 28 – Robinson returns to Cairo.
March 12 – Robinson and Smith leave Cairo for Suez.

In his letter from Smyrna, Smith was referring to Church Missionary Society (CMS) American emissary John J. Robertson who was stationed in Syra, an island in the Cyclades, located in the centre of the Aegean Sea about 150 km southeast of Athens. Syra functioned as a central point for the maritime transportation between Malta, Piraeus, Constantinople, and Alexandria. Whereas 'Athens was occupied by the Protestant Episcopal Church of America,' the CMS chose Syra and Smyrna as their main stations in the Greek east.[116] Notwithstanding, in 1838, Robertson convinced his superiors that Constantinople was a more suitable seat for the mission, and the station was moved to the Ottoman capital.[117]

Robinson embarked from Trieste on December 1, 1837, and after exploring Athens, visited the island of Syra and 'our kind American friends, Dr Robertson and his family.'[118] It is not clear where Robinson received Smith's first letters, in Syra, as Smith writes in his letter, or in Athens, as Robinson writes in the book. Robinson's earliest letter, included in the impressive collection of his letters in Eli Smith's papers, is dated Alexandria, January 3, 1838. As noted above, Smith's letters of July 11 and September 13 of the previous year have not been found to date, and there is no mention of former letters! Consequently, Williams' supposition that 'Smith and Robinson must have been in correspondence and decided upon 1837–38 as the propitious moment,'[119] while very logical, lacks proof in the existing letters! Smith writes in his November 10, 1837 letter, 'it gave me more pleasure than I can ex-press to learn that Providence has at last put it in your power to set out upon your long desired visit to

[116] Robinson (1841), I, 17; Stock (1899), 350; Oliver (2014), 131–32, 274–76, Syra and its centrality are mentioned repeatedly, e.g. 233–34.

[117] Stock (1899), 350; Kawerau (1958), 212, 567–68, 576–77; Oliver (2014), 274–80.

[118] Robinson (1841), I, 17.

[119] Williams (1999), 205.

Palestine.'[120] Smith indicates here, quite clearly, that he only now discovered that his teacher and friend was already on his way! Smith might have heard rumours of Robinson's impending arrival through other means, such as oral communications from missionaries traveling to and from Lebanon, but his letter appears to tell the story. This letter, dated November 10, 1837, is probably one of the most detailed documents describing Smith's whereabouts as well as his difficulties and hesitations concerning the planned expedition. He admits that it seems impossible for him to keep his promise from 1832 and accompany Robinson! Smith writes:

> Since our meeting and conversation in Andover, it has never been lost sight of by me, though I've been worried that your problematic health will prevent you from actualizing it. The interest I felt was not altogether destitute of a personal character. I have never forgotten my conditional engagement to be your travelling companion, but have always hoped and intended to be in readiness to journey with you, whenever Providence might lead your steps this way. It is a great disappointment to me that such an arrangement is now wholly out of my power. I need no arguments to show the importance of your object, nor that what little I might contribute towards its accomplishment would not be at all a diversion from the great aim of every Christian's heart, the advancement of the course of divine truth. The particular department of biblical geography is one in which I have long felt a deeper interest than in almost any other branch of biblical literation, & it is one I am more & more convinced in which greater improvements are to be made.

This is only the opening, and it seems most appropriate to present Smith's predicament in his own words:

> But what can I do? I am charged by my mission with a work to which I am absolutely tied for months to come, to carry on upon (?) the accomplishment of which some of our most important operations are suspended. I have on hand the making of a new & improved format of Arabic type which my brothers have deemed of so much consequence as to (?) their sending me away to own my other labors for a year or more with the chance of its calling me ever to the U. States & until it is done, not only my own missionary labors are suspended, but all the (?) of the mission which are connected with the press must move on very slowly.[121]

[120] Smith to Robinson, Smyrna, 10.11.1837, ESP, ABC 60 (105).

[121] Smith to Robinson, Smyrna, 10.11.1837, ESP, ABC 60 (105), both citations.

The important task forcing Smith to suspend all other obligations was the Arabic printing press. The Arabic print revolution is a focus of modern research; however, the exact role and importance of the missionary printing houses is disputed. The ABCFM followed in the footsteps of the CMS printing, which, in 1826, began publishing books from its Beirut house.[122] The ABCFM appointed the printer Hallock 'assistant missionary for the mechanical operation of the press,' and Hallock went with Smith to Malta in 1826. Later, he went to Smyrna on a special mission to prepare the types for the Arabic print. Though the story of the Arabic printing house has been studied extensively, it is helpful to refer here to some points that had direct influence on Robinson and Smith's travel and research.[123]

The work on the printing press is mentioned in various letters dating as early as 1834, describing clearly many obstacles. In spring 1836, Smith confides in his close friend Bird, 'the fact is, I have got tired of wearing myself out, & spending money, in trying to work this press, in so faulty a condition. If I am to spend my life in connection with it, I must be able to accomplish something.'[124] A letter from December 1837, containing 'a somewhat detailed account of my proceedings in reference to the Arabic type' is decidedly important and informative, detailing the difficulties of the initiative. Hallock, writes Smith, is at this very moment 'cutting the punches here and needs me by him from morning to night.' Smith also had to be ready at a moment's notice to travel to the US for various needs connected with the complex task.[125] The ABCFM had full confidence in Hallock and spent a considerable amount of money on the project. One condition was that the prepared types undergo full testing (possibly in the US, the location is unclear) before becoming an operational business. The fact that the work was carried out in Hallock's house made it possible for Smith to oversee it daily, and because it was an ABCFM financed project Anderson was also heavily involved. Smith gave his models of the type to Hallock:

> Just at that time, Mr Anderson's letter respecting retrenchments came to hand, putting it out of my power to go through Europe & rendering it highly desirable if not necessary (…) that the work should be done in

[122] Ayalon (2016), 25–27, and sources there.

[123] Coakley (2003); Tibawi (1966), 42, 49. For the history of steel engraving cf. Hunnisett (1998), 240–64. Cf. also Auji (2016).

[124] Smith to Bird, Beyroot, 28.4.1836, IBP-YSM, MS 82 Box 1 folder 13.

[125] Smith to the Brethren of the Syrian Mission, Smyrna, 15.12.1837, ESP, ABC 60 (105/7).

> our own foundry. Considering, also, how near the whole thing then was to testing & the amount of specific corrections & dissections I should have to suggest, I confess that I shrink from the responsibility of making a contract with a strange man, while there was the possibility, upon avoiding it. I also doubted whether any men without Mr Hallock's machine would be able to imitate my models so as to avoid the necessity of making the combinations (…). [Hallock] had spent a great deal of time on the work, was wholly devoted to it; & (?) with it perhaps more exclusively than he ever was with any new thing before that time in his life. I have evidence that he denied himself & his family & his other business the doing of a great many things, which he wished much to attend to. – In all these circumstances, brothers I wanted your counsel, which I could not have. And without it, I have acted according to the best of my judgment, after much anxiety, at the expense of many hours of sleep, with much prayer I hope you will approve of the course I took. The task is great, much greater than I at first supposed.[126]

In this rather convoluted letter to his missionary brothers, Smith reveals his deliberation over how to proceed with cutting of the punches and casting of the types for the Arabic printing press. His commitment to finishing the printing press led Smith to abandon the idea of accompanying Robinson, wishing the latter 'a rich harvest' and 'repaying for all your fatigues'! Smith's November 1837 letter to Robinson includes a long list of practical suggestions: Robinson must first visit Egypt, and then Palestine in April and May, the best months for travel, and include a visit to Petra. This itinerary would also avoid the two-week quarantine on travellers arriving in Jaffa by boat. In Jerusalem, Robinson could expect the help of the missionaries, of whom Smith recommends Nicolayson, 'a thorough scholar & can give you much information.'[127]

As the key Protestant missionary in Jerusalem, naturally, Smith assumed that Nicolayson was the best person to help Robinson upon his arrival in Jerusalem. 'I,' he adds, 'should stay here, in Smyrna, probably until January [of 1838], and would like very much to meet you here. You can do it by taking a boat from Syra, but this would delay you in a week or even two, and I hardly expect you to do that.' Yet, this was only the beginning of a drama revealed in their correspondence that intensified as 1837 drew to a close and Robinson approached the East. From the letter content, it is clear that there were other letters that did not sur-

[126] Smith to the Brethren of the Syrian Mission, Smyrna, 15.12.1837, ESP, ABC 60 (105/7).

[127] Smith to Robinson, Smyrna, 10.11.1837, ESP, ABC 60 (105).

vive. However, those that did, mainly in Smith's archive, are sufficient to understand the unfolding events. Still in Smyrna, Smith writes to Robinson on December 8, taking advantage of the presence of a French steamer, 'though I have not much to say, in addition to what my last contained.' He mentions Robinson's letter dated Berlin, October 21 and writes further, 'I cannot describe the pleasure it would give me, to see you, if you'll be able to come to Smyrna. (...) yet having travelled a great deal in Palestine, my hints might not be entirely useless.'[128]

At this stage, it seems that Robinson's plan and arrival were somewhat of a secret. Smith chose to announce Robinson's travel invitation only in his letter 'to the Brethren of the Syrian Mission' dated December 15, while Robinson was touring Athens, the same letter whose primary purpose was to describe the situation with the Arabic types. In the letter he adds, I am 'sending you the enclosed correspondence between Prof. Robinson & myself.'[129] Three days later, Smith writes to Whiting, then in Jerusalem, telling him that Robinson 'is now at Athens on his way to Egypt & Palestine for the purpose of a Charting & verifying parts from a biblical geography.'[130] This sentence hints at the fact that Robinson and Smith had not previously discussed the details, nor even the objectives of the expedition, since Robinson previously had mentioned verifying Biblical geography, but not 'charting,' though the idea might have been brought up in their 1832 meeting. This is also the first hint that Smith was about to reverse his decision. He adds that Robinson, accompanied by two other travellers, 'strongly [insists] upon having my company.' As for Hallock, Smith concludes that in about a month his work will reach 'as good advantage without as with (me?) & I am than to leave & join the party at Cairo about last of Feb.'[131]

Understanding that he would be able to join the expedition, Smith reacted quickly, and that very same day sent the decisive letter to Robinson stating, 'that the postponement of the time proposed for me to join you in your present journey, promised to put it within my power to comply with your invitation.' The letter, which awaited Robinson in Syra, announces that due to the change in Smith and Hallock's plans, Smith will join Robinson in Cairo when he returns from his planned visit to

[128] Smith to Robinson, Smyrna, 8.12.1837, ESP, ABC 60 (105/7).

[129] Smith to the Brethren of the Syrian Mission, Smyrna, 15.12.1837, ESP, ABC 60 (105/7).

[130] Smith to Whiting, Smyrna, 18.12.1837, ESP, ABC 60 (105/7).

[131] Smith to Robinson, Smyrna, 18.12.1837, 19.12.1837, ESP, ABC 60 (105/7).

Upper Egypt. Smith is also very practical: 'In the meantime, please keep me informed of whatever I need to know respecting your movements & wishes & tell me in what way I can serve you.'[132] Robinson wrote in the *Biblical Researches* that he was inclined to accept Smith's persistent invitation to join him in Smyrna and travel together to Egypt. But since Smith could not leave immediately and there was a quarantine of three weeks on visitors arriving in Smyrna from Alexandria, they decided that Smith should join him in Cairo in late February.[133]

The second issue influencing Smith's decision to join Robinson was financing. It seems that in one of the unrecovered letters Robinson took it upon himself, a fact that reassured Smith and made joining the expedition possible for him. After receiving Robinson's letter dated December 9 from Athens, the very next day, December 19, Smith writes in a return letter, 'Your last [letter] has induced me to review my decision,' he writes, noting that he will ask for Anderson's approval. Bearing in mind, Smith continues, that 'the defraying of my expenses also without drawing upon funds of the mission; which you had not mentioned before, happens to be a material consideration, mainly for the Board of the ABCFM.' Smith adds that Hallock 'sends his respects to you, will do everything he can, to get his work advanced so that I can leave in (?) to meet you in Cairo.' Smith hoped to leave Smyrna on January 19, expecting to reach Cairo six days later, which he actually did. He also repeats his letter from the previous day, 'I need not tell you how much satisfaction it offers me to be able to make this arrangement, so as not to disappoint you, nor be disappointed myself. I only fear your expectations of assistance from me are far too highly raised, ever to be realized. I beg you will do me the favor, to lower them among degrees before I reach you (...).'[134] When in Egypt, he adds, 'we both should expect and use the help of Consul Gliddon in Alexandria. Another letter will go to Missionary Thomson in Beirut, who will help in establishing the needed contacts for us through the journey (...).'[135]

In addition to the scientific mission, Smith also hoped 'to find many openings for missionary labor, & I trust you will allow me to act the missionary wherever I have opportunity.' The letter ends with the same

[132] Smith to Robinson, Smyrna, 18.12.1837, ESP, ABC 60 (105/7).

[133] Smith to Robinson, Smyrna, 8.12.1837, ESP, ABC 60 (105/7); Robinson (1841), I, 15–16.

[134] Smith to Robinson, Smyrna, 18.12.1837, 19.12.1837, ESP, ABC 60 (105/7).

[135] Smith to Robinson, Smyrna, 18.12.1837, ESP, ABC 60 (105/7).

request as the letter of the previous day, 'please let me know in what way I can serve you before leaving here.'[136]

Smith did not lose any time in pursuing his missionary obligation. From Alexandria, he sent a letter to Robertson in Syra encouraging him to come and look into the possibility of establishing a CMS station. Smith had obtained 'some statistical information' and was certain of the need and prospects of such a move. Only some days later, when he reached Cairo where he was hosted by local CMS missionary Lieder, Smith learned that the idea had already been brought to the CMS committee and summarily dismissed. Still, he writes to Robertson, 'It is clear to us all, that Alexandria ought no longer to be left without a missionary.'[137]

Robinson, now in Alexandria, replied to Smith on January 3, before leaving for Cairo. His letter is practical, detailing the necessary expedition preparations. Robinson instructs Smith that should he arrive in Cairo before Robinson returns from Upper Egypt, to 'feel yourself authorized to make at once all the needful preparations for the journey,' because 'we much all look to you as the leader & purveyor of our caravan, we all depend on you to be our guide and leader during the travel!'[138] Robinson also encourages Smith to enlist the help of the consuls and missionaries to hire a recommended guide for Mt. Sinai, and to make 'all such functionary arrangements as may seem expedient.' As for equipment, the primary concern was to acquire a compass, which Robinson was unable to find anywhere during his travels, a rather strange fact since his route took him through London and major German cities. The best thing, writes Robinson, would be to have two compasses. He had a small compass from ABCFM missionary Jonas King (1792–1869) whom he met in Athens, but 'it is small & not convenient.' In addition to the compass, they needed 'a pocket sextant' or any other 'not too enormously expensive' instrument to ascertain latitudes. Robinson assumed that Smith could use such a device, as he could not. Robinson concludes with some logistical comments, mainly concerning letter exchange, and notes the fact that, while travelling in Upper Egypt, 'my companions in travel are well (…). In this strange land we feel already very much the want of your experience & guidance; & often speak of it, how pleasant it would be if Br. Smith were with us now.'[139]

136 Smith to Robinson, Smyrna, 19.12.1837, ESP, ABC 60 (105/7).

137 Smith to Robertson, On the Nile, 22.2.1838, Cairo, 28.2.1838, ESP, ABC 60 (105/8).

138 Robinson to Smith, Alexandria, 3.1.1838, ESP, ABC 60 (63).

139 Robinson to Smith, Alexandria, 3.1.1838, ESP, ABC 60 (63). Robinson (1841), I, 9: 'We found a welcome home in the hospitable mansions of Messrs. King and Hill.'

As for measurements, Robinson carried a copy of Laborde's large map of Arabia Petraea. Laborde, later *Conservateur des Antiques* at the Louvre, visited Petra in 1828 together with the French engineer and Orient traveller Louis Maurice Adolphe Linant de Bellefonds (1799–1883), director of public works under Mehemet 'Ali.[140] Robinson used Laborde's map of Arabia Petraea extensively during his travels in Sinai and mentioned it extensively in the *Biblical Researches*. Kiepert used it as well in constructing his maps.[141]

Smith wrote to Robinson again on January 6, answering the latter's letter of December 24 (two weeks from Athens to Smyrna!), yet another letter that was not recovered. Although his work was not advancing as rapidly as expected, Smith notes that he still intends to leave Smyrna on January 19. He wasted no time in purchasing some needed equipment items, 'a Turkish portmanteau a piece of painted canvass to defend your bed from the rain & from the dampness of the ground & also a pocket compass.' From Beirut, he ordered for delivery to Alexandria 'a circumferentor or surveyor's compass,' a measuring tape of 60 feet and a thermometer. Smith asks Robinson if he acquired a barometer, useful for measuring heights, though 'very liable to break in carrying,' as reported by other travellers. He also notes that a sextant is obtainable in Smyrna, but not an artificial horizon or chronometer, without which it is impossible to make 'celestial observations on land.'[142]

In the next paragraph, Smith describes the printed and cartographic material he can contribute to the expedition. First, the *Palinarus* map, a map of Palestine that he has among his possessions in Beirut, noting that he 'will order it to meet us at Jerusalem.' The *Palinarus*, an East India Company ship under Captain Robert Moresby (1794–1863, some sources claim 1854), was engaged for the purpose of making a detailed

Graduating from Andover in 1819, King, professor of oriental languages and literature at Amherst College, worked in Syria for the ABCFM between 1823–5. In 1831, he went to Athens and remained there as a successful but controversial ABCFM missionary until his death. See Phillips (1939), 137–44; Oliver (2014), 96–105. For detail on the measuring instruments see chap. 4.d.

140 Laborde (1830); (1836). Cf. Dawson, W. R. (1995), 256–57; Kurz (1998), 68–69; Pfullmann (2001), 176–78, 286; Linan de Bellefonds (2001).

141 Robinson (1841), III, 1st Appendix, 25–38; Kiepert (1841), mainly 33–34. For Robinson's rather long and extremely favourable review, see Robinson (1837). In 1841, Laborde published his geographic commentary to the event of the exodus from Egypt and the Book of Numbers (Laborde (1841)), relying extensively on former sources and using illuminating maps; Robinson is not mentioned!

142 Smith to Robinson, Smyrna, 6.1.1838, ESP, ABC 60 (105). For the instruments, see also chap. 4.d.

survey 'of the northern parts of that Sea' (the first systematic measurement of the Red Sea), until that time an enigma to western seafarers. In January 1833, they reached and described 'Akabah and the island Jezirat Pharoun (Pharaoh's Isle). The map published in August 1836 does not include Palestine but it encompasses Sinai.[143] In his *Memoir*, Kiepert states that the map is the only existing authority for 'the mountains on the western coast of the Gulf of Suez, and along the eastern shore of the Gulf of 'Akabah.'[144]

To Alexandria, Smith forwarded pioneer British surveyor Major James Rennell's (1742–1830) 'very learned work,' the *Comparative Geography of Western Asia*, accompanied by an atlas of maps and published in 1831.[145] Robinson did use Rennell's publication and mentions it in his book.[146] Smith apologizes in the letter for not being able to locate a copy of Laborde's map, yet, as noted, Robinson had one in his possession. On the other hand, Smith promises to bring 'the last word' in the cartography of Jerusalem, Frederick Catherwood's (1799–1854) map, published in London in 1835 after his 'measuring in disguise' of the Temple Mount.[147] It should not be surprising that in Section VII of the *Biblical Researches*, 'Jerusalem: Topography and Antiquities,' there are eighteen references to Catherwood and his map. Robinson, who met him in New York, writes, 'I must here expect the notices kindly communicated to me since my return by Mr Catherwood.'[148] Robinson's archive holds a copy of the map, and Kiepert's 'Plan of Jerusalem' added to the *Biblical Researches* is 'sketched from Sieber and Catherwood.'[149] Smith asks Robinson to search in Cairo for the 'late work on the Egyptians by E. Lane.' The leading Arabic scholar Edward William Lane (1801–76) had been living in Egypt from 1825 and published this book in 1836.[150] The last item in Smith's list is that mentioned above, the 'list in Arabic of a large part of the towns, villages & ruins in Palestine, east of the Jordan &

[143] Moresby (1836). Cf. Dawson, L. S. (1885), I, 127–28, II, 195 ('an admirable survey'); Searight (2003); Henze (2011), 3, 535–36 (citation).

[144] Kiepert (1841), 34.

[145] Rennell (1831). Cf. Edney (1997); Henze (2011), 4, 578–80.

[146] Robinson (1841), I, 75, 277.

[147] Catherwood (1835). Cf. Ben-Arieh (1973); (1974); Dawson, W. R. (1995), 86; Davis, J. (1996), 59–71. For detail see chap. 6.

[148] Robinson (1841), I, 379 n. 2.

[149] Robinson (1844); Kiepert (1841a). See also Kiepert's comments, (1841), 39 n. 1.

[150] Lane (1836); Dawson, W. R. (1995), 235; Thompson (1996); Roper (1998a).

north of Damascus as far as Hamah' that he left in Beirut and had forwarded to him.

In early 1838, Reverend Story Hebard (1802–41), graduate of Amherst and Andover, writes this in response to Smith's joining the expedition:

> In view of all the circumstances of the case, I believe you acted wisely in leaving Smyrna, to your Prof. Robinson in his contemplated tour in Egypt & the Holy Land, & care, from my heart, bid you God Speed – You will thus be enabled to render very important aid to the Prof in collecting materials for his work, & at the same time, be engaged in direct missionary labor – Thus you will emphatically kill two birds with one stone.[151]

Robinson wrote his next letter of instructions in Cairo, before leaving for Upper Egypt.[152] This letter repeats much of the previous one, yet its additional information clarifies this rather hasty preparatory period. Robinson notes that he left sufficient money for Smith with the two local CMS missionaries, Kruse and Lieder, and asks that Smith take all responsibility for the needed preparations. There are seven items on Robinson's list: a water bin, tent ('with or without a floor – cloth, as you think desirable'), wooden carrying boxes, 'a second servant to be looked for, if you think one necessary,' a general guide, the necessary experienced guide for travelling in Sinai, and 'a firman [from Muhammed 'Ali] for Syria,' a royal mandate or decree which will be 'good enough to enable us to travel around the Dead Sea, & even part of this land,' for which consul Campbell 'has offered his best services.' Colonel Patrick Campbell (1779–1857), Royal Artillery officer and General-Consul to Egypt (1833–40), was an experienced diplomat, serving earlier in the British legation in Columbia.[153]

Smith was also entrusted with obtaining information regarding Petra, one of the greatest uncertainties of the expedition: recommendations for the best guides, route, and mode of travel to the city, as well as knowledge of anticipated problems and hazards. Two groups had succeeded in exploring Petra the previous year, the Germans headed

[151] Hebard to Smith, Beirut, 24.1.1838, ESP, ABC 60 (33).

[152] Robinson to Smith, Cairo, 19.1.1838, ESP, ABC 60 (63). All following citations are from that letter.

[153] Bindoff (1934), 38–39; Dawson, W. R. (1995), 82; Eliav (1997), 23–25 (his role in nominating a consul to Jerusalem). He is mentioned by many of the visitors and travellers. For his activity in Egypt, see Kutluoğlu (1998), 99–129.

by Prof. Gotthilf Heinrich von Schubert (1780–1860) and the English headed by Alexander William Crawford Lindsay, 25th Earl of Crawford, 8th Earl of Balcarres (1812–80). Schubert was one of the leading German natural historians, a distinguished scholar whose work combined the influences of Romanticism with the so-called German *Naturphilosophie*. In 1836–37, he went on a long pilgrimage-expedition accompanied by a painter and two of his young students who were charged with the scientific work, reaching Palestine after travelling in Egypt, Sinai, and Petra. The expedition was extremely fruitful, yielding natural history collections and numerous publications, including a diary, three-volume book, and various other scientific works.[154] Lindsay, student at Trinity College, Dublin, 'British nobleman, traveller, and writer on art,' held two earlships and spent most of his time building a magnificent library and in travel. In 1836–37, he travelled in Egypt and Palestine accompanied by friends; his letters were published in 1838.[155] Robinson adds that Dr Walne showed him letters from Lindsay and his companions describing their journey, and suggests that Smith try to obtain them.

In the end, Robinson and Smith did not take the well-travelled eastern Arabah route from 'Akabah, but were the first westerners to go through the Western Negev.[156] They reached Petra only later, from Jerusalem through Hebron and the Dead Sea. Accordingly, Schubert's and Lindsay's itineraries were of less use to them. Counting the number of references for the various traveller's accounts in chapter XII of *Biblical Researches*, the chapter narrating this leg of the expedition is telling.[157] Lindsay's letters are referenced only six times and Schubert's 16 times. In comparison, Laborde is referenced 29 times, the young British officers-adventurers Charles Leonard Irby (1789–1845) and James Mangles (1786–1867) 36 times,[158] and Burckhardt – 56 times!

[154] Schubert, G. H. (1839). Cf. Goren (1997); (2003), 117–24.

[155] Lindsay (1838); (1838a). Cf. ODNB Lindsay; Henze (2011), 5, 92–93.

[156] Robinson (1841), I, 244–46 (decision, reasons), 254–325 (description: From 'Akabah to Jerusalem). Cf. Goren (2015).

[157] Robinson (1841), II, 463–638.

[158] Naval captains Irby and Mangles belonged to a group of young British officers including Willian John Banks (1786–1855) and Thomas Legh (1793–1857, who also travelled with Buckingham), who toured the East in 1816–18. The story of their travels was published in 1823: Irby (1823). Cf. Shepherd (1987), 46–75; Dawson, W. R. (1995), 29; Henze (2011), 1, 687 (Irby), 3, 178–79 (Legh).

Smith was delayed for 16 days in a lazaretto in Syra before being able to continue to Egypt.[159] In his letter to Robertson in Syra, written while sailing on the Nile from Alexandria to Cairo, Smith notes that he was only able to set out for Egypt on February 16 and describes his journey. As the consul was absent, he was received 'with great hospitality' by Gliddon's son-in-law Alexander Ted: 'Immediately on my arrival he made arrangements for worship on the Sabbath, & by sending around cards, a congregation of some by a quite forty men assembled, including the whole consul's family, & others of the most respectable inhabitants.' Smith officiated at two services on that day. His primary message to Robertson, noted above, was that he should 'think of Alexandria as a Missionary station.'[160] Smith did not miss an opportunity for missionary activity wherever he visited. His letters to fellow missionaries were not only about this shared passion; he also added details useful in reconstructing the expedition preparations. Smith's exact itinerary was not clear to his missionary brethren; however, Hallock did send him a letter to Beirut on February 10, hoping he would receive it on time.[161]

Once in Cairo, it seems that with the help of locals, Smith did quite a good job of organizing the entire list of equipment requested by Robinson. Robinson returned to Cairo and met Smith there as planned. They remained in Cairo for twelve more days of preparations and short excursions to monuments around the city. Keeping to the aforementioned list in Robinson's letter, they hired two servants and purchased water-skins, provisions for a month, a tent, canvas to put on the floor under their mattresses, wooden boxes, firearms for demonstrative purpose only, and comfortable Arab clothing. They received a firmân from the Pasha through the intervention of Gliddon (the father) and not the British consulate, and letters of recommendation for the Governor of 'Akabah and the heads of the Sinai Convent.[162] Smith was not satisfied with the work of the British consulate: 'Though the vice consul Mr Walne has been very polite, there seems to be, however, in the consulate a delay in attending to the wishes of travelers, which gives occasion for complains.'[163] They chose the guide who accompanied Laborde and

[159] Smith to Anderson, Cairo, 1.3.1838, ESP, ABC 60 (105).

[160] Smith to Robertson, on the Nile, 22.2.1838, ESP, ABC 60 (105/8).

[161] Hallock to Smith, Smyrna, 10.2.1838, ESP, ABC 60 (32).

[162] Robinson (1841), I, 51.

[163] Smith to Alexander Ted [Gliddon's son-in-law], Cairo, 10.3.1838, ESP, ABC 60 (105).

their caravan included three dromedaries and five camels. They soon discovered that Bedouin camels are more slender than Egyptian camels and carry only about two-thirds of the weight, so they hired another camel.[164] Learning from experience and having the flexibility to make changes and improvements along the way characterized Robinson and Smith's approach throughout the journey. For example, they eventually realized that mattresses were not needed for travelling in Palestine and replaced them with easier to carry blankets, and abandoned the wooden boxes in favor of 'small sacks and larger saddle-bags of hair-cloths.'[165]

In contrast with the letters brought here dealing mainly with more technical issues, the development of the expedition motivations, goals, and itinerary have been dealt with extensively by every scholar interested in Robinson and his achievements. Robinson's own words in his introduction to the *Biblical Researches*, cited by many, are quite clear, 'my first motive had been simply the gratification of personal feelings' and, being a Congregationalist New Englander, he added that 'the scenes of the Bible had made a deep impression upon my mind (...).'[166]

Researchers typically write the introduction to their study upon completion of the manuscript and Robinson was no different. The full manuscript of the *Biblical Researches* is housed in the Burke Library at the Union Theological Seminary, New York.[167] In many places in the manuscript, Robinson noted the date a page was written, usually in its upper corner, though he seems only to have begun this habit late in 1839. The original introduction is signed 'Berlin, Aug. 30. 1840' and appears in the manuscript at the end of the first section (before Robinson and Smith leave Cairo).[168] 'Through the good providence of God,' writes Robinson to Smith, 'I was enabled to complete my [*sic*] in Berlin about the 1st of Sept. & on the 6th of that month left by way of Halle & Hamburg to this place [London].'[169] The introduction may not be describing Robinson's thoughts and plans before entering the Sinai Peninsula; rather, it seems to conflate pre-expedition expectations with post-expedition results. Understanding Robin-

164 Robinson (1841), I, 49–55.

165 Robinson (1841), I, 49–50.

166 Robinson (1841), I, 46–47. Cf. Davis, M. (1977), 7; Williams (1997).

167 Marked ERP-UTS, Edward Robinson Papers, 1836–38. Series 1. For the introduction: Box 1.

168 Robinson (1841), I, 46–48.

169 Robinson to Smith, London, 3.10.1840, ESP, ABC 60 (63).

son's goals and expectations is important for evaluating his scientific achievements. Did he anticipate potential findings and even fruits of the expedition? Did he have any concrete plans beyond simply walking the Holy Land, Bible in hand, imagining the unfolding of sacred events?

'I entered upon it without the slightest anticipation of the results to which we were providentially led,' continues Robinson in his introduction. The entire endeavour was a natural outcome of his New England education, beginning in early childhood. Robinson, eager to explain and present his motives, devotes a long description to the tremendous impact of his 'biblical childhood' in New England, where biblical names became part of his 'earliest recollections and holiest feelings.' It seems to have been a simple, natural process that developed in his adult years into a strong desire to visit this sacred part of the world in person.[170]

Still, he adds, these emotions were connected to a scientific motive, his long-held desire to prepare a study on biblical geography. He believed that the study should be based on personal observation; existing traveller's descriptions and other written sources lacked sufficient data. Yes, Robinson wanted to observe the actual country, which had been revealed before his eyes 'beyond the barren hillsides and dismal valleys,' in the words of Moshe Davis.[171] Robinson's education and theology led him to develop a different approach to the study of the country from most other American travellers of his era, as described in detail by numerous scholars.[172] In modern research, as early as 1970, Yehoshua Ben-Arieh most likely deserves credit as the first to name Robinson a central pillar in the 'rediscovery' of the Holy Land.[173]

Robinson's stated expectations were quite low, 'I never thought of adding any thing to the former stock of knowledge (...). I never dreamed of any thing like discoveries in the field.' These claims contradict, somewhat, the desire he expresses in the same introduction to prepare 'a work on Biblical Geography' in order 'to satisfy myself by personal observation as to many points on which I could find no information in the books of our travelers.' Here Robinson mentions Schubert, who 'had just preceded us' by about a year and a half, as someone 'we could hope to add

[170] Robinson (1841), I, 46.

[171] Davis, M. (1977), 14.

[172] As examples for 'Americans and the Holy Land': Davis, M. (1977); Shepherd (1987); Lipman (1989); Vogel (1993); Klatzker (1997); Oren (2007); Travis (2012).

[173] Ben-Arieh (1972a); (1979), 85–91 ('The Foundation of Palestine Exploration').

nothing to what he and others had observed.'[174] Less than a year later, in his letter to the president of the Royal Geographical Society thanking him for the gold medal he received, Robinson once again writes:

> My journey in the east was undertaken in weakness & fear & much trembling; I instead went first to satisfy an ardent longing to visit those scenes (...) & then to satisfy myself with regard to some few points round the Holy City, as to which I could gain no satisfaction from the works of former travelers. But I did not dream of discovery, nor of any special enlargement of the limits of Sacred Geography.[175]

Did Robinson really set out on the expedition without larger goals? In the introduction, he continues, 'the change in our aims, occurred as we advanced in Sinai and saw how much had been undescribed and hitherto unknown, names long forgotten, so we determined to do what we could with our limited means towards supplying this deficiency.'[176] The actual decision to write a book resulting from the expedition came only later:

> It is only since my return, that I became aware of the value of the materials thus collected, in a geographical point of view, from the judgment passed on them by eminent geographers; and I look back with painful regret on the circumstances, which prevented me from taking along more perfect instruments, and from obtaining a more exact knowledge of the observations necessary for the trigonometrical construction of a map.[177]

These words are quite clear: the decision regarding a publication was made only upon their return from the East and after recommendations from 'eminent geographers,'[178] but some of the letters question this assertion!

These thoughts, one should bear in mind, were written in retrospect; Robinson began writing the manuscript in early 1839. Both travellers kept journals, probably from the beginning of the expedition. The fact that they kept separate journals, writing down 'every thing we wished to record,' pencilling notes on the way and writing them out in full the

[174] Robinson (1841), I, 46–47. Robinson used to separate these words: 'any thing,' 'some thing,' 'every thing,' etc.

[175] Robinson to Hamilton, New York, ?.?.1842, ERP, I, 4.

[176] Robinson (1841), I, 46–47. For Schubert, who travelled from September 1836 to July 1837, see above, note 197.

[177] Robinson (1841), I, 48.

[178] See chapter 4.b.3 and 4.

same evening but without comparing them, by itself, is evidence of a plan to write a travelogue. Moreover, Smith meticulously recorded bearings from the beginning of the expedition, as well as the Arabic names for all artefacts and geographic features (places and ruins, streams, wadis, mountains and valleys, etc.), strengthening the theory for a prior plan, albeit a general one, for a publication of the expedition.[179]

It is difficult to find in the letters an explicit plan for writing a book, but there are some hints, beginning with the instruments they took with them. Again, in the introduction, Robinson writes that given these 'dis-expectations' of theirs, they hardly took any instruments, 'except an ordinary surveyor's and two pocket compasses, a thermometer, telescopes, and measuring tapes.' However, in the above-cited letter to Smith, he requests a 'pocket sextant, or any other instrument (not too enormously expensive) by which it may be possible to ascertain the latitude of places N. of Akaba.'[180] In October of 1838, Robinson, still recovering from his illness in Vienna, writes asking Smith whether he could complete his itinerary, as 'it would be of use to me at Halle.'[181] Quite clearly, he already had the book in mind and they had discussed it previously, hence Robinson does not explain his request for the diary. Upon reaching Halle and meeting with his scientific and social acquaintances, Robinson writes to Smith, 'They all agree in urging the propriety & importance of handwriting an account of our journey and also the lists of Arabic names, at least in the German edition. I shall take them to Berlin, & get the opinion of the geographers there, - & if it then be the best & print them, will transmit them to you at Leipzig for corrections etc.'[182]

Yes, the decision was made in late 1838, and the book was on its way!

[179] Robinson (1841), I, 47.

[180] Robinson (1841), I, 47; Robinson to Smith, Alexandria, Egypt, 3.1.1838, ESP, ABC 60 (63).

[181] Robinson to Smith, Vienna, 24.10.1838, ESP, ABC 60 (63).

[182] Robinson to Smith, Berlin, 14.11.1838, ESP, ABC 60 (63).

CHAPTER 3
THE 1838 EXPEDITION: ITINERARY AND DEVELOPMENT

3.a. Financing the expedition

One of the questions that should be raised with regard to such a pioneering enterprise is how it was financed, both in terms of raising the necessary funds and the logistics of their provision throughout the journey. It appears that Robinson had the capability to finance the entire expedition personally. His correspondence, as well as letters by Smith, provides clues to his sources for financing such a complex and expensive undertaking. The question of his financing nearly four years in Europe is also of interest, but will not be dealt with here other than to cite Robinson, reporting from Halle, that 'I find all my expenses here, exclusive of clothes, books, and traveling, and including instruction, lectures, &c, (...) is the rate of less than two hundred dollars per annum.'[1]

Robinson's inheritance from his first wife included a farm in Clinton. He hired Othniel Williams (1787–1832), a lawyer and agent in the village, to locate buyers and secure the transaction, and his son and successor, Othniel S. Williams (1813–80), to close the deal. In November 1830, Robinson writes that Williams, who 'I understand is now treasurer of the College' [from 1828], owed him money and was not answering his letters.[2] Robinson's letter dated February 24, 1832, describes in detail and rather argumentatively Williams' problematic handling of the matter, leading the former to declare that 'I am not anxious to sell the

[1] Hodge (citing Robinson's letter) to his wife, Paris, 21.12.1826, Hodge (1880), 111.

[2] Robinson to Norton, 13.11.1830 and 11.12.1830, summary, typed, of Edward Robinson to Rev. Asahel S. Norton, ERP, 1816, Robinson- 1; cf. Williams (1999), 110, 195; Weigel Williams (2007), 40.

land in this way.' Robinson discusses the issue of the interest on delayed payments and the fact that his (second) wife was confined to Andover with a small child, making it impossible to travel to Boston to conclude the sale. In the end, the land sale was successful when a Clinton group headed by John Powell purchased the plot for $2,500, Robinson's asking price.[3]

In his biography of Robinson, Jay Williams notes that after leaving Andover in 1834 and moving to Boston, Robinson 'remained essentially unemployed, working feverishly on more publications.' He investigated Robinson's sources of income for financing the upcoming expedition as well as his livelihood during this period. As the time of departure drew nearer, in addition to the aforementioned Clinton property, Robinson succeeded in selling his father's property in Twinsburg, Ohio, which together yielded a fair sum of money. Williams notes further that Robinson was also able to calculate into the expedition budget his salary from the UTS, which had been paid in advance for the first year of his leave of absence.[4]

Robinson's financial situation was sufficiently strong to withstand a nationwide financial crisis that hit in 1837. 'The panic of 1837 sweeps the nation,' write both Williams and Weigel Williams, describing the 'inflationary spiral [that] enveloped the nation' precisely when the Robinsons were preparing to set out for Europe. Banks closed, businesses failed, unemployment rose sharply, and the economy 'became very depressed and remained in that depressed state for five or six years.' Weigel Williams scrutinized the letters to learn how the Robinson family had been affected by the crisis. Robinson received the final payment for the property in Twinsburg, Ohio eight months late, in January 1837.[5] It seems, however, that Robinson had sufficient means for his planned one-year absence from the US, including his expedition to the East and his family's stay in Germany.

It was impractical and unwise for Robinson to travel with large sums of money. He had to find methods for sending advance sums to selected destinations. Before leaving the US, and in Europe during his stay prior to the expedition, Robinson arranged for money to be sent to dependable Europeans in exchange for letters of credit. Two letters

3 Robinson to John Powell, Andover, 24.2.1832, ERP, 1816 Robinson- 1.

4 Williams, (1999), 195–96; various letters in the Hamilton College Library collection, e.g., Robinson to John Powell, Andover, 24.2.1832, ERP, 1816 Robinson- 1.

5 Williams (1999), 207; Weigel Williams (2007), 103–4.

sent by Joshua B. Bacon of Messrs. Perkins, Bacon & Co. in London, to Robinson in Berlin, shed light on the process.[6] The correspondence, sections of which are barely legible, deal with the letters of credit that Bacon issued to Robinson. They reveal that a certain 'Mr Jacob' sent from Halle 'a draft in my favor,' as Bacon writes, of a significant sum of money. The Robinsons had arrived in Berlin after visiting Talvj's relatives in Stettin. 'Mr Jacob' is most likely Talvj's only brother, Adolph von Jakob.[7] Two other people are mentioned in the letters, both unidentified to date. A 'Mr Page' to whom Robinson wrote when still in London ('Mr Robinson left a letter for Mr Page that he did not now require the cash he could take it'), and someone with the initials R. K., to whom Page 'has since written to R K requesting a settlement which under the circumstances R K has yielded to.'[8] Robinson sent a letter from Halle to Bacon on October 16, which included 'a draft in my favor of £245.6.0,' signed by Jacob with the notation 'a greater sum at a shorter term was not to be had.' This letter reached London a month later and caused some misunderstanding. Nonetheless, after various fees, Robinson was able to receive a 'letter of credit for £240.'[9]

From other letters, we learn that Robinson and Smith did, of course, carry with them a significant amount of coin for paying their guides, and for their mounts, pack animals, food, and unforeseen needs on their journey. In Jerusalem, which they reached on April 14, 1838, they stayed on the premises of ABCFM missionary John F. Lanneau (1809–67), who had joined Whiting and his wife there in the spring of 1836. Lanneau stayed in Jerusalem until 1839, and helped Robinson and Smith with logistical arrangements throughout the expedition.[10] On May 17, they left for their long journey to Gaza, Hebron, Wadi Musa and back,

[6] Joshua B. Bacon in London to ER in Berlin, 2 and 17.11.1837, ERP H.3, Family letters to E.R. 1838–40. The firm printed books, bank notes and postage stamps ('The Black Penny' 1840). It was established in 1819 when Jakob Perkins (1766–1849), a genius 'engraver on steel,' moved from Boston to London, forming the firm 'Perkins, Fairman and Heath.' Joshua Butters Bacon (1790–1863; Perkins' son-in-law) bought Heath's interest in 1829 and formed 'Perkins & Bacon.'

[7] Weigel Williams (2007), 105; Williams (1999), 212. Bacon, an Englishman, probably used 'Jacob' instead of 'Jakob.'

[8] Joshua B. Bacon in London to ER in Berlin, 2.11.1837, ERP H.3, Family letters to E.R. 1838–40.

[9] Joshua B. Bacon in London to ER in Berlin, 17.11.1837, ERP H.3, Family letters to E.R. 1838–40.

[10] Bird (1972), 317; Taylor, A.R. (1957), 82–83; Abu-Gazaleh (1990), 24–25; Kawerau (1958), 256–57.

returning to Jerusalem only on June 9. Shortly before their return, Lanneau updated Smith writing, 'According to your injunction I opened your chest and took out the bag of money referred to.'[11] It is unclear whether they carried this bag of coins with them all the way from Egypt. Its contents are quite interesting, but probably standard for European travellers in the East. Not all the coins are mentioned by name in the letter; they included '21 Spanish dollars (...) 2 gold pieces (...) 21 piastres in small change and 3 small gold pieces (...) the whole amounting to 2827.20, two thousand, eight hundred and nine ½ piastres.' A portion of this money was intended for Whiting, who received 1100 piastres to organize the escorts, guides, and pack animals for the next leg of the trip, from Jerusalem north and onward to Beirut. In Jerusalem, Lanneau was able to change the different coins to 'standard money' totalling, at that time, '1727.20 seventeen hundred and 9½ piastres which are subject to your orders.'[12] According to Robinson, '100 Piastres were equivalent in 1838 to about 1 Pound Sterling.'[13] Hence, Robinson's £240 were equal to 24,000 Piastres, an amount that, when compared with the 2,827 Piastres totalled above, helps explain (even without being able to calculate all expenses) how Robinson was able to self-finance the expedition.

Lanneau then describes the missionaries' difficult financial situation, 'we are very much straitened in our money concern, and we know not when we shall be relieved,' before returning to the issue of the expedition's future financing. Apparently, Robinson and Smith planned to reach Damascus, as Lanneau asks Smith to 'tell Dr Robinson that Mr Kilber will write today to Mr Giustiniani of Damascus directing him to pay any order from the Dr up to 10.000 piastres.' The letter attests to Robinson's preparation of a rather large sum of money ('for 50,000') that he might draw from in Damascus. The immediate difficulty was that his foresight 'would not in all probability do the Dr any good in Jerusalem,' as 'it would be next to impossible to get a bill on Damascus for any sum however small, cashed here at present.' Moreover, the possibility of handling the transaction personally no longer existed, since 'the Damascus merchants who were here at the feast [Easter] have all left

[11] Lanneau to Smith, Jerusalem, 5.6.1838, ESP, ABC 60 (49).

[12] Lanneau to Smith, Jerusalem, 5.6.1838, ESP, ABC 60 (49). For the coins, their value and purchasing power, see Luncz (1980); Rubinstein (1997), 18–21, a very detailed study of the coins used in different periods in the nineteenth century. A Spanish dollar ('Rial Amud') was, in 1896, equal to 25 Piastres; 'Gold Pieces' probably referred to the US gold dollar, 130 Piastres in Jerusalem (Luncz (1980), 14).

[13] Robinson (1841), I, xvi.

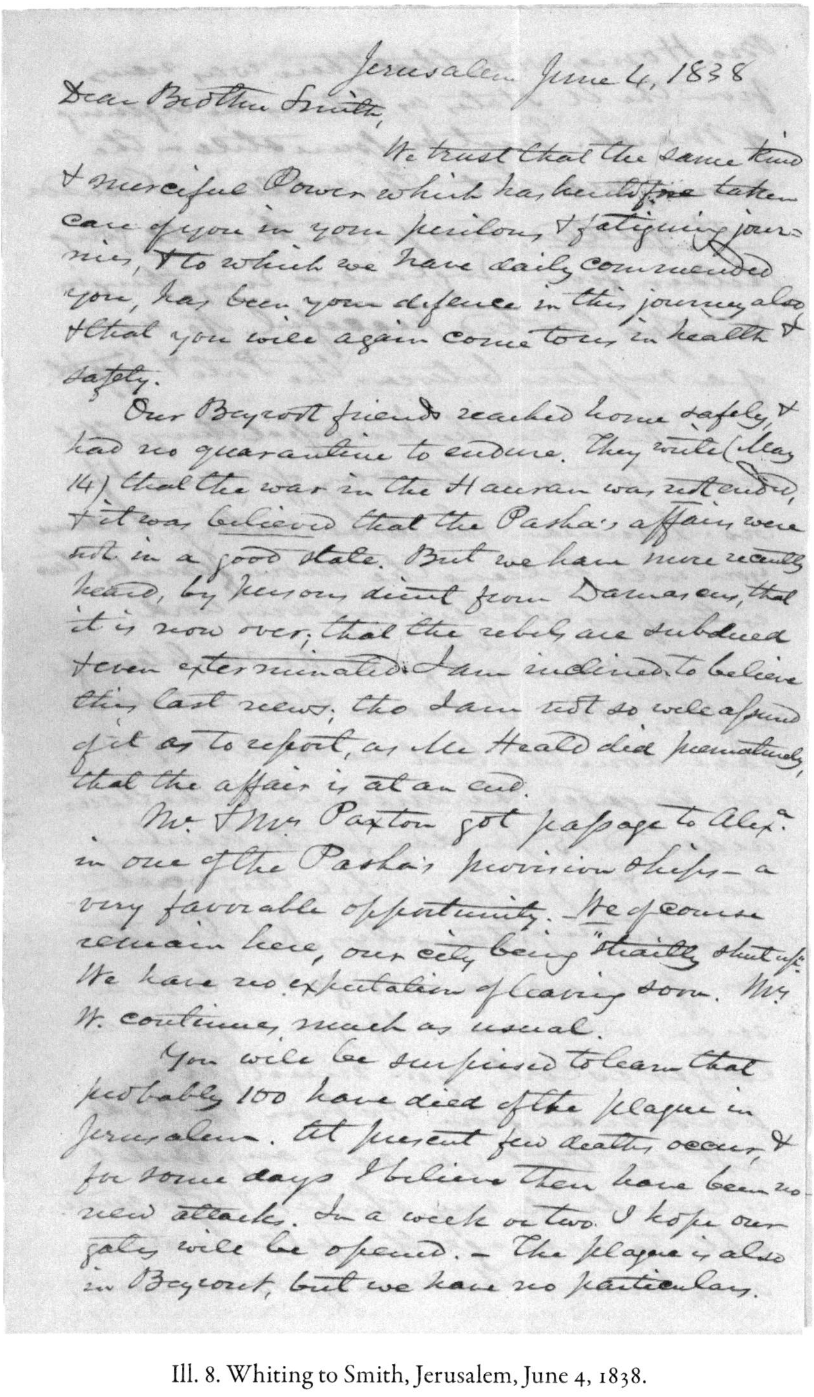

Jerusalem June 4, 1838
Dear Brother Smith,

We trust that the same kind & merciful Power which has heretofore taken care of you in your perilous & fatiguing journies, & to which we have daily commended you, has been your defence in this journey also & that you will again come to us in health & safety.

Our Beyroot friends reached home safely, & had no quarantine to endure. They write (May 14) that the war in the Hauran was not ended, & it was believed that the Pasha's affairs were not in a good state. But we have more recently heard, by persons direct from Damascus, that it is now over, that the rebels are subdued & even exterminated. I am inclined to believe this last news, tho I am not so well assured of it as to report, as Mr Heald did prematurely, that the affair is at an end.

Mr & Mrs Paxton got passage to Alexa in one of the Pasha's provision ships – a very favorable opportunity. We of course remain here, our city being "strictly shut up." We have no expectation of leaving soon. Mrs W. continues much as usual.

You will be surprised to learn that probably 100 have died of the plague in Jerusalem. At present few deaths occur, & for some days I believe there have been no new attacks. In a week or two I hope our gates will be opened. – The plague is also in Beyroot, but we have no particulars.

Ill. 8. Whiting to Smith, Jerusalem, June 4, 1838.

long ago and there is at other times no pecuniary transactions between the two cities.' In any event, these merchants carried only rather small amounts, 'perhaps seldom have one tenth the amount specified in Kilber's letter of credit on their command.'[14]

Lanneau's letter confirms Robinson's meticulous financial planning of the expedition. Dangerous yet unavoidable, Robinson carried enough money in local coin to pay for goods and services en route. To ensure his cash flow for local services, and for larger sums or prepaid services, Robinson arranged for letters of credit with European merchants or bankers residing in major cities on his expedition route. Initially, Robinson thought he could depend upon such services in Cairo and Jerusalem. However, there was no such agent in Jerusalem and he had to place his hopes on Damascus, which due to security conditions as the expedition developed they were unable to reach.[15] Lanneau's letter also demonstrates the extent to which Robinson and Smith's missionary brotherhood was essential to the smooth execution of financing the expedition.

A dependable source for calculating the actual expedition costs does exist. The source is a letter written by Smith in May 1840, to an anonymous recipient preparing to travel to the East who asked for information. They planned to meet in the near future, so Smith, 'much pleased to learn that you contemplate such a tour as that which you have sketched in your letter,' agrees to provide 'cheerfully all the information in my power,' but only briefly at this point.[16] Smith writes that sailing by steamer 'between Beyroot & Constantinople' costs 'somewhat more than $50' and from Constantinople up the Danube to Vienna, all expenses included, about $70. He then lists the following expedition expenses:

A servant – interpreter, from $10 to $12 a month.

About $10 a camel from Cairo to Akabah, & 9 more enough for 3 of us & 2 servants.

From Hebron to Petra & back, with 5 camels cost us about $70.

Generally, 75 cents a day for a mule, needing 3 or 4 of them.

Smith does not include other expenses such as equipment and food. For a delegation of three, for three and half months (although one can surmise that Adger paid his own expenses), costs totalled about $400

[14] Lanneau to Smith, Jerusalem, 5.6.1838, ESP, ABC 60 (49). Giustiniani belonged to the famous Venetian family and worked in Damascus, see Klein (1999).

[15] For this change of route see section 3.b.

[16] Smith to ?, Northford, CT, 4.5.1840, ESP, ABC 60 (105\12).

including the steamers. In this letter, Smith recommends setting aside three weeks for preparations and six weeks for travel in Palestine, including Damascus and Baalbek. He notes that 'the journey of the Danube from Cos̃ple to Vienna, consumes about a month.' However, after their difficult experience on the return trip, he cautions that 'in August & September [it] exposes one to sickness.'[17] In 1838, one US Dollar was equivalent to 20 Piastres,[18] so at the sum of $400, a total of about 8,000 Piastres per person was needed to finance such an expedition. The average daily expense for the first trip is unknown; however, 'during the second trip was under $5,' about 100 to 150 Piasters.[19]

3.b. Impressions en route: excerpts from the journals of Robinson and Smith

It was a scene which had often taken possession of my youthful imagination; but which I had not dared to hope would ever be realized.

Robinson (1839), 401, journeying in the desert.

Indeed, during the whole journey in the East, although surrounded by war, pestilence, and quarantines, we were enabled to pass through them all without harm or hindrance; without being detained from these causes even for an hour.

Robinson (1841), I, 369.

Itinerary:

1837
July 17 – Embark from New York
August 5 – Liverpool, London, Oxford, Antwerp, Brussels, Cologne, Frankfurt, Halle, Berlin
November 13 – From Berlin to Halle, Vienna, Trieste
December 1 – Corfu, Patras, Modon, Piraeus
December 6–25 – Athens
December 30 – Alexandria

17 Smith to ?, Northford, CT, 4.5.1840, ESP, ABC 60 (105\12).

18 Robinson (1841), I, xvi.

19 Bliss (1907), 196; Luncz (1980), 14.

1838
January 5 – Sail to Cairo and Thebes
February 28 – Return to Cairo; Smith arrived February 25
March 12–29 – Cairo, Suez, 'Ayûn Mûsa, Sŭrâbît el-Khâdim, Convent
March 29 – April 5 – Travel to 'Akabah
April 5–14 – Travel via the western route to Hebron and Jerusalem
April 15 – May 3 – Jerusalem
May 4–6 – Excursion from Jerusalem to Bethel and surroundings
May 7 – Jerusalem
May 8–15 – Excursion to 'Ain Jidy, Dead Sea, the Jordan, and surroundings
May 16 – Jerusalem
May 17–25 – Jerusalem to Gaza and Hebron
May 26 – June 5 – Hebron to Wady Musa (Petra) and back
June 6–9 – Hebron to Ramleh and Jerusalem
June 10–13 – In Jerusalem (outside Damascus Gate)
June 14–18 – Jerusalem to Nazareth and Mount Tabor
June 19–21 – Mount Tabor via Lake of Tiberias to Safed
June 22–27 – Safed via Tyre and Sidon to Beirut

The *raison d'être* of Robinson's various expedition manuscripts, including, of course, the three volume *Biblical Researches*, was publication of scientific findings. He began publishing immediately, shortly after reaching Berlin.[20] Yet, despite this focus, Robinson and Smith were no less interested in human relations, native culture, politics, and religion. They were also passionate Holy Land travellers. They journaled and wrote letters filled with 'non-scientific' information on a daily basis. Robinson's and Smith's reports hold numerous descriptions of diverse experiences and events of their journey, from the mundane to the quite harrowing. They both understood that these accounts were integral to the narrative and essential to their readers. For some, they hoped the anecdotal elements would liven the rather 'dry' academic texts. For readers planning an excursion to the destinations described, these texts would supply much useful information. Descriptions of their guides, fellow travellers, local tribes, and religious groups and sects, as well as local authorities, are but a few examples of the vast amount of what one might define as non-scientific, technical or anecdotal information. Issues such as travel logistics, daily itineraries, decision-making processes, explana-

20 Robinson (1839); (1839a); (1839b). See chap. 4.b.

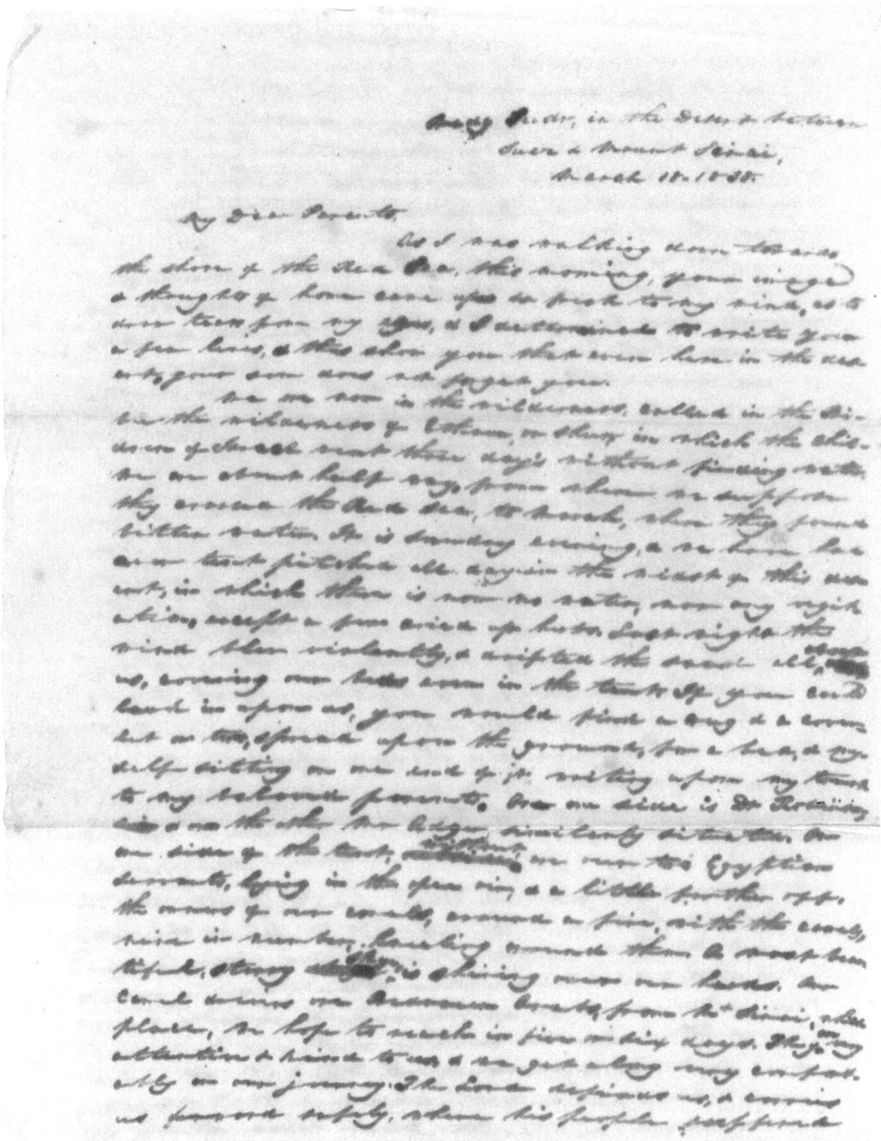

Wady Sudr, in the Desert between
Suez & Mount Sinai,
March 18. 1838.

My Dear Parents,

Ill. 9. Smith's letter from Sinai to 'my dear parents,' Wady Sudr, in the Desert between Suez and Mount Sinai, March 18, 1838.

tions of terms unfamiliar to the European and American reader, discussions of the preferred routes, special events, and unexpected confrontations are interwoven repeatedly in the texts.[21]

Hence, we learn that the party that left Cairo included three Americans, two Egyptian servants, five Arab Towara (Bedouin 'sons of the desert' who had the exclusive right to guide travellers to Mount Sinai), and nine camels.[22] In a letter to his parents written from the desert, Smith attempts to describe the incredibly different reality and atmosphere they experienced in the desert daily:

> If you could land in upon us, you would find a rug & a coverlet or two, spread upon the ground, for a bed, & myself sitting on one end of it writing upon my knees to my beloved parents. Over one side is Dr Robinson, & on the other Mr Adger similarly situated. On one side of the tent, (?) are our two Egyptian servants, lying in the open air, & a little further off, the owners of our camels, around a fire, with the camels, nine in number, howling around them. A most beautiful, sky is shining over our heads. Our Camel drivers are Bedaween Arabs, from Mt. Sinai, which place we hope to reach in five or six days. They are very attentive & kind to us & we get along very comfortably on our journey.[23]

How interesting it is to travel these scenes from history, Smith adds, in an attempt to bring Moses, Jesus, and Paul alive for his aging parents. He expresses his hope to send the letter from the Sinai convent, which they expected to reach the following day. They planned to rest there a couple of days before climbing the mountains. Robinson's control of 'some modern Greek' enabled him to converse with Father Neophytus, the Superior of St Catharine Monastery. In this way, they found particular favour in the eyes of the 'good old Superior,' who gave them gifts, including 'a pair of sandals worn by the Bedawîn of the peninsula, made from the thick skin of a fish which is caught in the Red Sea.' Robinson even succeeded in identifying the fish, which had been described by the natural historian Christian Gottfried Ehrenberg (1795–1876) from Berlin. In 1823–24, Ehrenberg studied the flora and fauna of the Red

[21] Bliss (1907), 203–04, who read only the book, takes a different approach: 'We are bound to admit, however, that Robinson yielded to some of the temptations to which the editor of his own journals is objected.'

[22] Robinson (1839), 401.

[23] Smith to parents, Wady Sudr, in the desert between Suez and Mount Sinai, 18.3.1838, ESP, ABC 60 (105/8) 1838 Jan.-June. For Adger see chap. 5.c.

Sea region together with his colleague Wilhelm Friedrich Hemprich (1796–1825).[24]

Another letter, written nine days later, describes an exceedingly difficult two-day walk for which the route is presented in detail:

> Yesterday morning we set out for an excursion to the tops of Sinai & St Catherine. (...) The highest point of Sinai is not far from 2000 feet above the convent, & the ascent a great part of that way, is by steps. (...) The peaks we were seeking are above us, a giant mountain, of one naked solid rock, only cracked here & there by some convulsive of nature. Even our old Bedaween guide said it were impossible to reach its top summit. We determined to make the attempt & stripped ourselves of all unnecessary appendages.[25]

For the European and American reader, Robinson and Smith describe *curiositates*, natural otherness, such as the character of the desert, the vision of a mirage, and the phenomenon of the wady, which 'in the desert, means a shallow bed, through which the waters of the rainy season are carried off.'[26]

Moreover, to help his readers, Robinson included a short 'reader's guide' to the *Biblical Researches*[27] with the following examples: (1) 'the rules for the PRONUNCIATION of Arabic Names as written in Roman letters;' (2) 'the measure of distance' is recorded by hours, and in order to commute time to distance, one must know the kind of animal used, camel, horse, etc., and the nature of the ground, for which Robinson added a small explanatory table; (3) heights are recorded in French feet; (4) measures of land; (5) measures of corn; (6) weights; and (7) currency including the value of piastres.

In further correspondence, Smith supplies a vivid description of the daily schedule they developed upon leaving Egypt for the Sinai and held constant throughout the expedition:

> The course of affairs with us on ordinary days is as follows. In the morning I wake at 4 o'clock & call my companions & our domestics & Arabs. By the time we are dressed, the domestics have our leather table

[24] Robinson (1841), I, 169–71. Cf. Henze (2011), 2, 156–58; Goren (2003), 58–62, and the sources cited therein.

[25] Smith to Hoadly, Mount Sinai, 27.3.1838, ESP, ABC 60 (105/8) 1838 Jan.-June.

[26] Robinson (1839), 401–02; (1841), I, 55–61.

[27] Robinson (1841), I, preface, xv–vi.

> spread on the ground, with some bread & butter & a little dried fruit. Then we eat, drinking cold water; our Arabs, in the meantime taking their frugal need having finished, we pack our things, our tent is taken down & the camels loaded. This last is generally done with much noise & scolding among the Arabs. By sunrise we are mounted & we do not stop again until near sunset, generally making a day's journey of 10 or 11 hours. Then we are all fatigued, our tent is pitched, we spread our carpets & blankets on the ground, & lie down to rest a little, while dinner is preparing. This, from day to day, is the same thing, boiled rice, with a little salted butter poured over it & some dried fruit.

Robinson and Smith, devout Presbyterians, were very strict in observing their Sabbath rest:

> This is Sabbath evening, & we have had a day of rest. The Sabbath is never more welcome than in the desert. We pitch our tent on Saturday evening & do not break up our encampments until Monday morning. It gives us time to rest. We read the word of God & have opportunities for social & private devotions. And on Monday morning we start again refreshed in body & in spirit.[28]

They accomplished an enormous amount of travel in three and half months while omitting Sunday's, which adds to the wonder and unbelievable achievement of their expedition!

Robinson's and Smith's deep emotional connection to the Biblical text and events resonates throughout their writing. Letters written throughout the expedition include long and detailed geographical descriptions connected to different scriptural traditions. Robinson's reverence was stirred early on at the 'evening devotions' in Suez, where after a long day of searching for the exact location of the crossing of the Red Sea, 'we read and felt in its full force, the magnificent triumphal song of Moses,' citing Exodus 15:1. About Jerusalem, Robinson writes, 'our journey *to* Palestine was now complete; and our researches and travels *in* Palestine were to begin,' and it is 'as if the realization of a former dream.'[29]

On April 29, 1838, two weeks after arriving in Jerusalem, Smith wrote a single page letter from the city. The report, relatively brief for Smith, begins in Beersheba. He writes, 'we were very much interested' in uninhabited houses, the wells ('the first good water' since 'Akabah), the

[28] Smith to Master L. Ives Hoadly, Wady el Ghudaghud, 8.4.1838, ESP, ABC 60 (105/8) 1838 Jan.-June (both citations).

[29] Robinson (1839), 410.

place where Abraham and Isaac lived and, of course, in Hebron, though it was impossible to enter 'where is supposed to be the tombs of the Patriarchs.'[30] Another letter, written after the expedition, is a particularly wonderful example of Smith's writing; unluckily, the document is partially torn:

> Since I wrote you from Jerusalem, we have added much to our geographic (?) discoveries. The tract between Jerusa[lem and] Gaza, we have discovered pretty through (...) near which David killed Goliath, Beth-shemesh, Zorah the birth place of Samson, Timnath his wife's place, Ekron, Eleutheropolis (...) in Gedor & Adoraim, besides Ajalon & the valley of Ajalon between Jaffa & Jerusalem. We have surveyed the region between Hebron & Petra in the W to Edom going to the latter place by way of the southern extremity of the Dead Sea & returning direct on this route we found the site of Arad, examined a Mount of salt at the south of the Sea, five miles long, with perpendicular precipices 80 or 70 feet high, visited a place which we are satisfied was Kadesh, where Moses brought water out of a rock, & ascended the very Mount which the children of Israel did, contrary to the command of Moses (...). Then in coming from Jerusalem Northward, we discovered Shiloh, Samaria, & Jezrael.[31]

In their correspondence, both Robinson and Smith highlight the prior experience of their expedition guides in leading Europeans through different sections of the country. 'We found bedawy guides generally very reserved of first,' Smith writes, but this obstacle was easily overcome 'after two or three days acquaintance.' Smith also opines that 'within their own territory they know the name of every mountain & valley & almost every stone (...), yet at the distance of two or three days journey their information becomes very vague & imperfect.'[32] This made it difficult to extract any useful and trustworthy information concerning regions outside their immediate vicinity.

However, in contrast to Smith's general impression, deeper investigation reveals a slightly different picture. Besharah, who had travelled with Laborde, guided their group from Cairo, and Tuweileb, Rüppell's guide and 'a man of more experience and authority than Beshârah,' led

[30] Smith to ?, Jerusalem, 29.4.1838, ESP, ABC 60, 1838.

[31] Smith to Hoadly, Alexandria, 13.7.1838, ESP, ABC 60 (105/9) 1838 Jul.-Dec. Kadesh turned out to be one of Robinson's most incorrect identifications!

[32] Smith to Homes, Smyrna, 4.8.1838, ESP, ABC 60 (105/5) 1838.

them from the convent to 'Akabah.[33] They carried a letter of recommendation from the governor of Suez for the governor residing in the castle at 'Akabah ('an oblong quadrangle of high thick walls, with a tower or bastion at each of the four corners'[34]) asking him to 'provide them with what camels they need, they paying the hire,' and 'extend to them protection & assistance in all that they need & charge the Arabs who take them to serve them to their content.'[35] In 'Akabah, they planned to hire for Wadi Musa and Petra the services of the 'faithless and mean-spirited' Sheikh Hussein of the Alawin tribe, 'a lawless tribe, standing in no good repute among their neighbours.' Originally intending to take the 'regular' eastern route, due to uneasy, even dangerous circumstances, they changed course and took the western road to Gaza and Hebron along the Haj route, described as 'terrible wilderness.' Contributing to their decision to take the safer, western route was the preciousness of every expedition day and the knowledge that 'it would require at least four days time' to bring Hussein to 'Akabah, and 'such a loss of time was insupportable.'[36]

Ill. 10. 'View of the Fortress of Accaba,' by Laborde.

[33] Robinson (1839), 401, 405; Robinson (1841), I, 216. Differences in the name spelling in original. Edward Wilhelm Peter Simon Rüppell (1794–1884), see ADB Rüppell; Goren (2003), 57–60.

[34] Robinson (1841), I, 241, added (note 1) that a view of the castle 'is given by Rüppell and by Laborde.' Cf. Pringle (2009).

[35] Kasim to the Governor of 'Akabah, n.d., ESP, ABC 60 (137); cf. Robinson (1841), I, 242. For discussions concerning different Bedouin tribes in Sinai, see Robinson (1841), I, 197–213.

[36] Robinson (1839), 406; (1841), II, 244–45; cf. Goren (2015).

For the difficult and dangerous trip from Jerusalem to the Jordan River and the Dead Sea, in place of soldiers, Robinson and Smith preferred employing the Sheikh of the Ta'amra Bedouin tribe with four of his men: 'In one and in all our intercourse with the Taamrah, we discovered hardly any of that prejudice against Christianity which is so characteristic a trait of Mohammedanian; & I found a freedom in conversing with them on religious topics, which I have never experienced in intercourse with any other followers of the Arabian prophet.'[37]

In Petra, they narrowly escaped confrontation with the local Sheikh, Abu Zeitûn.[38] Robinson recounts that the old Sheikh, 'who caused so much difficulty to Mr Bankes and his party in 1817,'[39] demanded a rather large tribute 'for the privilege of visiting the territory.' When they refused, he threatened to bar them from continuing to Mt. Hor, the traditional site of Aaron's tomb. At the drama's height, 'swords were drawn and banished,' resulting in them forgoing the climb up Mt. Hor and leaving Wady Musa after visiting Petra, even though the Sheikh, fearing Ibrahim Pasha's reaction upon hearing the story, withdrew all his demands.

Robinson and Smith also documented the Americans whom they encountered on their way, making special note, of course, of the missionaries who assisted them. Robinson first met missionaries in Cairo. Later, in Jerusalem, Missionary Lanneau graciously hosted the travellers for three weeks in his house 'which was one of the better class (...). It furnished indeed one of the most desirable and healthful residences in the city.'[40] They arrived in Jerusalem on April 14, the timing of which must have been pre-planned. As Easter Sunday 1838 fell on the 15th,[41] they were able to attend the Jerusalem ceremonies and, most importantly, participate in the annual meeting of the ABCFM Near East missionaries held in Jerusalem that year. Knowing their itinerary, Smith may have

[37] Robinson (1839), 417; (1841), II, 153–54; Smith to Homes, Smyrna, 4.3.1838, ESP, ABC 60 (105/5), 1838.

[38] Robinson (1839), 425; for greater detail see Robinson (1841), II, 535–47; cf. Williams' description of the incident, (1999), 248–49.

[39] The difficulties faced by travellers wishing to visit Petra, mainly due to rivalry between tribes and their leaders, has been described, *inter alia*, by Irby (1823), 120–21 (with his companions Mangles and Bankes), and by Laborde (1836), 144–47, where he cites the Bedouin guide's version to Bankes's story.

[40] Robinson (1839), 410; (1841), I, 327.

[41] Good Friday on 14.4, Easter Saturday 15.4, Easter Sunday 16.4. Cf. Robinson (1841), I, 329–31 about Easter Sunday.

arranged for the annual meeting to be held in Jerusalem rather than in one of the other ABCFM Near Eastern centres.

The missionaries came to Jerusalem with their families, 'like the Hebrews of old, at the time of the Passover, to worship in this place, and to consult together on the best measures of promoting the great work in which they were engaged.' Eight ABCFM missionaries attended from the stations in Jerusalem, Beirut, Cyprus, and Constantinople.[42] In his report to secretary Anderson after the expedition, Smith describes the gathering as 'not the least interesting event in our journey (...), when I had the pleasure of meeting all my brothers, except brother Pease of Cyprus.' He notes some rather impressive achievements: 'I was thus enabled to preach the annual sermon, to read an essay on missions among the Mohammedans in Turkey, & present for the apparition of the mission, a treatise in the Office a work of the Holy Spirit, for publication in Arabic, all which had been assigned to me at the previous meeting, & had been composed during my service in Beyroot.'[43] The missionaries also discussed progress of the Arabic printing press and recommended that Smith travel to Europe at the expedition's end, most likely to Germany, to work on production of the Arabic types before continuing to the US.[44] The gathering took place until the outbreak of the plague in Jerusalem.[45]

The Jerusalem friends accompanied the travellers on selected routes. On their first excursion from Jerusalem to the mountainous area northeast of the city about which Smith commented, 'a region which, so far as I know, has never been visited by any Frank traveler,' their party was augmented by ABCFM missionary Lanneau, LJS missionary Nicolayson, and visitor Paxton.[46] Robinson could not resist writing a sarcastic comment in the *Biblical Researches* about Paxton, whom the ABCFM had refused to accept into its service. In his book, published in 1839,

[42] Robinson (1841), I, 327–28.

[43] Smith to Anderson, Smyrna, 9.8.1838, ESP, ABC 60 (105/9) 1838 Jul.-Dec. Lorenzo M. Pease served in Larnaka, Cyprus from 1835 (Kawerau (1958), 286).

[44] See in detail chap. 5.a. Interestingly, Nicolayson was in Jerusalem at that time and refers to American missionaries in his journals (Nicolayson (1836), 10–12, 17), but does not mention this conference!

[45] Robinson (1841), I, 333.

[46] Robinson (1841), II, 107. See Smith's letter to Paxton, Smyrna, 24.1.1837, ESP, ABC 60 (105/5), 1837. For a later (1846), bitter discussion between both concerning Paxton's 'calling upon me to correct certain statements, in a late communication of mine to the N. Y. Observer,' see Smith to Paxton, Boston, 8.10.1846, ESP, ABC 60 (105/5) 1846.

Paxton makes no mention of travelling with Robinson and Smith. Yet, in a letter appearing in the book that Paxton dates to October 1836, he mentions place names that he surely only learned for the first time while accompanying Robinson and Smith, 'thus leaving upon the mind of the reader the impression, that he became acquainted with them two years earlier,' prior to the expedition.[47] No further comment is needed.

Letters sent by Jerusalem missionaries Whiting and Lanneau provide a fascinating glimpse into the plague that struck Jerusalem and its implications for the expedition. Both missionaries were asked to make arrangements for supplying the group with 'seven spirited mules' for their ensuing journey from Jerusalem to the Galilee and then to Damascus. Robinson, Smith, and their party returned to Jerusalem on June 9, but the city was shut down due to the plague. They had to pitch their tent near the Damascus Gate, where they were joined by both missionaries and their families, who received special permission to quarantine the group at Lanneau's house.[48] Lanneau's long and detailed letter to both Robinson and Smith, dated June 5, 1838, ten days after their initial departure from Jerusalem, is packed with information about the quarantine. 'I embrace this the first opportunity I have had since your departure (...). We have now been closely shut up for 19 days by a strict quarantine (...) how many have fallen victims to the plague it is difficult to ascertain.' The letter continues with an expression of the safeguarding of the expedition party, 'We were rejoiced that a kind providence had watched over you during your journey and brought you this far in your way in health and safety.' Lanneau then provides detailed instructions for the timing of their arrival in Jerusalem ('by Friday afternoon, so that we may join you early on Sunday morning') and where to pitch their tent outside the city ('somewhere in the grove of olives on the north as near to the city as practicable'), concluding with the reassurance that 'Brother W.[hiting] will see that every thing be done according to your directions.'[49]

Robinson's opinion of the Roman Catholics in Jerusalem is quite interesting. Bliss writes that his approach was 'a scepticism not negative but brilliantly positive,' but as for the Holy Sepulchre, 'here speaks the

[47] Robinson (1841), II, 107, n. 1. Paxton (1839), letter XV (Nazareth, 17.10.1836), 169.

[48] Robinson (1841), III, 68–73, 1st Appendix, 80; Lanneau to Robinson and Smith, Jerusalem, 5.6.1838, ESP, ABC 60 (49).

[49] Lanneau to Robinson and Smith, Jerusalem, 5.6.1838, ESP, ABC 60 (49). See also Whiting to Robinson, Jerusalem, 4.6.1838, ESP, ABC 60 (91).

Puritan, not the Explorer.'[50] This paragraph from the *Biblical Researches*, though a bit long, is worth citing, as it bears evidence of his extreme criticism. After observing the 'high day' of the Passover festival, Robinson conveys his impression that 'the whole scene was to a Protestant painful and revolting.' He does not stop with this censure but adds these rather harsh and sarcastic comments:

> I was struck with the splendor of their [the Latins] robes, stiff with embroidery of silver and gold (...); but I was not less struck with the vulgar and unmeaning visage that peered out from these costly vestments. The wearers looked more like ordinary ruffians, than like ministers of the cross of Christ. Indeed there is reason to believe, that the Latin monks in Palestine are actually for the most part ignorant and often illiterate men (...) who come or are sent hither as into a sort of exile (...). There was hardly a face among all those before us, that could be called intelligent.[51]

Although some European Catholics joined the procession, they appeared to Robinson 'ashamed of their companions.' He found it necessary to explain that these remarks should not be viewed as stemming from 'any spirit of prejudice against the Roman church or her clergy.'[52]

Naturally, Robinson also appears to have made a point of meeting the American consuls stationed along their route. His first encounters were in Egypt with consuls John Gliddon and son George Robins Gliddon.[53] An interesting document in his archive, written in Arabic and signed by the 'Consular Agent of the American State in Tyre,' had been written as verification of his intention to continue to Beirut.[54] There was no American consular-agent serving in Jerusalem at the time, and the first Americans he met after leaving Egypt were the consular-agents in Tyre (Ya'kôb 'Akkâd, a Greek-Catholic) and Saida (Ibrahîm Nŭkhly, 'a wealthy Greek-Catholic').[55] Robinson's original passport for the expedition, issued in Washington on June 21, 1837, contains various stamps,

50 Bliss (1907), 213–14.

51 Robinson (1841), I, 330–31. See also Bliss (1907), 213–14, citing Robinson's 'obvious reluctance' with regard to the Roman Catholic and Eastern clergy and their 'obnoxious' ceremonies.

52 Robinson (1841), I, 331.

53 See in detail chapter 2.b.

54 Tyrus, 12.6.1848, ERP, J.2. My gratitude to Prof. Mustafa Abbassi for the translation. See Ill. 7, p. 41.

55 Robinson (1841), III, 393–94, 416–17.

including a stamp dated July 9, 1838, of Salonika-born businessman Jasper Chasseaud (1780–1857) who served as American consul in Beirut from 1836–50, and one dated September 3, 1838, of John P. Brown, first American consul general in Constantinople and later the US consul in Berlin.[56]

The expedition party also met European travellers en route, mainly during their stay in Jerusalem. Robinson mentions Duke Maximilian of Bavaria (Maximillian Josef, 1808–88) and his 'somewhat numerous suite' who stayed in the city during the same period, though it is not clear whether they met. The Bavarians were not spared from the plague.[57] They did meet 'Mr [Jules de] de Bertou, a Frenchman,' who had just returned from an excursion to Petra and 'Akabah, and spent an evening with him listening to his travel experiences.[58] Upon returning to Jerusalem on June 5, they called at two English travellers who had pitched their tent nearby at the city gate. They were the British churchman and historian William Hodge Mill (1792–1853), first principal of Calcutta's Bishop's College, and the Spanish Colonel Joseph de Hezata (José Hezeta Zenea, 1788–1862), who returned from India via Egypt, and sailed with them from Beirut to Alexandria and Smyrna.[59]

The general geo-political situation had a rather strong influence on the travellers and their expedition; where relevant, Robinson describes the climate in the *Biblical Researches*. Arriving in Jerusalem, he recalls being 'struck with the comparatively few people we met.' Only the bazaars and the main streets were relatively crowded, yet outside the city it was quite deserted.[60] The vacant city was a consequence of Ibrahim

[56] Robinson Passport, ERP, ERP, J.2. For Chasseaud: Kark (1994), 81–87, 342; Oliver (2014), 82–83. Brown: Kark (1994), 79, 340.

[57] Robinson (1841), I, 370. For the Duke and his visit see Maximilian (1978); Goren (2009), 106–10.

[58] Robinson (1841), I, 370; II, 589, 601 n. 2; 'Remarks on Bertou,' Robinson (1841), II, 659–69; Bertou (1838); (1839). For detail on Robinson and Bertou, see chap. 4.c; Goren (2011), passim.

[59] Robinson (1841), II, 637–38, III, 448–49. Mill, graduate of Cambridge, resided in India from 1820–38 when he returned to Europe; he was a leading scholar in the Bengal Asiatic Society (ODNB Mill). Hezata, a Havana born Spanish naval officer, was secretary of the Spanish War Office in 1820. In 1823, he was forced to take refuge in Gibraltar, subsequently returned to Spain and then escaped to England in early 1826. Hezata received a senior commission to the West Indies and arrived in India in April 1829, where he devoted himself to the cultivation and manufacture of indigo. In 1837, he embarked on a four-month voyage in the Red Sea to Egypt, traversing the Sinai to the Holy Land, and back to Spain (Gil Novales (2019)).

[60] Robinson (1841), I, 362–63.

Pasha's first campaign, in 1834, defeating the Peasants' Revolt. Ibrahim Pasha's 1838 campaigns against the rebelling Druze took place at the time of their visit.[61] Robinson, who heard rumours of the rebellion before leaving Cairo, described upon arriving in Jerusalem its effect on the city, 'full of rumors (...) the unquiet spirit of the Land began to rouse themselves.' For some time, it was doubtful whether they would be able to continue travelling. Soon, however, it became clear that Ibrahim Pasha was well on his way to quashing the Druze rebellion and in Jerusalem reigned 'the same feeling of security as in England and Germany.'[62] Yet, 'as we were to have a specimen of all the evils to which the oriental world is exposed, a few days after our arrival in the Holy City, rumours of the plague began to be circulated.' For the American traveller, it was 'a state of things such as I had never anticipated, and which I shall never forget. Men's life seemed to hang in doubt before them. No one knew what to do or whither to turn himself.' Nonetheless, the party continued exploring the empty streets, 'and a kind of Providence preserved us from the dangers by which we were surrounded.'[63]

Upon reaching Tiberias, Robinson and Smith learned that the rebellion was still waging in the north, thwarting their plan to proceed directly to Damascus. Instead, they reached Safed, then 'still little more than a heap of ruins' due to the earthquake of January 1, 1837. The party had not intended to reach Safed at all but to study the sources of the Jordan River: 'it had been our plan to proceed from Tiberias along the lake northwards, and so by the Jordan and the shore of el-Hûleh to Bâniâs.' They then planned to travel by way of 'Hâsbeiya and Râsheiya to Damascus' and continue to Ba'albek and only then to Beirut.[64] But they were compelled to seek haven in Safed where they would decide whether to continue to the Hula Valley and the Banias. 'Our aim in coming hither, was to obtain intelligence respecting the safety of the roads to Damascus,' Robinson writes of Safed. It was a problematic decision, as initially it was difficult to gather information, with 'every one being silent for fear of the government.' They even considered waiting for a

[61] For campaign details in contemporary and modern studies, see Perrier (1841); Prokesch-Osten (1877); Anderson, M. S. (1966), 88–109; Fahmy (1997); Kutluoğlu (1998). For Robinson's description: Robinson (1841), II, 461; Nicolayson's impression: Nicolayson (1838); Bird's description: Bird's journal, IBP-YSM, Box 2, Folder 25, 105–17.

[62] Robinson (1839), 414; (1841), I, 365–67.

[63] Robinson (1841), I, 367–69.

[64] Robinson (1841), III, 274.

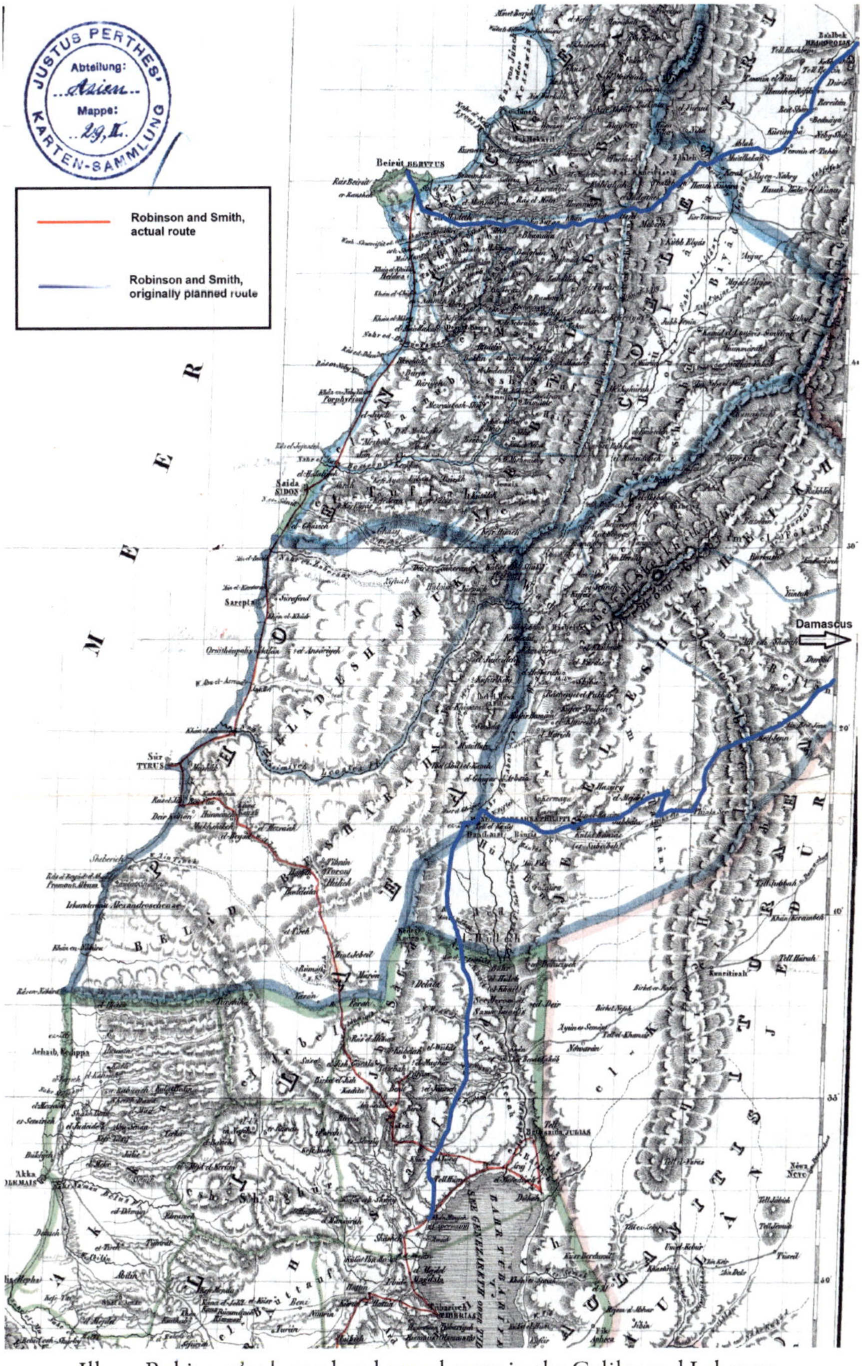

Ill. 11. Robinson's planned and actual route in the Galilee and Lebanon.

large caravan from Acre to Damascus, but slowly intimations against taking that route won over.[65] Instead, one of their muleteers, a resident of Safed, offered to take them 'to a spot about an hour N. E. of the town, where he said we might have an extensive prospect.' The party travelled to the mountain summit of Benît, today Yavnit (+897 m), 'commanding a splendid view over the whole basin and the surrounding region.'[66] Even more astonishing is Robinson's detailed description of the whole area including the sources of the Jordan and Mount Hermon, compiled from what he observed in a mere few hours and in constellation with all his sources.[67]

Ill. 12. Landscape of Hula Valley, Sources of the Jordan, Golan, Mount Hermon.

At that point, the security threat in and around the Upper Galilee due to the Druze rebellion forced Robinson to change the expedition route: 'the account became more threatening; and we were compelled to turn our faces towards Beirout by the way of Tyre and Sidon.'[68] Still lingering under the impression of continued instability and uncertain-

[65] Robinson (1841), III, 336–38.

[66] Robinson (1841), III, 339.

[67] Robinson (1841), III, 339–64.

[68] Robinson (1839), 429–30; for the longer discussion see Robinson (1841), III, 274–75 (Tiberias), 336–39 (Safed); Bird's journal, IBP-YSM, Box 2, Folder 25, 125–30.

ty, Robinson writes in the *Biblical Researches*: 'Such was, and such is, this land of wars and rumours of wars! Reports fly current from mouth to mouth, of which no one can learn, whether they are true or false; suspense, agitation, and anxiety prevail continually; without the possibility of ascertaining whether or not there actually exists the slightest cause of inquietude.'[69] These matters, needless to say, had nothing to do with Biblical Geography. Still, one might expect that the letters would reveal even more information concerning such dramatic events. Unfortunately, neither Robinson nor Smith wrote many letters during the expedition, lacking both the time and the possibility of delivering them.

The earthquake devastation in Safed, as well as in Tiberias, also left a very strong imprint.[70] Robinson devotes much space in the *Biblical Researches* to the aid given by a joint expedition of representatives of the two Protestant missionary societies in Beirut. They were the first to reach the devastated city, on January 18, a long two weeks after the disaster hit. The delegation included ABCFM missionary William McLure Thomson and CMS Jerusalem missionary Erasmus Scott Calman (1796–1889), an aid to Nicolayson who, he noted, 'has fully supplied my place' in Beirut.[71] Arriving in Beirut in early 1836, Nicolayson met Calman, and upon 'not finding a vessel for Jaffa, was induced to accompany the Rev. Mr Elliott in a journey overland to Jerusalem' in March 1836. Nicolayson's report supplies important information concerning his travels, places, and people.[72] Interestingly, the report in the *JI* has no reference to the four days that Calman spent in Safed.[73] Thomson and Calman wrote relatively long reports of the devastation in Safed (Calman's is longer and much more detailed). Robinson, who met Thomson in Jerusalem and again in Beirut, received his report and published it in his book.[74]

69 Robinson (1841), III, 275.

70 Robinson (1841), III, 253–56 (Tiberias, also mentioning the 1759 earthquake), 321–23 (Safed).

71 Smith to Bird, Beirut?, 1831, IBP-YSM, Box 1.

72 Nicolayson (1837); (1837a). Reverend Charles Boileau Elliott (1803–75), English travel writer. His travels: Elliott (1839), for Nicolayson and the missionaries, II, 104–07, 192.

73 Mordechai Eliav found Calman's report and published it in Hebrew, adding many details about the different delegations following the earthquake (Eliav (1996)). Elliott met Calman in Beirut: Elliott (1839), II, 99–102. Cf. also Smith to Bird, Smyrna, 4.8.1838, ESP, ABC 60 (105/5) 1838.

74 Robinson (1841), III, 471–75.

The ABCFM missionaries in Lebanon were deeply involved in the events surrounding the 1838 Druze revolt. Smith, whose task in life as a missionary was always primary, concerned himself with the unfolding developments even while engrossed in post-expedition research. Smith, now in the US and disconnected from the events, received a report from Thomson, who was heading the Beirut station. In his letter, Thomson responds to Smith's discussion of the revolt's expansion, the movements of the Pasha, and the possibility that conversion to Christianity might safeguard the Druze from the Egyptians: 'In regard to the plan concerning the Druzes which is drawn out at some length at your letter,' Thomson replies that he did not expect a significant number of converts to Christianity.[75]

As noted, the *Biblical Researches* contains numerous interesting and important descriptions that have no direct connection to toponymy and scriptural geography. Many of them deal with everyday life, agriculture and agricultural methods in different locations along the route, confrontations with local authorities, whether Ottoman or resident leaders and dignitaries, and the customs and habits of local people and sects. Robinson and Smith were particularly fascinated by the Samaritans in Nablus (Nâbulus by Robinson), noting their history, religion, and customs.[76] There are various records of flora and fauna, one outstanding example being their description of the camel, with a long discussion of its purchase and rental prices, uses, and advantages and disadvantages.[77] Another outstanding report deals with arms and gunpowder production of the Jehâlîn, the Bedouin tribe from the Hebron area that provided expedition guides to the adventurous Petra expedition. Egyptian authorities at that time officially collected arms and prohibited the production of gunpowder, but Robinson reported that 'the Jehâlîn have not been disarmed; they have still their old muskets with match-locks; and make their own gunpowder.' They obtained the needed components, the nitre [potassium nitrate] 'from the dust of the ruined villages in their country' and the sulphur, naturally, from the shores of the Dead Sea.[78]

Given the attention to other topics, a curious omission from both their correspondence and publications is any mention of the personal equipment they carried en route. There is one hint, a list of Smith's be-

75 Thomson to Smith, Beirut, 28.12.1838, ESP, ABC 60 (83).

76 Robinson (1841), III, 96–133.

77 Robinson (1841), II, 630–35.

78 Robinson (1841), III, 469, and 221–22 concerning minerals around the Dead Sea.

longings that he carried with him on his return from Constantinople to Vienna and Berlin. One would expect this list to be different from the list of items he carried on the expedition, even though he expected that both legs of travel would entail a rather long period of wandering. Upon entering the Habsburg Empire from the Ottoman Empire, he received a document from Austrian customs.[79] The document contains a description of his person: Smith, 36 years-old, average stature, brown hair, beard and eyebrows, gray eyes. It also lists the items he carried in addition to the clothes he wore: '*2 Stöcke* [sticks], *2 Fracke* [tioxido], *4 Kittel* [overalls], *6 wollene Unterhosen* [underpants], *3 pr Handschuh* [gloves], *3 pr Hosen* [trousers], *3 wollene Hemden* [shirts], *2 pr Stiefel* [boots], *2 pr Schuh* [shoes], *1 Schreibzeug* [writing material],' and again '*8 Hemden* [shirts], *2 Handtücher* [towels], *2 Nachtmützen* [nightcaps], *4 Sacktücher* [sackings], *1 … tierspritze* [??], *5 Halstüchrl* [scarfs], *5 Krägen* [collars] *2 Halsbinder* [neck tie] *1 lederner Koffer* [leather suitcase], *10 pr Strümpfe* [socks].' He also transported ten packets of manuscripts and thirteen books. Were the packets of manuscripts the daily diaries he wrote during the expedition and later gave to Robinson for the writing of the *Biblical Researches*?

3.c. The post-expedition journey: Robinson and Smith to Germany

Robinson, Smith, and their companions reached Beirut, the final station of their journey, on Thursday, June 26, 1838, three and a half months after their departure from Cairo. In Beirut, Robinson had a slight attack of the chronic illness that plagued him from time to time, but soon recovered enough to continue.[80] His travel plans from Beirut were straightforward. His wife and two children awaited him in Halle. Robinson intended to go to Halle and maintain his original plan to begin his teaching and other duties at UTS in the fall.[81] Heading eastward at

[79] Sanitäts-Fehde für Personen, 3.9.1838, ESP, ABC 60 (137). See chapter 3.c. See ill. 13, p. 92.

[80] In his youth, Robinson began suffering from 'upper respiratory disorders' (Williams (1999), 64) and much later from Bright's disease, acute or chronic nephritis, a kidney disease after Richard Bright (1789–1858) and cataracts (Robinson, M. A. (1862–64), 157). Robinson's illness is discussed as motivation for his resigning from Andover in 1833 (Weigel Williams (2007), 86–87).

[81] As mentioned in various Robinson letters, e.g. Robinson to Norton, New York 1.7.1837, ERP, ER 1.12.

the start of the expedition, Robinson had 'set off from Berlin on the 13th Nov. by way of Halle,' travelling through Vienna to Trieste ('I had chosen the route by Trieste as the shortest'), and then sailed through Athens and Syra to Alexandria.[82] Now in Beirut at the end of his journey, he decided to sail up the Danube, taking a direct route back to Germany.

Smith's travel plans were far from clear. Originally intending to go directly to the States, he now meant to actualize a plan deferred given the timing of the expedition with Robinson. As noted, prior to the expedition, Smith was preoccupied with preparing and producing the Arabic types for a printing press in Beirut where they planned to print an Arabic translation of the Bible. The Arabic types became a determining factor in Smith's post-expedition travel plans. The story of the Arabic printing press is well published.[83] The short account here, illustrated by excerpts from Smith's correspondence, explains the basis for Smith's travel decisions that contributed not only to the success of the printing press, but to the writing of the *Biblical Researches* as well.

The decision to print an Arabic translation of the Bible had roots in both the aims of the ABCFM and the emerging legacy of foreign language publications initiated by Andover scholars. As early as the 1820s, Moses Stuart, Robinson's teacher and tutor at Andover, printed books with Hebrew letters and 'as early as 1829 he had at his command fonts of type for eleven oriental languages and dialects.'[84] Robinson conveys this detail even earlier in a letter dated June 15, 1828, to 'Messrs Breitkoff, Harteli [Härtel] and Co., Leipzig.' This firm, world renowned for its publication of sheet music, an influential journal and books on music, and manufacture of musical instruments, also published theological works and operated a large foundry and printing house. The firm had shipped a box of types to Robinson in the US that unfortunately 'was totally lost on the passage between N. York & Boston.' In the same letter, Robinson requests the Leipzig firm 'to make anew for me <u>all the types contained in the first box</u>, i.e. all the types that I ordered from you <u>except</u> the smallest Hebrew, which was sent in January in the <u>second box</u>,' and orders Greek, Coptic, Armenian, Ethiopian, and Samaritan capital letters.[85]

82 Robinson (1841), I, 3–19; Receipt for a parcel sent by Smith with the Austrian Lloyd to John M. Adger, Smyrna, Syra, 10.2.1838, ESP, ABC 60 (128).

83 Hallock (1929); Glass (1997), esp. 14–30; Roper (1998), 58–62; Coakley (2003).

84 Kitto (1853), 313.

85 Robinson to Breitkopf, Härtel and Co., Dresden, 15.6.1828, ERP, ER 1.6.

In 1834, four years before the expedition, the task of completing the Arabic printing press had been put in the hands of punchcutter Homan Hallock under the supervision and responsibility of Smith. Hallock, son of a Massachusetts Congregational family and apprenticed to printers 'Flagg and Gould' in Andover, 'had the capability, almost alone of American printers at the time, of dealing with Oriental languages.' Secretary Rufus Anderson recruited him to take over the printing office of the ABCFM's board in Malta, which he reached in December 1826, later marrying and entering the ABCFM service as a missionary. In 1833, the printing office and equipment were moved from Malta to Smyrna where the Hallock family settled, and the Arabic printing material was sent in 1834 with Smith to Beirut.[86]

Smith's archive holds numerous letters concerning the production of the types.[87] Hallock's detailed letters foresee many of the personal, technical, and financial obstacles that would hamper his efforts and successful completion of the Arabic printing press in the coming years:

> I offer you no apology for repeating my mind freely. So much of my wisdom as is worth taking, accept. At the remainder be angry or pleased, as the case may require, only let your anger bear only as long (for your own sake) as the scriptures allow. – I have many things, on at least a matter in parvo, to say to the man who shall take the place of responsible printer in connexion with your press; and in relation to him I will venture to add, and then to wish, that he will say less and do much to help you, if you will say less and do nothing to help him.[88]

The project certainly did not advance as expected. In April 1836, an already disillusioned Smith writes to his fellow missionary Isaac Bird:

> I know not how the plan for getting the new Arabic types, may strike you, & regret very much you could not have been here [Beirut], to give your advice. The fact is, I have got tired of wearing myself out, & expending money, in trying to work this press, in so faulty a condition. If I am to spend my life in connection with it, I must be able to accomplish something.[89]

[86] Coakley (2003), 19–21; (2013), 550–65, also in reference to Anderson's role in advancing Arabic printing.

[87] 1834 letters: Hallock to Smith, Smyrna, 20.4.1834 (sends first samples) and 28.4.1834 (exact prices), 28.5.1834, ESP, ABC 60 (32) (the most detailed and technical letters). Extensive correspondence followed, mainly from 1837.

[88] Hallock to Smith, Smyrna, 28.4.1834, ESP, ABC 60 (32).

[89] Smith to Bird, Beirut, 28.4.1836, IBP-YSM, Box 1.

Following the shipwreck and after spending two months with the missionaries in Constantinople where he 'produced a new set of 1,800 drawings,'[90] Smith returned to Smyrna in January 1837, bringing them to Hallock. He updates fellow missionary Whiting in Jerusalem by letter, 'I have just returned from a visit of two months at Constantinople, endeavouring to make models for our Arabic font of type,' and contemplates whether Hallock will be able to produce the Arabic types successfully. Hallock's slow progress in cutting the punches of the Arabic typeface designed by him prevented a disappointed Smith from returning quickly to Syria, which he describes with longing, 'I love that land. My heart is there, so much of is as yet remains on earth.'[91] The situation changed little by the end of that year. To his fellow missionary Daniel Temple serving with Hallock in Smyrna,[92] Bird writes from Andover that 'I have wanted very much to know about the success of Bro Hallock in cutting the Arabic Type. We only hear that he is likely to succeed.' Temple (1789–1851), one of the earliest ABCFM missionaries, had moved the printing house from Boston to Malta in 1821. In 1826, he was assisted by Eli Smith, serving in his first ABCFM post. Bird also hoped to see Smith soon on American shores.[93] Four days later, Smith, now in Smyrna, wrote his above-mentioned letter to the brethren of the Syrian mission unveiling his plan to accompany Robinson. In the same letter, he devotes a significant portion to 'a somewhat detailed account of my proceedings in reference to the Arabic type.'[94]

Immediately before the expedition, Smith ruminated on his responsibility for the Arabic press and the factors that convinced him to leave the work in Hallock's hands and disappear for the expedition duration with Robinson. Smith's reasoning is apparent from this letter: 'The amount our Society had expended in establishing a foundry here, & the confidence they had referred in Mr Hallock by putting it into his hands. (...) When returning from the excursion, I will still be able to control the work in Hallock's house, and if it will be accomplished, could the necessity of my crossing the Atlantic be evident.' Smith further explains

[90] Leavy (1993), 13.

[91] Smith to Whiting, Smyrna, 10.1.1837, ESP, ABC 60 (105/5). See Leavy (1993), 11–13.

[92] Temple (1855); Anderson, R. (1872), I, 15, 101, 139–40; Kawerau (1958), 188–89, 204–05, 208 (Temple's wife died in 1827), 365–66; Leavy (1993), 7.

[93] Bird to Temple, Andover, 8.12.1837, IBP-YSM, Box 1.

[94] Smith to Brethren of the Syrian Mission, Smyrna, 16.12.1837, ESP, ABC 60 (105/7).

that he has a letter from Anderson instructing that the work should be done in their own foundry:

> Considering, also how near the whole thing then was to reexperiment & the amount of specific corrections & dissections I should have to suggest, I confess that I shrink from the responsibility of making a contract with a strange man, which then was the possibility of avoiding it. I also doubted, whether any men without Mr Hallock's machine would be able to imitate my models so as to avoid the necessity of making the combinations. (…) the work is a great one, much greater than I at first supposed. It would take time anywhere.[95]

Smith remained absorbed with the printing press on his way to Egypt to meet Robinson for the start of their expedition. Anderson, also very troubled by the lack of progress, had offered to send George (Christopher) Hurter, former CMS missionary born in Malta to a Swiss father and English mother. Hurter wished 'to get employment under the board as printer' and Anderson suggested Hurter given that Hallock, Temple, and Smith were well acquainted with him. Smith responds to Anderson, 'concerning his acquaintance with local languages, I can recommend him, though there might be a problem with his health. (…) His departure need not be hurried until we get our type finished. In the mean time he can be more profitably employed in improving himself in the arts.'[96]

After reaching Jerusalem in April 1838, Smith writes to Hallock in Smyrna that 'in reference to my future movements, my brethren have concluded that it will possibly be best for me to go to Germany to get the matrices fitted up.' Smith adds that because Robinson had paid his expenses to date, he was expected to accompany him, most likely to Germany, and could only stay in Smyrna for a few days.[97] On May 28, Hallock responds that he plans on finishing the punches in time for Smith's arrival in Smyrna so that Smith can 'remove the whole to America without delay.'[98]

[95] Smith to Brethren of the Syrian Mission, Smyrna, 16.12.1837, ESP, ABC 60 (105/7), both citations, cited partially in chap. 2.d. We could not trace Anderson's letter.

[96] Smith to Anderson, Syra, 31.1.1838, ESP ABC 60 (105/8). Hurter reached Beirut in 1841 as the first qualified printer there and served until 1864: Leavy (1993); Roper (1998), 60; Auji (2016), 73–74.

[97] Smith to Hallock, Jerusalem, 25.4.1838, ESP, ABC 60 (105/8).

[98] Hallock to Smith, Smyrna, 28.5.1838, ESP, ABC 60 (32). The letter contains an exact counting of exactly 240 punches: Ready or almost ready for driving - - 79 + 8 = 87

In the end, upon reaching Beirut, Smith decided to sail first to Alexandria. Writing on July 12, Smith repeats that at the start of the expedition he intended to sail directly to the US upon its conclusion, but his 'brother missionaries were still in favour of my going to Germany,' a decision he finalized only upon reaching Smyrna on August 4.[99] Even then Smith wavers, still uncertain considering the financing of his next moves. Moreover, he expresses, 'I never was in a greater quandary than I am about my types. I absolutely know not what to do. I suppose I shall go to Germany. But I shall take the step with much hesitation.'[100] Even much later, Smith continued to receive anxious and even disheartening letters from fellow missionaries complaining about the highly unsatisfactory progress with production of the types.[101] In the meantime, having arrived in Constantinople from Beirut, Robinson writes to Smith:

> I was in hopes to have heard definitely of you accompanying me to Germany; but I can well enter into all your doubts; & also into your wishes & the motives for going directly to the U. States. As to the mere matter of the bound of types, my belief is that you can get them done much more readily & cheaply in Germany; – whether better or not I cannot of course decide; – but whether these motives are sufficient to overbalance the general virtues for proceeding to America, you can judge much better than I.[102]

Smith chose Leipzig, explaining his decision to execute the next stage of the project in Germany in a rather detailed letter to Anderson sent from Smyrna, dated August 9, 1838.[103] 'We did return safe and highly satisfied with our results,' he opens, 'and plan to publish our journal; I also had the opportunity to meet my brother missionaries.' The remainder of the

Made and finished -43
Do but not finished -51
Made & not to be finished at present - - - - - - -26
Fraction ready for driving - - - - - - - - - - - - - - - -10
Remaining to be made, embracing numerals & dble accts 23.

[99] Smith to 'my dear Brother,' Alexandria, 12.7.1838, Smith to Hoadly, Alexandria, 13.7.1838, Smith to Homes, Smyrna, 4.8.1838, ESP, ABC 60 (105/5).

[100] Smith to 'my dear Brother,' Alexandria, 12.7.1838, ESP, ABC 60 (105/5). See also Robinson to Smith, Constantinople, 7.8.1838, ESP, ABC 60 (63).

[101] See the letters from missionary Homes in Constantinople, 12.11.1838, 22.1.1839, ESP, ABC 60 (37).

[102] Robinson to Smith, Constantinople, 7.8.1838, ESP, ABC 60 (63).

[103] Smith to Anderson, Smyrna, 9.8.1838, ESP, ABC 60 (105/9). Following citations are also from this letter; see also Recent Intelligence (1838a).

letter, however, is dedicated to one issue: the types. Smith notes that his decision not to remain with Hallock in Smyrna turned out to be justified, as 'the length of time [he] has consumed upon it, shows clearly how undesirable it would have been for me to remain here.' Subsequently, Smith presents for Anderson's authorization the 'further steps which I propose to take in order to secure the completion of the works,' namely the production of the types, which had been approved by consensus ('not one expressed any doubt') at a general meeting of the missionaries. 'Acknowledging my great responsibility, it had not been an easy decision,' he writes, and 'it is not until since my arrival here, that I have made up my own mind on the subject.'

Clearly, until this point, Smith had been in favour of going directly to the States, but now he hoped to receive permission to travel to Germany. He notes that among other reasons, the conviction that mechanical difficulties impeded accomplishing the task in the East 'finally decided me in favor of going to Germany.' He details: 'Being to the shape of the letters, & their uniting with one another, not only the making of the punches, but the fitting up of the matrices of this species of types is peculiarly difficult. This difficulty is very much increased, in the format we are making, by some peculiarities introduced to inform Arabic typography.' Smith writes further that Hallock had only 'little acquaintance with the business,' the actual casting and producing of the types, and that 'I have never been able to rely fully upon his opinions.'

Apparently, the solution was not to be found in the US. Germany was the leader in Arabic typography and casting of types, offering the services 'of the best workers, & also the best advices.' Smith had one person in mind, 'probably there is none superior to it [his establishment]': Karl Christian Philipp Tauchnitz (1798–1884) who, in 1836, inherited his father's (Karl Christoph Traugott T., 1761–1836) printing business, established in 1796. The Tauchnitz script foundry and printing and publication house also concentrated in 'improving the ancient scripts, the Russian, Greek as well as the Oriental writings.'[104] Robinson had met Tauchnitz Jr. in 1828, describing him then as 'scientifically & practically acquainted with the subject, is a fine man & an ardent friend of Missions.'[105] Smith writes further that since Robinson offered to finance

[104] Smith to Anderson, Smyrna, 9.8.1838, ESP, ABC 60 (105/9) 1838 Jul.-Dec. Arabic scripts: 'Proben aus der Schriftgiesserei von Karl Tauchnitz in Leipzig [trials from the script foundry of Karl Tauchnitz in Leipzig], 1925,' Schwarz (1924), 5.

[105] Weigel Williams (2009), 282–83. For both father and son Tauchnitz and their role in Arabic typography, cf. Schwarz (1924), 19; Glass (1997), 20–24 (unfortunately

Smith's travel to Berlin, the only additional cost to the ABCFM would be Smith's trip from Germany to the States, and the work in Germany was expected to be less expensive. He also notes that Robinson suggested that Tauchnitz might 'do the work, or a part of it, as a contribution to the missionary causes.' Finally, Smith emphasizes his desired outcome, 'having a format of type from my matrices, of which, there would be no chance in the U. S.'[106] While in Leipzig, Smith stayed in Tauchnitz's house and improved his German.[107] Production of the punches was not Smith's sole aim; he was also considering the future:

> We must calculate for a long time to come, to do our own printing, & if so, we must have ultimately, at least two more parts of the same shape as this, & if possible, as long to serve as (studies?). How & when shall we get these parts? (…) Is it not, therefore well that we shall (have?) at the content, by personal inquiry, the whole state of Arabic typography in Germany, prone where more than from any other place, we are likely to receive assistance.

Having made his decision, Smith joined Robinson in Constantinople to embark on their course up the Danube. 'I trust,' he adds in his letter to Anderson, 'that the committee, - yourself, will (…) approve of my courses.' Smith, of course, needed the ABCFM to supply him with the necessary financing and added Hallock's estimation for the cost of the types in Germany. With Smith on his way to Leipzig, Hallock continued his work in Smyrna, remaining in contact with him.[108]

It is quite clear that Smith received permission to travel to Germany only after acting upon his decision. Robinson arrived in the Ottoman capital on July 30, a week before Smith wrote to Anderson while still in Smyrna! Smith joined Robinson, as the latter wrote, 'on the morning of August 13th (…) in order to accompany me to Germany; whither he had been commissioned to repair, in order to procure the casting of new founts of Arabic types, for the press at Beirût.' Upon receiving authorization several months later, Smith quickly thanks Anderson: 'These letters were a great relief to me, & I feel very thankful to you & the Committee for the favorable view you & they have taken of my pre-

contains some inaccuracies); Liebrenz (2014), 37; Mienert (2017), though the book deals with another Tauchnitz printing house.

106 Smith to Anderson, Smyrna, 9.8.1838, ESP, ABC 60 (105/9). Subsequent citations are from this letter.

107 Talvj to Kopitar, Berlin, 14.12.1838, Milović (1941), 99.

108 For example, Hallock to Smith, Smyrna, 29.7.1839, ESP, ABC 60 (32).

sent course. It has relieved me of a weight of responsibility which I had long felt resting very heavily upon me. I now know that you are satisfied; & it remains only for me to satisfy myself.'[109]

On the very day that Smith arrived in Constantinople, Robinson and Smith embarked on a Danube steamboat. They were quarantined for ten days at Orsowa (today southeastern Romania), at the end of which they received the above mentioned *Sanitäts Fehde für Personen* (rules of health for persons) signed by the *k. k. Contumaz-Amte* (Imperial quarantine office) validating their release 'in full health.' They continued on their way, spending two nights in 'Gasthof vom Jägerhorn' in Pesth and arrived in Vienna on September 13.[110]

The Robinson family had arrived in Germany in September 1837, Talvj and the children spending the winter in Stettin with her brother and sister and her husband. They then moved to Berlin, from where she writes on March 23, 1838, that they were planning to continue to Dresden the following week where they would await her husband.[111] But things were not to be that smooth; this was only the beginning of a rather dramatic narrative. 'Among the immense marshes which stretch along the Danube in this quarter,' writes Robinson, 'I probably inhaled the poison of an intermittent fever.' He was not the only one to be affected:[112]

> At Vienna where we had intended to stop but four or five days, I was detained more than a month, which (?) by the sickness of Dr Robinson. Among the low ground & marshes, through which the Danube runs in Hungary, he was sickened with an intermittent fever, of the form known in those parts under the name of (?) fever. Three other of our companions had the same fever, making four of nine, the number of which our party consisted when we left Constantinople. Dr Robinson succeeded in reaching Vienna, & after a few days became convalescent. But a relapse soon brought him to the borders of the grave.[113]

[109] Smith to Anderson, Leipzig, 9.1.1839, ESP, ABC 60 (105) 1839.

[110] Smith to Anderson, Leipzig, 1.11.1838, ESP, ABC 60 (105/9); Invoice, 9–11.9.1838, ESP, ABC 60 (128); Document: ESP, ABC 60 (137). Robinson (1841), III, 448–51; cf. Williams (1999), 259. For the quarantine system on the border between the Ottoman and Habsburg Empires, established in the 1829 Treaty of Adrianople to fight the dangers of epidemics such as the plague, see Sechel (2011); Promitzer (2011); Mehedinţi (2012); Cotoi (2016). My deepest gratitude to Dr Ines Grigorescu of the Institute of Geography, Romanian Academy, for the relevant references.

[111] Talvj to P. I. Klöppen, Berlin, 23.3.1838, Krause (2001a), 294–95 (sent with Robinson on his way to the East); Weigel Williams (2007), 105–06.

[112] Robinson (1841), III, 451.

[113] Smith to Anderson, Leipzig, 1.11.1838, ESP, ABC 60 (105/9).

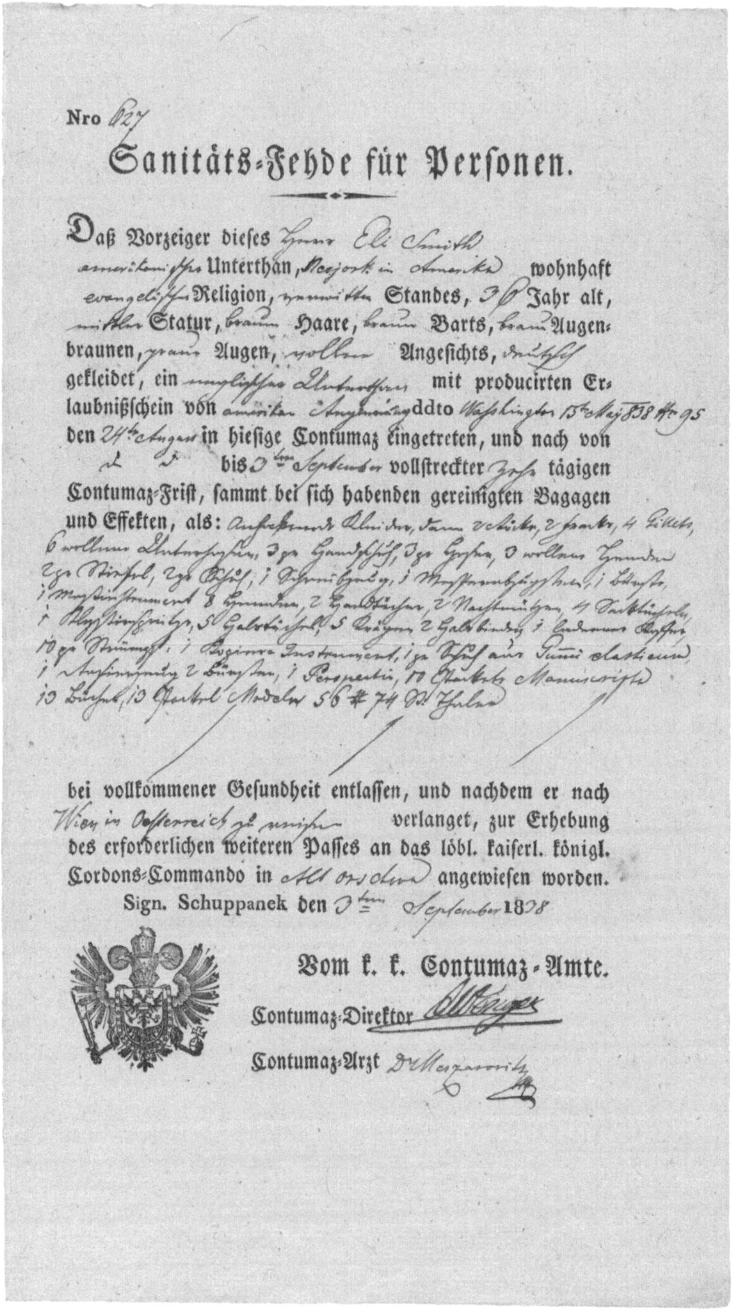

Nro 627

Sanitäts-Fehde für Personen.

Daß Vorzeiger dieses Herr Eli Smith amerikanischer Unterthan, Newyork in Amerika wohnhaft evangelischer Religion, [illegible] Standes, 36 Jahr alt, mittler Statur, braun Haare, braun Barts, braun Augenbraunen, grau Augen, vollen Angesichts, deutsch gekleidet, ein [illegible] mit producirten Erlaubnißschein von [illegible] ddto Washington 15 May 1838 Nr 95 den 24t August in hiesige Contumaz eingetreten, und nach von d. d. bis 3ten September vollstreckter zehn tägigen Contumaz-Frist, sammt bei sich habenden gereinigten Bagagen und Effekten, als: [illegible]

bei vollkommener Gesundheit entlassen, und nachdem er nach Wien in Oesterreich zu reisen verlanget, zur Erhebung des erforderlichen weiteren Passes an das löbl. kaiserl. königl. Cordons-Commando in Alt Orsova angewiesen worden.

Sign. Schuppanek den 3ten September 1838

Vom k. k. Contumaz-Amte.

Contumaz-Direktor [illegible]

Contumaz-Arzt [illegible]

Ill. 13. Sanitäts-Fehde für Personen [rules of health for persons] by the *k. k. Contumaz-Amte* [Imperial quarantine office], for Mr Eli Smith, September 3, 1838.

In the *Biblical Researches*, Robinson writes sparingly about his brush with death, adding at the end of volume III that 'our voyage up the Danube has already been described by one of the Party, Dr E. Zachariä (...). He had travelled (...) in search of manuscripts of the Byzantine law.'[114] Carl Edward Zachariä von Lingenthal (1812–94) was a German jurist and expert in Roman and Byzantine law. During his travels in 1837–38, he met Robinson twice, first in Vienna on their way to the East and then in Constantinople on their return. His travels brought him, as detailed in his book's title, through 'Wien, Venedig, Florenz, Rom, Neapel, Malta, Sicilien und Griechenland,' to 'Saloniki, dem Berg Athos, Konstantinopel und Trapezunt.'[115]

Robinson's party sailed by Varna, Constanta, and Tulcea to Galatz and Braila, where they spent the first Danube night on board, leaving the entire party feeling sick. Changing boats twice, each time to a smaller vessel, they sailed to Orsowa through the 'Iron Gates,' a 134-kilometre gorge on the Danube, but had to walk part of the way while the boat was towed by 'twenty oxen' upstream. In the Austrian border city Orsowa, they were quarantined for ten days, then continued in a carriage and again by boat, reaching Pest on September 9. The party left some of its sick travellers behind and continued to Vienna, which they reached on September 13. The voyage from Constantinople lasted 32 days.[116]

What was intended as a short stop in Vienna turned into a long and rather difficult ordeal. On September 26 and 27, two weeks after reaching Vienna, Smith sent two very alarming letters to Talvj in Dresden in which he writes: 'you will suffer nothing to delay your coming as speedily as possible. – Sympathizing with you most deeply in the feelings this Sickness of your excellent husband must occasion you, & praying that our fears may not be realized.'[117] The address at which the letters were initially received was Dresden; yet on the second letter's envelope it was crossed off and changed to Vienna. Talvj, as will become clear, was already on her way to the Austrian capital!

Smith detailed the illness in an exhaustive, 19-page letter to Robinson written before they left Vienna, another highly interesting document that has never been published: 'My Dear Sir, your request that I would

[114] Robinson (1841), III, 451 n. 1.

[115] Zachariae (1840); ADB Zachariae. For a detailed description of an 1836 sailing on the Danube from Vienna to Galaz, see Elliott (1839), I, 21–98.

[116] Zachariae (1840), 322–34.

[117] Smith to Talvj, Vienna, 27.9.1838, ERP, H. 2; found also in Robinson's archive and cited by Weigel Williams (2007), 106–07. We could not locate the first letter.

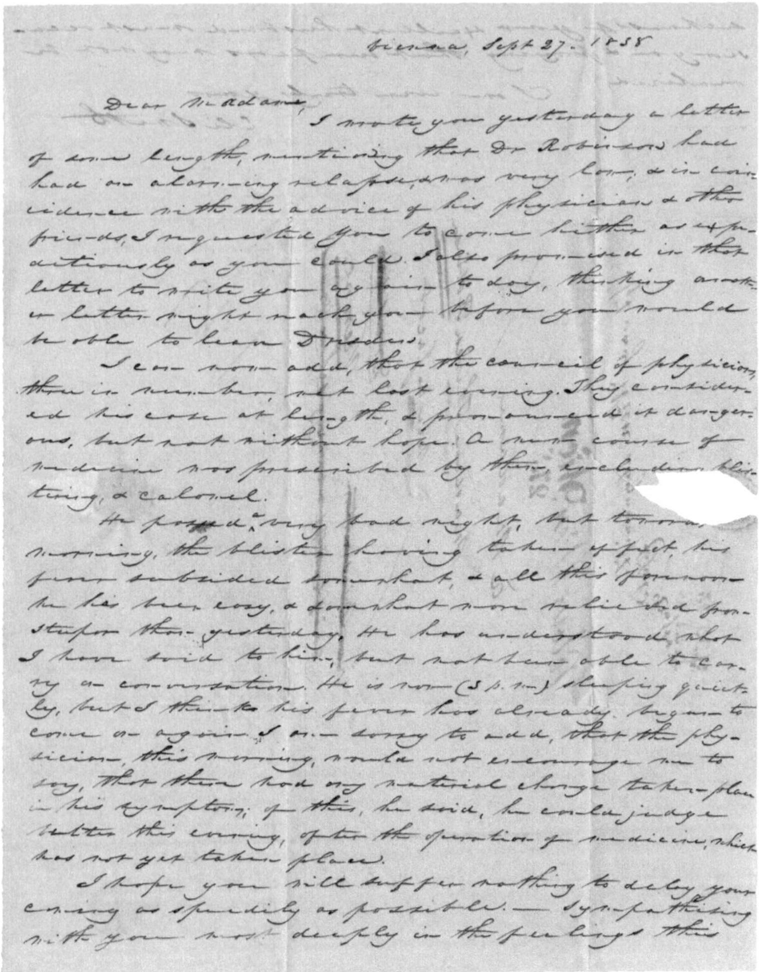

Vienna, Sept 27. 1838

Dear Madame,

I wrote you yesterday a letter of some length, mentioning that Dr Robinson had had an alarming relapse, & was very low, & in coincidence with the advice of his physician & other friends, I requested you to come hither as expeditiously as you could. I also promised in that letter to write you again to day, thinking another letter might reach you before you would be able to leave Dresden.

I can now add, that the council of physicians, three in number, met last evening. They considered his case at length, & pronounced it dangerous, but not without hope. A new course of medicine was prescribed by them, including blistering, & calomel.

He passed a very bad night, but toward morning, the blister having taken effect, his fever subsided somewhat, & all this forenoon he has been easy, & somewhat more relieved from stupor than yesterday. He has understood what I have said to him, but not been able to carry on conversation. He is now (3 p.m.) sleeping quietly, but I think his fever has already began to come on again. I am sorry to add, that the physician, this morning, would not encourage me to say, that there had any material change taken place in his symptoms; of this, he said, he could judge better this evening, after the operation of medicine, which has not yet taken place.

I hope you will suffer nothing to delay your coming as speedily as possible. — Sympathizing with you most deeply in the feelings this

Ill. 14. Letter to Thérèse Robinson née de Jacob in Dresden, from Eli Smith, Vienna, September 27, 1838.

draw up a brief act of your late sickness, clearing the days of which you have no recollection, enforces upon me a painful duty; but a duty far less painful, than I expected to be called upon to perform, when I took the few notes, which now furnish me the data for this communication.'[118]

Naturally, at the first signs of illness they sent for a physician, but Robinson's fever worsened and he entered a state of 'an almost constant

[118] Smith to Robinson, Vienna, 11.10.1838, ESP, ABC 60 (105/9).

drowsiness.' They could not find a nurse to treat him around the clock, and were at first totally ignored by the American consul as well as the ambassador.[119] Smith did have names of two of Robinson's acquaintances, Dr Julius and Mr Kopitar, though 'the residence of the latter I was ignorant of.' The Slovenian linguist Jernej Barthol (Bartholomäus) Kopitar (1780–1844), Talvj's acquaintance and friend, served as librarian and later administrator ('Custos') of the Imperial Library in Vienna.[120] Talvj had alerted him of Robinson's expected arrival in her July 1838 letter.[121]

In Vienna, a letter from Smith to Dr Julius went unanswered as he was away, and the night was terrible:

> I passed a bad night. Your fever continued without intermission. You were restless; unable to speak, & evidently without reason. Constant attention was necessary, even to pack the bed-covering on you, especially such covering as your bed had; & yet deprived of rest as I had been the night before, & after a day of too much anxiety, I was not in a state to give you the attention you needed.[122]

A local physician was called who wrote prescriptions. In the evening, his fever still very high, and unable to find a servant, Smith once again cared for Robinson during the night. The next day, the attending physician 'seemed to flatter himself that he had again got the better of the disease, & said that a nurse was no longer necessary.' Dr Julius was still away, and when the local physician returned on his daily visit, 'he found you unable to converse, & helpless, & the fever ever higher than yesterday. He now seemed to apprehend the worst & it was the first time he had intimated to me that there was danger.' Now Smith had to update Talvj, whose Dresden address he received from Kopitar. In the evening, Smith summoned three physicians, '& then they called me in to make known their conclusion.' Smith once again attended to Robinson throughout the night: '(...) for though I would gladly have slept a little,

119 Findling (1989), 368: Only six months earlier, on February 8, 1838, the US appointed its first diplomatic envoy to Vienna, Henry A. Muhlenberg (1782–1844, DAB Muhlenberg), who served until 1840. The consul (1829–1852?) was John George Schwarz (Smith spelled his name Schwartz).

120 Krause (2001a), 265–97 (letters); Milović (1941), Einleitung. On Kopitar: BLKO Kopitar; NDB Kopitar; Merchiers (2007). See chap. 5.b.

121 Weigel Williams (2007), 106.

122 Smith to Robinson, Vienna, 11.10.1838, ESP, ABC 60 (105/9). All subsequent citations from the same document.

I could not, until near morning. Such a night I hope never again to pass. I felt that I was a stranger in a strange place, & the only peace I had, helpless & deprived of reason by disease, which would soon bring Mine to the grave. (...) But in the morning the good providence of God made my way plain (...).'

Recognizing the seriousness of the situation, Smith decided to ask for 'advice from the [US] consul, if not as a friend, yet as a magistrate, before taking any further steps.' This time Schwarz cooperated and 'entered at once with much friendly sympathy, into your case,' and 'urged the calling of a council of physicians as soon as possible.' The disease, however, was worsening and did not await these developments: '(...) you could understand what was said to you, better than in the night, & could answer yes & no to my questions. But a single sentence, of yourself, you could not form & though there was repeatedly something you risked & tried to say, you were unable, in a single instance, to make me understand what it was.'

They succeeded in finding a nurse so that Smith could at least move to the adjoining room, while 'leaving the door open, which communicated directly with the one you were in.' A second physician council was convened, more medicines dispensed, and another day passed; the situation continued to be 'very dangerous,' but now at least 'not without hope.' On Thursday, the fourth day of the illness, Robinson was 'entirely helpless' with a high fever; however, Consul Schwarz called several times a day and was of great assistance to Smith who had lost confidence in the first physician. In the evening, 'the fever was now higher than last evening, & the coma worse. You could not be induced to open your mouth, in order to show your tongue, or to take medicine, or even to spit (...).' On Friday morning, the physician informed Smith: 'I must profuse myself for your departure; & seemed pleased when I told him that I had had religious connection, & had prayed with you. I had already spoken to the consul to make arrangements for the farewell, in case our fear was realized; & he had informed me, that he must also take an inventory of your effects.'

'Painful as it is in the extreme,' Smith adds, 'I believe there is hardly anything as profitable to the Christian, as to wait at the bedside of a dying friend.' But in the afternoon, they started noticing the first signs of improvement. Saturday morning 'found you still alive, but the same alarming symptom continued, (?) increased, though alternated with short success of sleep & quietness.' Robinson started communicating and his first inquiry was about physicians. 'I answered it, but added what was my own deep impression at the time, that our hope was now in God, & not in physicians.' Robinson asked to pray with Smith, who revealed that Mrs Robinson was on her way. On Sunday, the physicians 'seemed to see in your symptoms, even more danger than I had done,' and doubled the medicine. One of them, before leaving, told Smith that they 'revised only the bear possibility of your surviving.' Upon his return on Monday, the physician 'announced the victory gained.' The healing was quick and 'every symptom of this kind entirely disappeared':

> I cannot close, without assuring you, how heartily I join with you in thankfulness to God for your recovery. (...) allow me to say that I thanked God, during your sickness, that I was with you, & could do anything to assist you. (...) my prayer is, that I may have, at least, one kind & faithful friend by my bedside. Such a friend as I would wish for myself, I have tried to be to you. Your life has been spared.

Smith summarizes the event in a letter to Anderson:

> We were strangers, in a hotel, in a strange place. I was his only friend, & I devoted myself to him night & day. God helped me, in procuring physicians & nurses, & raised up a friend, & gave affect to medicine. So that when his wife arrived, for whom I had written, when no one had any expectation of his living, I had the unspeakable pleasure of presenting him to her, not only better, out of danger.[123]

What exactly was the nature of Robinson's illness? Eran Dolev, a leading Israeli expert in the history of medicine and internal diseases, has suggested that Robinson probably suffered from malaria. Europe in that period was home to many diseases that were transferred by microbes and malaria was prominent among them. The Delta of the Danube was a well-known source of the disease, which actually existed in Hungary until the 1950s. Clinically, it was a severe, remittent fever malady, where the fever is followed by a stupor. As for the remedies mentioned by Smith, Sulphur medication has, of course, no effect. When the fever rose, the Austrian physicians decided to treat with calomel, a substance used in that period against diseases characterized by high fever.[124]

[123] Smith to Anderson, Leipzig, 1.11.1838, ESP, ABC 60 (105/5).

[124] See Ellioston (1839), Lectures, published in the same year, lecture xxxi: Treatment of fevers; lecture xxxii: Remittent fever, pp. 211–23. My gratitude to Prof. Eran Dolev for this reference.

Robinson was not the only victim of such a disease, nor the only American. After a long and adventurous excursion in Egypt, Philip Rhinelander (1815–39) and his New York friends left Constantinople for Vienna on July 1, 1839, arriving there on August 5. Letters of his companions describe the same symptoms. Rhinelander died on August 11 and was buried in Vienna, a fact corroborated in a letter of Consul Schwarz.[125]

During the initial period of Robinson's illness, Talvj stayed in Dresden, having moved there in May 1838. Shortly thereafter, she received a visit from her long-time correspondent and close acquaintance, the leading philologist and mythologist Jacob Grimm (1785–1863).[126] When Grimm and Kopitar looked for a translator of Slavian poems into German, it was the linguist Vuk Stefanović Karadžić (Karadjich, 1767–1864) who suggested Talvj, whom he first met in 1824.[127] The information concerning Talvj's involvement in her husband's recovery comes primarily from a letter dated September 21 sent from Dresden to Smith in Vienna, and from a short account of the experience that she wrote to Grimm in March 1840, beginning 'some months after I had the pleasure to welcome you, a storm-wind cast me away to Vienna.'[128] In her letter to Smith, Talvj refers to a letter from him (that we were unable to trace) that 'made a very melancholy impression upon me.' Weigel Williams assumes that she did receive this letter from Smith. However, we can only be certain of receipt of a former letter describing the difficulties Robinson and Smith faced on their way from Constantinople, and the fact that some of their group fell ill and were left behind in Pest.

Talvj writes that while no immediate danger seemed apparent, her instinct was otherwise: 'I cannot deny myself, that, as my husband's

[125] Oliver (2014), 285–304.

[126] Weigel Williams (2007), 108. It is important to note here 'Talvj's Slavian circle,' in which Karadžić was a central figure, together with Kopitar and Grimm, discussed in detail in chap. 5.b. (cf. Steig (1894)). Gabriella Schubert and Friedhilde Krause edited an outstanding collection of papers concerning Talvj's *Balkan-Slaven* connections and works (Schubert (2001)), including a biography titled 'Talvj, Deutsche und Südslaven' (14–18), and an intensive publication of Talvj's letters (Krause (2001a)). Most of the papers include long discussions of the three, their relations and collaboration (e.g. Dobrašinović (2001); Delić (2001); Ellenmeyer-Životić (2001)). Another important study is Merchiers ((2007), 222–87), where she discusses Kopitar and Karadžić's connections and network.

[127] Dobrašinović (2001), 69.

[128] Talvj to Grimm, Berlin, 16.3.1840, Schubert, G. (2001), 269–71 (original letter in Grimm's legacy, National Library, Berlin); Theresa Robinson to Smith, Dresden, 21.9.1838, ESP, ABC 60 (63); Weigel Williams (2007), 106–07.

complaint had been for some time increasing, there was reason to fear that it would become dangerous, or at least result in a protracted illness.' She relays her decision to go immediately to Vienna with her three children and their caretaker, a difficult four-day coach journey, and asks Smith to send an update on Robinson's health with his Vienna address to one of the intermediate stations, either Jglaw (today Jihlava, Czech Republic) or Znaym (today Znojmo, Czech Republic): 'You can probably ascertain at the post office in Vienna where I can receive your letters. I shall inquire at both places. (…) May God be so gracious as to make this journey unnecessary! And may He give me strength to bear it, if it must be made!' This letter was sent on September 22, and later Talvj writes that she succeeded in doing the impossible; only ten hours after receiving the information they were already on their way.[129]

Talvj reached Vienna on September 27, when Smith sent his letter, which she naturally never received in Dresden. She arrived after the major crisis subsided and was happy to be able to help with Robinson's care, noting how painful the meeting was after a full year of separation.[130] Robinson's recovery, which took six weeks, delayed them in Vienna. His illness resulted in another big change in Robinson's life. When Robinson began getting sick in Andover, his physicians prescribed him a meatless diet. Now, after five years of vegetarianism, he started eating meat 'and his health improved perceptibly.'[131]

Until Robinson was able to continue to Germany, Smith spent time 'in the Imperial Library gathering much valuable information respecting oriental literature, from its (?) Librarians, & its rich stores of Oriental Manuscripts.' He most probably relied on Kopitar, searching for books on 'his' Oriental languages and obtaining the catalogues.[132] Smith then left Robinson in Vienna with Talvj, going first to the Saxonian city Meissen where he visited the Orientalist Gustav Lebrecht Flügel (1802–70) who supplied him with 'cards of introduction, of essential service to me here.' In a December 1838 letter, Smith thanks Flügel for 'the kind manner, in which you begged Mr Tauchnitz to undertake the work of fitting up & casting my type.'[133]

129 Theresa Robinson to Smith, Dresden, 21.9.1838, ESP, ABC 60 (63).

130 Theresa Robinson to Smith, Dresden, 21.9.1838, ESP, ABC 60 (63).

131 Robinson, M. A. (1862–64), 153.

132 Homes to Smith, Constantinople, 12.11.1838, ESP, ABC 60 (37).

133 Smith to Flügel, Leipzig, 17.12.1838, ESP, ABC 60 (105/9) 1838 Jul.-Dec.

Smith finally arrived at his destination, Leipzig, in late October 1838. As noted, he immediately started writing letters, first to Anderson:[134] 'I have since heard from him [Robinson], & am expecting him here after two or three days.' The remainder of the letter is a rather long report dealing almost solely with the possibility of missionary work among the Bedouin![135] On October 24, Robinson, still in Vienna, writes to Smith that as 'my health has been constantly improving (...). We are now making our preparations to leave for Prague tomorrow.'[136] After 'a not uncomfortable, though rather a late ride,' Robinson and his family arrived in Halle on November 14. He was received 'with open arms' by the local Orientalists and his long-time acquaintances, 'Gesenius, Roediger, & Tholuck,' and they were intensively busy with social duties. On the day of the letter, at one p. m., they took the mail-coach (*Personen-Post*) to Berlin, which was to reach its destination the following morning.[137] After a long and rather harrowing delay, both Smith and Robinson reached their destinations: Leipzig for the Arabic letter types, Berlin for the book.

Timetable in Vienna:
Arrival in Vienna: Thursday, September 13, 1838
Robinson's sickness becomes nearly fatal: Sunday, September 23, 1838
Smith's letters to Talvj: Wednesday and Thursday, September 26 and 27, 1838
Talvj reaches Vienna: Thursday, September 27, 1838
Robinson improves: Monday, October 1, 1838
Smith's report to Robinson about his illness: Thursday, October 11, 1838
Robinson and family leave Vienna: October 25, 1838
Robinson and family leave Halle: November 14, 1838 and reach Berlin November 15, 1838.[138]

[134] Smith to Anderson, Leipzig, 27.10.1838 and 1.11.1838, ESP, ABC 60 (105/9).

[135] Smith to Anderson(?), Vienna, 13.10.1838, Leipzig, 25.10.1838, ESP, ABC 60 (105/5). For detail, see chap. 5.a.

[136] Robinson to Smith, Vienna, 24.10.1838, ESP, ABC 60 (63).

[137] Robinson to Smith, Halle, 14.11.1838, ESP, ABC 60 (63).

[138] Robinson to Smith, Berlin, 23.11.1838, ESP, ABC 60 (105/9).

CHAPTER 4
IN BERLIN AND HALLE: WRITING THE *BIBLICAL RESEARCHES*

4.a. Return to NY or stay in Berlin? Robinson and his UTS superiors

'The Theological Seminary in this country may be regarded as one of the characteristic institutions of American Christianity. It is mainly the growth of our own soil during the present century.' Thus, opens George Lewis Prentiss (1816–1903) his 50th anniversary history of the UTS, originally named New York Theological Seminary. At these institutions, Americans of all beliefs are trained, Prentiss continues, 'While differing radically as to polity and doctrine, they are nearly all agreed in assessing the divine origin and claims of Christianity, the ruling authority of the Holy Scriptures...'[1] The idea for the establishment of the Union Theological Seminary arose in October 1835 and it opened its gates for instruction, initially on temporary premises, on December 5, 1836.[2] 'The pledge required from the professors, touching both doctrine and polity,' was 'I believe the Scriptures of the Old and New Testament to be the word of God, the only infallible rule of faith and practice...'[3]

The fact that Robinson prolonged his stay in Europe from the originally approved one to three years is noted in most of his biographies, short of investigating how he succeeded in extending his leave without losing his new position at UTS! Even Williams simply states that 'the

[1] Prentiss (1889), 3–4; Handy (1987), 1–7. Following Robinson, Prentiss studied from 1839–41 in Halle and Berlin. From 1851, he was Robinson's minister, and later faculty at the UTS. For the history of the UTS cf. DAB Prentiss; Handy (1987).

[2] Prentiss (1889), 24–25.

[3] Prentiss (1889), 43–44.

Seminary, to its good credit, exhibited supreme patience with its newly appointed but untried professor.[4] Archival material, including letters, primarily between Robinson and Smith, and UTS board protocols, shed light on Robinson's leave. The protocols, which appear not to include all board meetings and typically are limited to relatively technical reports (participants, started\ended with a prayer, accepted and raised communications, resolutions), provide an outline of events. However, absent are the discussions crucial to understanding the decision-making process at the seminary.

A resolution adopted at the December 20, 1836 board meeting, after Reverend George Howe (1802–83) had declined the position, states that the board would 'now proceed to the election of a Professor of Biblical Literature,' and 'upon counting the ballots it appeared that Edward Robinson, D.[octor of] D.[ivinity] was unanimously chosen.' The board, following regular procedure, appointed a committee of three members 'to confer with Professor Robinson on the subject of his appointment.' The board also decided upon an annual salary of $2,500.[5] The committee was very efficient and only two weeks later, on January 3, 1837, reported receiving 'a communication' from Robinson 'with a view to further correspondence.'[6]

Robinson received the invitation in December 1836, shortly after the first year of instruction began. In January, he went to New York 'to inquire into the state & prospects of the Seminary, before arriving to any decision.'[7] His aim was 'to obtain information, as to the plan and prospects of the Seminary,' and he received 'the most frank, and full communications.' He did not deliberate at length. On January 20, 1837, he announced by a rather long letter, entered in full in the protocol, his formal acceptance of the 'office of Professor of Biblical Literature in the seminary.'[8]

Prentiss located and cites this letter almost in full, deeming it 'one of the most important documents connected with the early history of the institution.' It includes an extensive discussion of Robinson's acceptance

[4] Williams (1999), 259.

[5] BP-UTS, 20.12.1836. Howe (DAB Howe), former professor at Dartmouth College and then professor for biblical literature at the Theological Seminary at Columbia, South Carolina, was the second to decline the offered position. See Williams (1999), 288. The D. D. defined a university education licensed to teach theological issues.

[6] BP-UTS, 3.1.1837.

[7] Robinson to Norton, Boston, 9.2.1837, ERP, ER 1.13.

[8] BP-UTS, 20.1.1837.

of the seminary's theological approach and his curricular suggestions. Robinson also raises a 'single point which is personal to myself,' his request to visit Europe with his family, where he would purchase books for the UTS library, and his desire to embark on his long-awaited journey to Palestine, which he notes 'with reference to the preparation of a Biblical Geography.' To this end, he requests a leave of absence 'from and after the close of the present academic year for a period not exceeding one academic year.'[9]

The letter's last chapter is a kind of compendium of the person Robinson, his world and beliefs:

> Should you deem it compatible with the interests of the Seminary, that I take the office under these conditions, I am ready to throw myself, head and soul into the work, and exert to the utmost, all the feeble powers, which God has given me trusting, that in cooperation with my respected colleagues, and with the blessing of God upon his own work an institution may be raised up, which by its happy influences upon the churches of this city and of our land, shall repay an hundredfold into the bosoms of its founders, the cares and exertions and sacrifices which they have been called to make in its behalf.[10]

In a letter dated February 9, 1837, Robinson writes to Norton, 'If Deo prospers our journey, we hope to return in the autumn of the next year; so that I can enter upon my duties at the seminary in the opening of the term in October.' The agreement was unmistakably clear: Robinson intended to fulfil it and present himself at the UTS in October 1838, ready to assume his teaching duties. Robinson then enumerates the unique prospects of the new institution:

> (...) its support & direction upon men who are the very serious of Christian enterprize & liberality in that city. They are (??) to do every thing which may seem important & expedient, to build up an institution which shall sustain in high character & excel a wide influence. (...) this seemed an opening in providence, with which I might enter with the hope of being useful; all my previous studies & pursuit, having prepared me for just such a station, if for any.[11]

9 BP-UTS, 20.1.1837; Prentiss (1889), 246–48. Cf. Williams' discussions, (1999), 205, 288–89.

10 BP-UTS, 20.1.1837.

11 Robinson to Norton, Boston, 9.2.1837, ERP, ER 1.13.

Robinson entered into his new position immediately and taught the first half of 1837. However, his impending absence was not simple to resolve. He was one of only three regular faculty members, others being replacements or auxiliary faculty.[12] The seat for biblical literature could not and should not remain vacant, and an interim needed to be found. In May 1837, the Reverend Ichabod S. Spencer (1798–1854) was elected to the position.[13] Spencer, Presbyterian preacher and author, became the minister of the Second Congregational Church in Brooklyn in 1832. From 1836–49, he served as member of the UTS Board of Directors, and from late 1836 to Robinson's return in 1840, as Professor Extraordinary of Biblical History. His appointment was extended on December 26, 1838, when it became clear that Robinson would not be returning as early as planned.[14]

Robinson's first request to extend his leave of absence was discussed in November 1838. The protocol is short, stating that the board received a communication from Robinson requesting an extension, and approving his stay in Germany until 'the close of the next academic year.' The board also adopted Robinson's suggestion that 'his salary be discontinued from the 1st of October last until his return,'[15] indicating that Robinson did receive his salary for the 1837–38 academic year, even though he did not teach. Another committee was appointed 'to procure a teacher in the place of Prof. Robinson.' The committee suggested splitting the position of language instructor, appointing one instructor for 'the German and Oriental languages,' and the second 'to give exegetical instruction in Hebrew & Greek & to act as a temporary librarian.'[16] The Hebrew instructor chosen was young Rabbi Isaac Nordheimer (1809–42), born and educated in Germany, who in 1835 emigrated to New York and served at the UTS from 1836 until his early death.[17]

Yet another academic year, 1838–39, came to a close, and Robinson knew he was not ready to return and begin teaching in October. Aware

[12] Union Theological Seminary (1850), 6–7; Williams (1999), 286–87.

[13] BP-UTS, 6.5.1837.

[14] Union Theological Seminary (1850), 4, 7 ('During the first years of the Seminary, members of the Board of Directors were annually appointed as *Professors Extraordinary*, to give lectures once a week'); BP-UTS, 26.12.1838.

[15] BP-UTS, 20.11.1838.

[16] BP-UTS, 20.11.1838, 17.12.1838.

[17] BP-UTS, 20.11.1838, 26.12.1838; DAB Nordheimer; Union Theological Seminary (1850), 7; Goldman (1990–1), 1–2, 7–8. Williams (1999), 287, 295, 301, stresses the cooperation between Robinson and 'his young Jewish assistant.'

that his absence was problematic, and knowing that Smith would soon be returning to New York, Robinson writes to Smith, asking him to negotiate with the UTS directors the possibility of a prolonged stay in Europe:[18] 'I am more & more every day persuaded of the impossibility of returning this autumn, whatever may be the consequences,' and that the seminary 'will doubtless only be the more pressing for my return.'[19] Smith had reached Leipzig in late October 1838. The first mention of his return plans can be found in a letter dated May 5, 1839. Two weeks later he writes of his intention to sail by steamer, shortening the trip by 'some 20 or 25 days,' leaving England before August 3.[20] In early June, still in Leipzig (this specific letter is tremendously difficult to read), Smith confirms his plan to visit Robinson briefly in Berlin. He continues, 'my route will be by way of Hamburg to London. By the way, can you tell me about how much money I shall need to take [with] me from Berlin to London?'[21]

Smith left Leipzig soon after posting this letter, travelled to Berlin (June 14) and Hamburg and then sailed to London. He left the British island on July 10, docked in New York on July 28, where he spent the next three months, reaching Boston only in early November. He stayed in the States, mainly in Boston and New York, until he left again for the East on April 27, 1841.[22] Robinson now had an agent who could meet with UTS decision-makers, participate in board meetings (though no 'guests' are ever mentioned in the protocols), and fight for their mutual cause: to allow Robinson to complete the book, without loss of his position. 'I feel very grateful to you, for your kind exertions in my behalf; I doubt not but that your representations, & the pressure of the times, will reconcile the Board to my prolonged absence,' writes Robinson in April 1840.[23]

In London, Smith met with UTS faculty members Reverends Thomas Harvey Skinner (1791–1871) and William Patton (1798–1879), who were exploiting their summer vacation for travelling in Europe. Both Skinner and Patton are listed as UTS directors from 1836 and 'Profes-

18 Robinson to Smith, Halle, Sept. 10.9.1839, ESP, ABC 60 (63).

19 Robinson to Smith, Berlin, 10.7.1839, ESP, ABC 60 (63).

20 Smith to Robinson, Leipzig, 5.5.1839, 18.5.1839, ESP, ABC 60 (105/11), 1839 Apr.-Dec.

21 Smith to Robinson, Leipzig, 3.6.1839, ESP, ABC 60 (105/11), 1839 Apr.-Dec.

22 Smith to Robinson, London, 8.7.1839, ESP, ABC 60 (105/11), 1839 Apr.-Dec.; Recent Intelligence (1838a); Kawerau (1958), 263–64.

23 Robinson to Smith, Berlin, 15.4.1840, ESP, ABC 60 (63).

sor Extraordinaries,' Skinner for sacred history from 1840, and Patton for pastoral theology from 1843.[24] Skinner was Robinson's colleague at Andover beginning in 1832 and a UTS co-founder. Patton was also a UTS co-founder, a strong abolitionist, and frequent visitor to Europe (14 visits between 1825 and 1879). The UTS Board of Directors in the years under discussion, 1837 to 1840, included Rev. Thomas McAuley (1778–1862), president, replaced in 1840 by Richard T. Haines; Zechariah Lewis, vice president; Eskine Mason, recorder; William M. Halsted, treasurer; nine clergy, and fourteen laymen.[25]

Smith conversed with Skinner and Patton and they naturally raised the matter of Robinson's absence. In a letter to Robinson, Smith relays the highlights: Skinner described the absence as a 'serious temperance' to the seminary, creating 'a feeling of disappointment' and 'considerable dissatisfaction in some questions.' Patton, even more outspoken, stressed 'the injury the seminary was suffering from your absence & the dissatisfaction it was occasioning.' Smith further recounts that he argued for extending the leave until Robinson successfully shepherded the book to printing and that he might return by March, and that Skinner and Patton suggested 'to converse fully with Dr Peters.' They believed that Reverend Absalom Peters (1793–1869), UTS director from 1836, editor of the *ABR* from 1838, and professor of pastoral theology and homiletics at the seminary, might be influential in convincing the board.[26]

Robinson's answer from Halle, dated September 10, mentions two letters from Smith, one from London (most likely dated July 8) and the other from New York (dated July 31). Robinson attests to now being relaxed and reassured, as he received letters which 'put my mind quite in rest.' He had received 'a very kind & friendly letter' from Peters, who considered Robinson's reasons for the desired extension, 'giving his opinion more fully & frankly than I have hoped.'[27] Robinson does not

[24] DAB Skinner; DAB Patton; Prentiss (1889), 254–59 (Skinner), 121–30 (Patton); Union Theological Seminary (1850), 3–7. In January 1841, Patton was the pastor of the church in which Robinson delivered his inaugural address (ERP, AL COLL R65 B43 1841: Edward Robinson, *The Bible and Its Literature; an Inaugural Address*, New-York: Office of the American Biblical Repository and The American Eclectic, 1841). Skinner, Robinson's 'old friend and colleague,' delivered his funeral eulogy (Williams (1999), 331–32).

[25] Union Theological Seminary (1850), 3–4; Handy (1987), 5–11.

[26] Smith to Robinson, London, 8.7.1839, ESP, ABC 60 (105/11); DAB Peters; Union Theological Seminary (1850), 3.

[27] Robinson to Smith, Halle, 10.9.1839, ESP, ABC 60 (63). The letter 'came by the Post over England, & reached me in Berlin in a few hours over 19 days from New York.'

reveal Peters' words and the letter was not located, but Robinson writes, quite frankly, that he had already decided to remain in Europe. To UTS President Thomas McAuley he wrote:

> I was forced to decide this, unless I was ready to give up the work; although my remaining involves a large pecuniary sacrifice, which the profits of the work will probably hardly cover. Yet having once begun, I wish to go on to the end in the same manner; although the progress will be slow; & I have my doubts whether under the most prosperous circumstances I shall be able to see New York before the summer.[28]

In his next letter, sent from Saratoga, NY, Smith recounts that he discussed the matter with McAuley '& others,' 'expressing the opinion that you had no other way than to remain until your work was done, & then they would be very glad to receive you again as professor at New York. (...) all your friends whom I have seen are most decided on the opinion that you ought to finish your work where you are.' McAuley's answer was simple: Robinson should not trouble himself with the subject, 'for they had dismissed it.' Smith understood, incorrectly, that they dismissed Robinson from the UTS, albeit temporarily, and as the board meeting had concluded, he did not have the opportunity to ask for an explanation.[29]

Perhaps Robinson was content with this answer; in any event, he was occupied with the book. Five months passed until he next wrote Smith in February 1840, that he hopes 'to see the end before May,' which meant giving up any intention of returning to the seminary by the coming summer. While unavoidable, he could not anticipate the directors' reaction.[30] From this letter, it is clear that Robinson was worried about his future at UTS. Robinson asks Smith, who previously had proven his powers of persuasion, to use his influence while still in New York to convince the directors to accept this last delay. For the first time, Robinson reviews the list of UTS directors and their views concerning his absence. Whereas Peters 'takes strong ground in my favour,' McAuley's 'sails are all in the wind, to be trained according as the wind shall blow hereafter.' As for the clerical directors, 'some might be disposed to think hard of my absence,' but people like Halsted and Haines 'could be led to look at the subject in its time light,' and then the others will follow. Robinson,

28 Robinson to Smith, Halle, 10.9.1839, ESP, ABC 60 (63).

29 Smith to Robinson, Saratoga, 17.9.1839, ESP, ABC 60 (105/11), 1839 Apr.-Dec.

30 Robinson to Smith, Berlin, 12.2.1840, ESP, ABC 60 (63).

adamant in his decision to finish the *Biblical Researches* before returning to New York, authorizes Peters confidentially 'to tender my resignation to the Board, should he clear the circumstances such as to render that step necessary & proper'![31] This letter was never put in use.

Robinson's absence hovered uneasily at the seminary. A letter from Robinson was read at the April 4, 1840 board meeting noting, 'that he cannot yet state the time when he shall be able to return.' It generated neither an official comment nor resolution.[32] Robinson's April 15th letter from Berlin attests to the fact that the board meeting protocols do not reflect the discussions held, which in Robinson's case were far from calm. Smith's reports are absent as well from the protocols; it is difficult to discern whether he even participated in the meetings. In a personal letter to Smith, it is remarkable that Robinson uses the word 'putsch' (spelling it 'Pfutschee') to describe Peters' acts, this time not in his favour, while excusing the latter's behavior: 'I also expressly stated my wish that Prof. Silliman or Hitchcock might be induced to take up the subject; & I am truly sorry (according to your report) that Dr Peters has so hastily let in a Pfutschee, instead of waiting for one of them.'[33] Robinson also admits that he was dependent upon his UTS salary, 'I am very sure that without a provision of that kind, I could not have remained in connection with the Seminary' and would have to look for another source of income. He continues, 'I have no friends in the seminary,' and then agrees to Smith's suggested next steps, 'I shall endeavor to act upon them.'[34]

In the same letter, Robinson raises another issue related to his delayed return to New York, a scientific debate that had become public. One of Robinson's early post-expedition publications was a long article, 'On the Dead Sea, and the Destruction of Sodom and Gomorrah,' published in 1840 in the *ABR*.[35] This paper received a critical review by a certain Dr Lee. 'I know nothing, & don't remember that I have heard

[31] Robinson to Smith, Berlin, 12.2.1840, ESP, ABC 60 (63). William M. Halsted, treasurer and director, and Richard Townley Haines (1795–1870), director, president in 1841, were partners in Halsted, Haines & Co., one of the largest dry-goods stores in the US.

[32] BP-UTS, 7.4.1840.

[33] Robinson to Smith, Berlin, 15.4.1840, ESP, ABC 60 (63). Benjamin Silliman (1779–1864), professor of chemistry and natural history at Yale from 1802, editor of the *American Journal of Science*, was also strongly connected with the ABCFM (DAB Silliman; Phillips (1939), 301).

[34] Robinson to Smith, Berlin, 15.4.1840, ESP, ABC 60 (63).

[35] Robinson (1840).

of him before,' Robinson writes and continues, 'if he wishes to get into notice by breaking a lance with me, I fear he will lose his trains.' Peters, the *ABR* editor (on the cover he writes 'conducted by'), failed to show the article in advance to his colleagues Silliman and Hitchcock, most likely due to lack of time before publication.[36] It appears from the correspondence, nonetheless, that the board was resigned to Robinson's delay, hoping that he would appear in New York in time to begin teaching in Autumn 1840.

Robinson finished his book-related tasks in Berlin on September 1, 1840. Five days later, the family set out through Halle to Hamburg, and from Hamburg sailed to London. In London, Robinson wrote his next letter to Smith, not knowing whether Smith was still 'in the United States, or already in Syria.' Actually, Smith left the US only on April 27, 1841; he even married (his second marriage, to Maria Ward Chapin, who died of dysentery within a year) in Rochester on March 9, 1841.[37] In the letter, Robinson justifies the protracted delay in completing the book, which 'arose from the very great amount of pictorial investigation and illustration which came up,' particularly for the routes to Petra and back, to the Samaritans in Nablus, Nazareth, Tabor and the Esdraelon Valley, and to Tiberias and the Sea of Galilee.[38] This letter contains no further references to his teaching. In London, Robinson naturally exploited his brief layover to reconnect with as many old friends and acquaintances as possible. This included Samuel Lee (1783–1852), English Orientalist and Cambridge professor of Arabic and Hebrew, who was also connected with the CMS.[39] Was this the same Lee who criticized Robinson's Dead Sea paper?[40]

Robinson stepped on American soil again only in November 1840. Back in New York, Robinson wrote to Smith in the closing days of 1840 about his teaching, 'I was glad to find that it [ancient Jerusalem] interested the students.' After a long delay, 'on January 20th, 1841, Edward

36 Lee (1840); Robinson to Smith, Berlin, 15.4.1840, ESP, ABC 60 (63); Robinson (1840), 24.

37 Robinson to Smith, London, 3.10.1840, ESP, ABC 60 (63); John Nevin Lewis to Smith, Geneseo N.Y., 8.3.1841, ESP, ABC 60, General letters L (53), excuses himself for not being able to attend the wedding. We were unable to determine whether Robinson attended. Cf. Finnie (1967), 201.

38 Robinson to Smith, London, 3.10.1840, ESP, ABC 60 (63).

39 Samuel Lee, near London, to Robinson in London, 6.10.1840, ERP, L. Correspondence to E.R. 1840–60.

40 Robinson did maintain contact with Lee and mentions him in a letter written on his way to his second expedition, Robinson to Smith, 8.9.1851, ESP, ABC 60 (63).

Robinson was officially inaugurated in his office as Professor of Christian Literature at Union Seminary.[41] However, Robinson's focus at that time remained the *Biblical Researches*, as the majority of his letter to Smith describes the final stages of the book's publication.[42]

4.b. The *Biblical Researches*: planning, writing, and distributing

It appears that this Holy City will yet, for a long time to come, offer a rich field for investigation, to which you yourself opened the way, showing the plan on which future operations must be conducted, and awakening the dry bones of antiquity to a new life.

Carl Ritter to Edward Robinson, Leipzig, August 4, 1847, read in the Second Meeting of the Ethnological Society of New York (Scientific Proceedings (1847)).

My husband recommends you on the best way, and indeed actually from Jerusalem, as he lives there more than here [Berlin]. [Mein Mann empfiehlt sich Ihnen bestens, und zwar eigentlich aus Jerusalem, denn er lebt mehr dort als hier]

Talvj to Kopitar, Berlin, August 12, 1839, Milović (1941), 101.

The three-volume *Biblical Researches in Palestine, Mount Sinai and Arabia Petraea in 1838* was published in 1841 in London, Boston, and Halle (by Murray, Crocker & Brewster, and *Weisenhaus* [Orphanage] respectively). The volumes of the Boston edition comprise nearly two thousand pages [(571+679+(475+246)].[43] The *Biblical Researches* is certainly one of the most detailed scientific undertakings originating from travels in Sinai and Palestine, the outcome of an expedition that lasted only three and a half months!

This section attempts to resolve various questions concerning this unique publication. The first section explores Robinson's decision-making process, addressing such questions as his pre-travel expectations, the origin and actualization of the idea for the book and whether he out-

[41] Williams (1999), 292; Weigel Williams (2007), 116–18. I tend not to agree with Wagner (1897), 13, who was probably the first to write that Robinson asked 'for three or four years (on his own account) to study the Holy Land there and then.'

[42] Robinson to Smith, New York, 28.12.1840, ESP, ABC 60 (63).

[43] Robinson (1841); (1841a); (1841b).

lined the publication in advance. Specifically, it considers the process resulting in the final contents and dimensions of the *Biblical Researches*. The second section of this chapter concentrates on the 22 months spent writing in Berlin. How did Robinson work? What were his connections and sources of information and who were his partners? Choosing publishers was also quite a task, as was locating or creating the types for non-English languages, mainly for Arabic. Some of these questions were addressed in an earlier book concentrating on the cartographical work accompanying the *Biblical Researches* and other publications.[44] Fortunately, Smith saved most of the letters he received during this period as well as copies of the ones he sent (on transparent paper, which makes them even more difficult to decode).

4.b.1. The preface: explaining the underlying scientific approach

Robinson's preface is important for understanding the process leading to the *Biblical Researches'* final form. Robinson wrote the preface, as is customary and expected, only after finishing the book. The manuscript is signed 'Berlin, Aug. 30. 1840,' and the date of publication was 'New-York [*sic*], June, 1841.'[45] The preface will be analyzed here in detail and later compared with the picture that can be deduced from the letters.

Ill. 15. Robinson's Gold Medal.

There is another source for reconstructing Robinson's path from expedition to publication. On May 23, 1842, the RGS awarded one of its two annual gold medals to Robinson. Accepting on Robinson's behalf, who was unable to attended the public ceremony, was Edward Everett (1794–

[44] Goren (2017).

[45] ERP-UTS, Boxes 1–4. Preface: Box 1, F 1. Cf. Robinson to Smith, New York, 14.4.1841, ESP, ABC 60 (63): the preface had been written in August 1840 in Berlin.

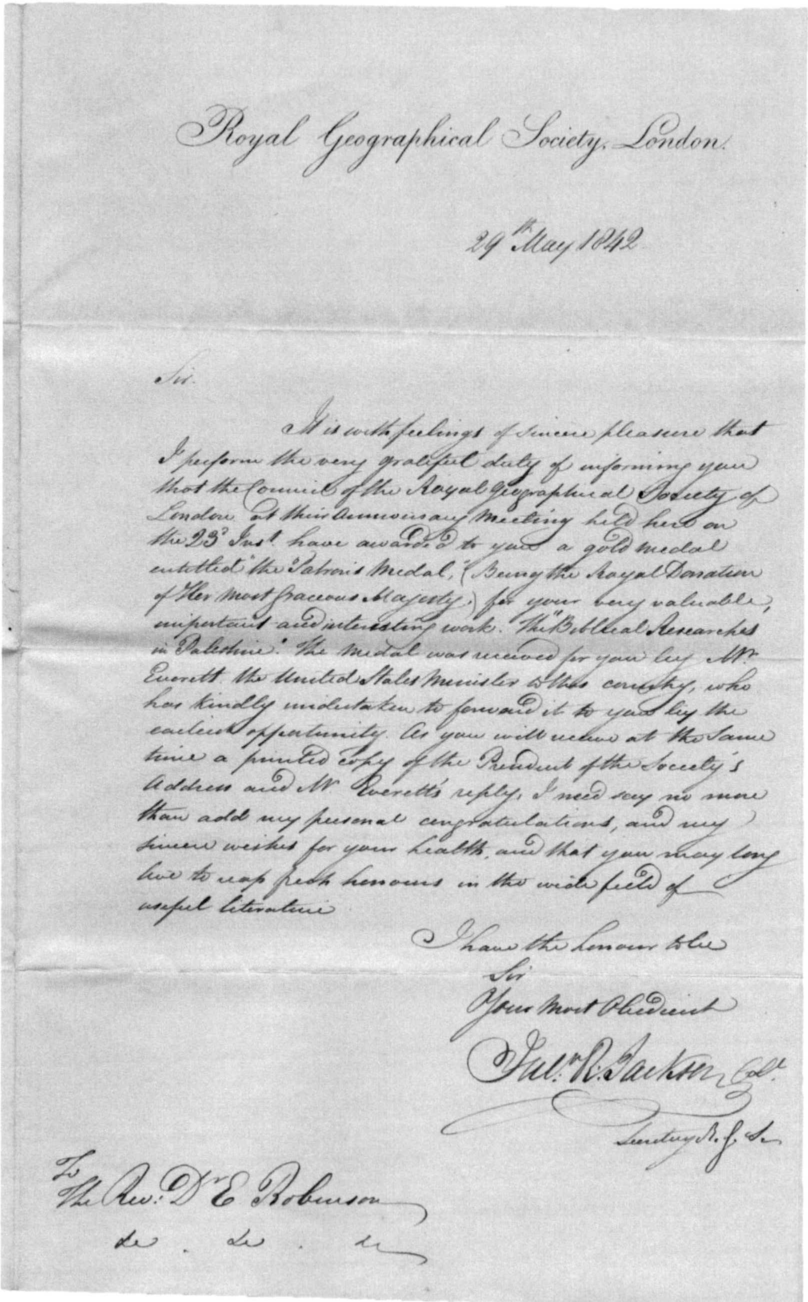

Royal Geographical Society, London.

29th May 1842

Sir.

It is with feelings of sincere pleasure that I perform the very grateful duty of informing you that the Council of the Royal Geographical Society of London at their Anniversary Meeting held here on the 23d Inst. have awarded to you a gold medal entitled the "Patron's Medal," (Being the Royal Donation of Her most Gracious Majesty,) for your very valuable, important and interesting work "The Biblical Researches in Palestine." The Medal was received for you by Mr. Everett the United States Minister to this country, who has kindly undertaken to forward it to you by the earliest opportunity. As you will receive at the same time a printed copy of the President of the Society's Address and Mr. Everett's reply, I need say no more than add my personal congratulations, and my sincere wishes for your health, and that you may long live to reap fresh honours in the wide field of useful literature.

I have the honour to be
Sir
Your most Obedient
Jul. R. Jackson Col.
Secretary R. G. S.

To
The Rev. Dr. E. Robinson
&c. &c. &c.

Ill. 16. J. R. Jackson's letter announcing Robinson's Gold Medal awarded by the RGS, London, May 24, 1842.

1865), US envoy in London.[46] Everett, born one day after Robinson, preceded Robinson and studied for two years in Göttingen, travelled intensively in Europe (1815–19), and taught ancient Greek literature at Harvard. He 'became one of America's foremost orators and politicians,' and developed a long-lasting friendship first with Stuart, and then with Robinson and later also with Talvj.[47] RGS President William Richard Hamilton (1777–1859)[48] delivered the address at the ceremony, Everett delivered the response in place of Robinson, and Robinson received his medal and copies of the speeches by mail. A thrilled Robinson sent Hamilton a letter of gratitude in which he describes, probably for the first time, his journey beginning with his departure from the US until returning with the finished manuscript.[49] 'This was to me wholly unanticipated; & of course on that acc[oun]t the less grateful, as coming from impartial judges, & that too in favour of a stranger & a foreigner,' writes Robinson to Ritter.[50]

The preface of the *Biblical Researches* is dedicated to what Robinson calls 'the form, in which the materials have been wrought up,' referring to the materials collected during the expedition.[51] He admits to having doubts concerning the form, as the original plan was to present the results of 'our researches in Palestine, without any reference to personal incidents.' His 1842 letter to Hamilton sheds light on the evolving project: 'On returning to Europe therefore, I was conscious of pursuing a large stock of new materials, but was still not aware of their value in a scientist aspect.'[52] Robinson changed his mind upon 'advice of friends,

[46] ERP, ERP, I, 1–4: Prudhoe to Robinson, Whitehall Gardens April 23.4.1842, transcript; W. R. Hamilton to Robinson, London, May 2.5.1842, transcript; J. R. Jackson to Robinson, RGS 29.5.1842; Edward Everett to Robinson, London 28.5.1842; Hamilton (1842) and Everett, print; drafts of Robinson's thank you letters; Robinson's letter to W. R. Hamilton, New York, ?.?.1842; ER's thank you to Everett; Robinson's thank you to Prudhoe. See DAB Everett; Findling (1989), 180.

[47] Brown, J. W. (1969), 34–59, including a detailed discussion of his connections with John Thornton Kirkland (1770–1840, DAB Kirkland, J. T.), president of Harvard College (1810–28), Stuart and Robinson; Williams (1999), 9 (citation), 72–73 (his attraction to Eliza Kirkland, Robinson's first wife), 202 (as anti-Masonic governor of Massachusetts), 298–99 (RGS medal); Oliver, (2014), 58–78.

[48] For Hamilton, see, e.g., ODNB Hamilton; Mill (1930), 35–40; Stoneman (1987), 155–79. For his long-time interest in Palestine exploration, cf. Kark (2011), 266–67.

[49] Robinson to Hamilton, New York, ?.?.1842, ERP, I, 4.

[50] Robinson to Ritter, SBB-PK, 31.12.1841, Nachl. 133 (Carl Ritter), K 5, Q – SCH, Bl. 35.

[51] Robinson (1841), I, Preface, v–vi.

[52] Robinson to Hamilton, New York, ?.?.1842, ERP, I, 4.

whose judgement I could not but place over my own,' referring to 'Ritter & others.' He writes further that it was Ritter's 'judgement which led me to give up pressing duties at home and remain for two years in the midst of the science and privileges of Berlin.'[53] 'The brightest period, connected with my journey to the East, – besides the journey itself,' he writes later to Ritter, 'is the time I spent in Berlin in the preparation of my materials, where I enjoyed such cherished intercourse with yourself & others.'[54] In his letter to Hamilton, Robinson not only expresses his gratitude to the RGS for the medal, but also for the society's role in the conception and development of the book, primarily 'the kind interest manifested in the subject by their late secretary, Capt. Washington.'[55]

The extensive correspondence between Robinson, Berghaus, and Washington regarding the production of the maps surely encouraged Robinson in his work as well.[56] The British naval officer and geographer John Washington (1800–63) was one of the central personalities in the world of geographical research and exploration during the mid-nineteenth century. After a nearly 25-year naval career, in 1836, he was appointed secretary of the RGS, of which he was also a co-founder. Subsequently, Washington decided to return to the navy 'to resume the more active duties of my profession.' In 1855, he replaced the legendary Francis Beaufort (1774–1857) as chief hydrographer of the navy.[57] Their many letters reveal the deep personal friendship and academic interchange between Washington and Robinson, as well as with Smith, Ritter, and the cartographers Berghaus and Kiepert. The extensive correspondence housed in the archives in Britain, Germany, and the USA (described in detail in the aforementioned study) evidence the substantial influence of their scholarly exchange over the maps on Robinson's decision-making process in Berlin.[58]

[53] Robinson (1841), I, Preface, v–vi; Robinson to Hamilton, New York, ?.?.1842, ERP, I, 4.

[54] Robinson to Ritter, New York, 31.12.1840, SBB-PK, Nachl. 133 (Carl Ritter), K 5, Q – SCH, Bl. 33.

[55] Robinson to Hamilton, New York, ?.?.1842, ERP, I, 4.

[56] See in detail Goren (2015); (2017); Herbert (1983), 67, 71, 97.

[57] Washington to Robinson, Shearwater, Harwich, 1 May 1842, ERP, L.2 Letters from J. Washington (citation); Robinson to Ritter, 2.11.1840, SBB-PK, NL 133: Carl Ritter, Bl. 31; Dawson, L. S. (1885), 93–111; ODNB Washington.

[58] See the discussion in Goren (2017), esp. 25–30. For the letters, e.g. in Robinson's letters to Smith, Berlin, 11.2.1839, London, 3.10.1840, ESP, ABC 60 (63); Smith to Robinson, London, 8.7.1839, ESP, ABC 60 (105) 1839; Smith to Washington, London, 20.10.1839, ESP, ABC 60 (101); Smith to Washington, Boston, 21.4.1840, ESP, ABC

This exchange also resulted in the personal narrative being 'everywhere interwoven' in the text. Robinson explains its twofold aspect in the preface: 'the manner in which the Promised Land unfolded itself to our eyes' and 'the processes by which we were led to the conclusions and opinions advanced in this work.' The advantages of this framework were clear: it would yield a considerable amount of new information concerning the historical topography of Palestine while enabling the reader to judge the 'general accuracy' of the writers' conclusions.[59]

Robinson then notes a second dilemma concerning 'the form of narrative to be adopted,' raising two possibilities used by Holy Land travellers. Richard Pococke (1704–65), British prelate and scientist who travelled in 1738, and Carsten Niebuhr, who visited the Holy Land briefly in 1766, arranged their accounts topically. Henry Maundrell (1665–1701), chaplain of the Aleppo factory of the Levant Company, who travelled in 1697, and Johann Ludwig Burckhardt, who travelled from 1810–12, each wrote a daily journal.[60] Robinson chose the daily journal format so that the reader could follow 'the process of inquiry and conviction in the traveller's own mind.'

During the expedition, both Robinson and Smith kept daily diaries. The need for such recordkeeping was probably clear, even if its ultimate purpose was undecided. Robinson and Smith undoubtedly exemplified the spirit of the period of evolving methods of geographical observation and research, with emphasis on accurate observation and simultaneous recording (though they recorded only every evening) necessary for producing a reliable field journal.[61]

However, the simple chronological diary designed 'merely to describe what we saw' underwent significant change. It was enlarged 'fourfold,' in Robinson's words, by the 'introduction of historical illustrations, and the discussion of various points relating to the historical topography of the Holy Land.' Robinson writing about historical topography in the framework of the scriptural narratives is systematic

60 (105/12) 1840; Washington in Harwick to Robinson in NY, 1.5.1842 ('received the book from Murray, it took time to read it – but now full of praising; met also Ritter, Humboldt, not to forget Kiepert'), ERP, L.2; Robinson to Washington, Berlin, 19.6.1839, RGS-IBG, RGS Corr. Block 1834–40, RGS/CB2/454: Berghaus to Washington, Potsdam, 30.4.1839, RGS-IBG, Corr Block 1834–40, RGS/CB2; Robinson to Ritter, 2.11.1840, SBB-PK, NL 133: Carl Ritter.

59 Robinson (1841), I, Preface, v–vi; and following citations.

60 Pococke (1745); Niebuhr, C. (1969); Maundrell (1703); Burckhardt (1822). All travelogues were intensively used by Robinson, as described in chap. 4.c.

61 See summary and sources by Withers (2013), 169–70.

and in accordance with his deterministic approach, his strongly held belief in the crucial role of physical geography in human development. Mere description rarely suffices; Robinson methodically discusses and answers questions that continually arose 'which I could not pass over without at least satisfying my own mind.' They led Robinson on long investigations, exchanging letters extensively, first with Smith but also with other scholars, and researching all available written sources. Some of the questions required long discourses. 'Most of these were topics relating to the geography of the Bible, and intimately connected with its interpretation,' he writes, adding that 'they have never been discussed by any one, who had himself visited the Holy Land.' Robinson's novel approach originated in two 'important points.' The first was the ecclesiastical tradition, 'the attempt made to point out, in most cases, not only what is truth and what is legendary tradition, but also to show how far the latter reaches back.' The second was the native tradition, the 'preservation of ancient names among the native population.'[62] Robinson was deeply invested in reviving what he termed the 'Holy Land's topographical tradition,'[63] '(...) to exhibit an historical review of the Sacred Geography of Palestine, since the times of the New Testament; pointing out under each place described, how far and in what period it has hitherto being known.'[64]

The examples chosen by Robinson for discussion in the preface were the location of Mount Horeb, 'the probable place of the giving of the law,' and Beit Jibrîn, 'the ancient Eleutheropolis' visited twice by different routes and known as one of Robinson's important identifications.[65] In order to demonstrate more explicitly what he termed the 'process of oblivion,' the course of forgetting and failing to retain ancient place names to all but the most local memory, Robinson examines three works from three different periods: the fourth-century *Onomasticon* by Church Father Eusebius (*c.* 330–400), a thirteenth-century work by Brocardus (Burchard of Mount Sion, *c.* 1283) and a seventeenth-century work by the Franciscan Franciscus Quaresmius (1583–1650). Carefully following the thread of their toponymy, 'the

[62] Robinson (1841), I, Preface, vi–vii.

[63] Robinson (1841), I, Preface, vii; For detailed discussion see Robinson (1841), I, 371–79, as preface to chap. VII: 'Jerusalem. Topography and Antiquities.'

[64] Robinson (1841), I, Preface, ix.

[65] Robinson (1841), I, Preface, vi. See: 'Sinai of the Old Testament,' Robinson (1841), I, 175–80; 'Eleutheropolis,' II, 404–20. The PEF scholars cite here Robinson's discussion as proof of its significance: Conder, C. R. (1883), 272–73.

topographical respect,' he identifies, quite easily, this process of disappearing names.[66]

Robinson's focus on the topography illuminates Smith's major contribution to the research. The orthography of Arabic names, 'both in Arabic and Roman letters,' a 'point which we gave particular attention,' is embroidered throughout the study. Most likely an outcome of their meeting in Andover in 1832, Smith prepared lists of Arabic place names according to Ottoman administrative regions. He obtained some of the names during his own travels while others were 'written by educated natives (...) according to the established rules of the language.'[67] At the general meeting held in Jerusalem of the missionaries serving in the East, attended by Robinson and Smith, they solved the problem of the lack of 'a regular system of orthography' in Latin letters, the 'romanization of Arabic names,' by adopting a system proposed in 1818 by the Boston lawyer and linguist John Pickering (1777–1846). Pickering 'exerted a special influence on both Therese and Edward' for a uniform orthography 'for the Indian languages of North America.' The system was already in use by the ABCFM and others, so Smith could readily adopt its principles for his Arabic orthography.[68] Pickering was a close friend of the Robinsons who, in 1846, was formally accepted as a member of the American Oriental Society founded by Pickering in 1842.[69]

During the expedition, they made use of the orthography of former travellers. Robinson singles out Burckhardt as 'the only Frank traveller in Syria, who has, to any extent, given us Arabic names written with Arabic letters.' Burckhardt was helpful for the regions of Sinai

[66] Robinson (1841), I, Preface, vii–ix. For the reference to the three works he mentions in his source list: Robinson (1841), III, 1st Appendix, 4, 9–10, 16. The years are as assigned by Robinson. Quaresmius served as Franciscan Holy Land Custos in 1516–20 and again from 1525–26; from 1627, he served as Rome's emissary to the Christian Churches. His book was written between 1616–26. A new critical edition of Burchard's text was recently republished, including a detailed biography (Bartlett, J. R. (2019)). Various scholars have published on the issue of toponymy and place-name reconstruction, see e.g. Soler (2014), 167–72.

[67] See further discussion in Chap. 5.a. Robinson's and Smith's archives hold vast material on this subject. See above, note 137.

[68] Robinson (1841), I, Preface, x; Smith (1841a); Weigel Williams (2007), 95 (citation). Cf. Phillips (1939), 306 (citation); DAB Pickering; Pickering (1820); Stevens (1956). Talvj translated his paper to German (*Indianische Sprachen*). For the missionaries' conference see Chap 3.b.

[69] Committee of the American Oriental Society to Robinson, Boston, 9.6.1846, ERP, 1816 Robinson- 2, ER 2.24; Williams (1999), 325–26.

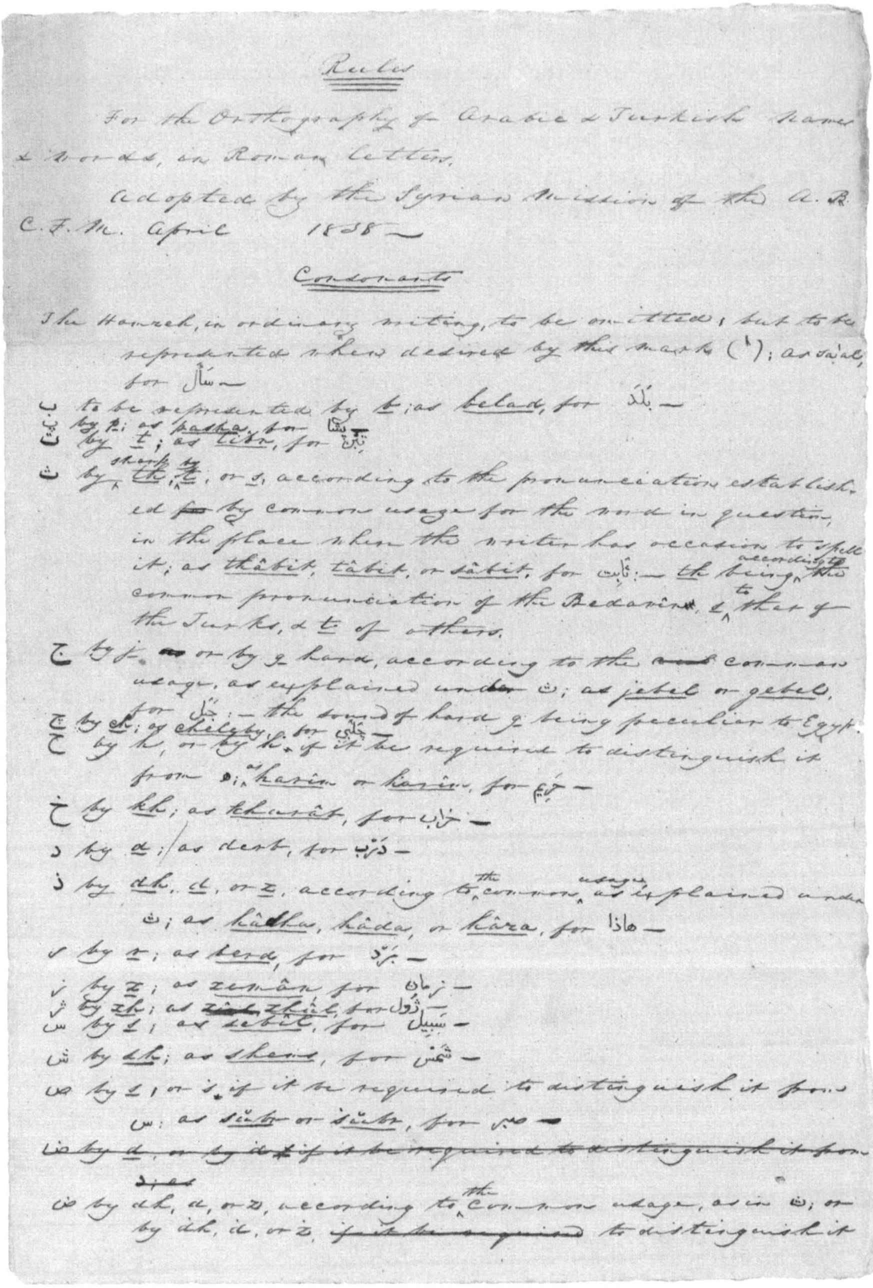

Rules

For the Orthography of Arabic & Turkish names & words, in Roman letters,

Adopted by the Syrian Mission of the A.B.C.F.M. April 1838 —

Consonants

The Hamzeh, in ordinary writing, to be omitted; but to be represented when desired by this mark (ʾ); as saʾal, for سأل —

ب to be represented by b; as belad, for بلد —
پ by p; as pasha, for پاشا
ت by t; as tibn, for تبن
ث by sharp th, t, or s, according to the pronunciation establish-ed ~~for~~ by common usage for the word in question, in the place when the writer has occasion to spell it; as thâbit, tâbit, or sâbit, for ثابت: — th being according to the common pronunciation of the Bedawin, & t to that of the Turks, & t of others.
ج by j, ~~or~~ or by g hard, according to the ~~word~~ common usage, as explained under ث; as jebel or gebel, for جبل; — the sound of hard g being peculiar to Egypt.
چ by ch; as chelby, for چلبي
ح by h, or by ḥ, if it be required to distinguish it from ه; as harîm or ḥarîm, for حريم —
خ by kh; as kharâb, for خراب —
د by d; as derb, for درب —
ذ by dh, d, or z, according to the common usage, as explained under ث; as hâdha, hâda, or hâza, for هاذا —
ر by r; as berd, for برد —
ز by z; as zemân for زمان —
ژ by zh; as ~~zhûl~~ zhûl, for ژول —
س by s; as sebîl, for سبيل —
ش by sh; as shems, for شمس —
ص by s, or ṣ, if it be required to distinguish it from س; as sabr or ṣabr, for صبر —
~~ض by d, or by ḍ if it be required to distinguish it from~~
ض by dh, d, or z, according to the common usage, as in ث; or by ḍh, ḍ, or ẓ, ~~[illegible]~~ to distinguish it

Ill. 17. Rules of the orthography of Arabic and Turkish names and words, in Roman letters. Adopted by the Syrian Mission of the ABCFM, April 1838, (Smith's?) handwritten comments.

and Petra and his orthography 'was found to be usually, though not always, correct.'[70] Robinson acknowledges in a footnote that Burckhardt was not their only source, but that the bilingual place names in the maps drawn by the French surveyors under Colonel Pierre Jacotin and published by him, and the report of the German Catholic Orientalist and biblical scholar Johann Martin Augustin Scholz, 'are so very incorrect, as to form no exception to the above remark.'[71] Smith, however, was able to depend on the help of his fellow missionaries. Bird sent him information concerning Mount Lebanon, and Whiting, who in summer 1839 stayed in the US, sent a list of 'the names which you request for Dr R.[obinson].'[72] The rest of Robinson's preface is dedicated primarily to the maps, whose production is detailed in a prior study.[73] In this context, Robinson explains two important objectives: his intent to eliminate the multiplicity of names characteristic of most Palestine maps and to discover accurate and authoritative sources. Robinson expresses his hope that 'very much has been gained in truth and correctness' and that their study would succeed in laying the groundwork for future research:[74] 'We wish it [the work] to be regarded merely as a beginning, a first attempt to lay open the treasures of Biblical Geography and History still remaining in the Holy Land,- treasures which have lain for ages unexplored, and had become so covered with the dust and rubbish of many centuries, that their very existence was forgotten.'[75]

4.b.2. Origin of the manuscript from inception to format

After 22 months of intensive work, Robinson summarizes in the preface the process that guided the writing of the manuscript. He also refers briefly to an 'original plan' for the manuscript. The letters, however, do not reveal such an early determination to write a book, certainly not in its final scope. The picture they reveal of the writing process is different in important respects.

70 Robinson (1841), I, Preface, ix. Smith, who closely checked Burckhardt's orthography, was not as satisfied, see note 136.

71 Robinson (1841), I, Preface, ix, n. 1; Jacotin (1810); Scholz (1823).

72 Whiting to Smith, Elmira, NY, 22.8.1839, ESP, ABC 60 (91); see note 97.

73 Goren (2017).

74 Robinson (1841), I, Preface, xii.

75 Robinson (1841), I, Preface, xiii. Cf. Robinson (1843a), as retrospect.

During the three and a half months of travel, on long nights and on Sundays, their Sabbath of rest, Robinson and Smith had plenty of time to consider publication of their discoveries. It became clear that the sizeable wealth of new information required nothing less than a book; consequently, in his first long letter-report to Anderson after the expedition, Smith writes from Smyrna:

> I am happy to be able to inform you that I arrived here safely, in company with Dr Robinson (...). Our journey was very successful, & brought us acquainted with much more information of a geographical nature, than we or even I expected. (...) The results of our observations, Dr R. will embody in the form of a Journal, for publication. For this purpose, I shall put into his hands my own notes, which are as full as I could (?). I have (nice compensation?) in the thought that our Journey will (thereby?) inform the geography of the Bible. I propose, and soon as practicable, to communicate to you, such information as I may be able, respecting the people with whom we have had intercourse, especially the Bedawin Arabs.[76]

Quite clearly, Smith touches upon two issues: the planned publication and locating potential communities for missionary purposes. Notwithstanding that the publication would be of tremendous benefit to the ABCFM board, he never forgot his primary role as missionary.[77] In furtherance of the first-mentioned goal, Smith carried his diaries with him and packed '50 packages of manuscripts' in his equipment, which had been checked following their quarantine upon entering the Habsburg Empire.[78]

According to their letters, at the time of Smith's first report to Anderson, they were still uncertain as to the framework and contents of the manuscript. As described in chapter 3.c., after recovering, Robinson left Vienna with his family on October 25, 1838, for Dresden, Leipzig, and Halle. Because the family stayed in each city longer than originally scheduled, they only reached Berlin on November 15. Talvj, writing from Berlin in early December, states that they were planning to stay there only for the winter, for six months. She also notes that her husband had

76 Smith to Anderson, Smyrna, 9.8.1838, ESP, ABC 60 (105/9).

77 This issue is discussed in chap. 5.a. See, for example, Smith's long report to Anderson on the possibilities of missionary work among the Sinai Bedouin, Vienna, 13.10.1838, and Leipzig, 25.10.1838, ESP, ABC 60 (105/5), 1838.

78 Sanitäts-Fehde für Personen, k. k. Contumaz-Amte, 3.9.1838, ESP, ABC 60 (137).

begun writing short descriptions of his travels that were published, and that they both intended to do some work in Berlin, but that the issue of a thick travelogue would be left for Schubert.[79] Gotthilf Heinrich von Schubert, a leading natural historian from Munich, travelled in the East in 1836–37, and his three-volume work was published in 1839.[80] As early as November 1837, Smith, who at the time thought he was unable to accompany Robinson, tried to reassure him by writing 'I hope you will be able to pass by (…). Professor Schubert & his wife with another lady in company have recently done it.'[81] Naturally, Robinson was familiar with Schubert's travels and publications. Eight months later, Talvj, unaware that they would be staying a much longer period in Berlin, writes that 'we are still here,' happy to spend another winter in her *Vaterland*.[82]

To Smith in Leipzig, Robinson writes that 'the orientalists here, Gesenius, Roediger, & Tholuck, received me with open arms, & seem truly to rejoice at the results of our journey in Palestine.'[83] Emil Rödiger (1801–74), who was to take a leading role in production of the German edition of the *Biblical Researches*, studied and taught in Halle. Following Gesenius' death, Rödiger completed some of his books and re-edited his Hebrew Language Thesaurus.[84] They all agree, Robinson continues, 'in urging the propriety and importance of handwriting an account of our journey.' Being Orientalists, they also urged Robinson to emphasize the study of Arabic names. Robinson, seeking a second opinion of scholars from a different field, went to Berlin 'to get the opinion of the geographers there' and to make a final decision.[85]

Robinson reached Berlin and did not waste time, beginning his extensive letter exchange with Smith. The latter's archive holds 18 letters from Robinson written between November 23, 1838 and April 15, 1840, and about the same number of letters from Smith to Robinson between December 4, 1838 and late 1839, when Smith was back in the USA.[86] Robinson's archive does not hold any of these letters, or any other letters

79 Talvj to Kopitar, Berlin, 14.12.1838, Milović (1941), 97–99.

80 Schubert (1839). Cf. Goren (2003), 117–24; (2011), 207–29. Robinson mentions Schubert most extensively in Chapter XV, which deals with his travel 'From Mount Tabor by the Lake of Tiberias to Safed.'

81 Smith to Robinson, Smyrna, 10.11.1837, ESP. ABC 60 (105) 1837.

82 Talvj to Kopitar, Berlin, 12.8.1839, Milović (1941), 99.

83 Robinson to Smith, Halle, 15.11.1838, ESP, ABC 60 (63).

84 ADB Rödiger.

85 Robinson to Smith, Halle, 15.11.1838, ESP, ABC 60 (63).

86 ESP, ABC 60 (63); ESP, ABC 60 (105/7–105/12).

exchanged between the two. How long did it take for a letter to cover the relatively short distance between Leipzig and Berlin? In a letter dated January 28, 1839, Smith writes that after returning from four days in Halle, he received both of Robinson's letters dated late December.[87] Hence, we can assume that the mail between the two cities took at least two or even three weeks' time.

They soon discovered that another technical problem hampered both their communication and publications: their handwriting. Both admitted that it was not their strong point, and soon enough they realized that their poor handwriting affected their work, particularly when sending documents to third parties. 'Pray take all the pains you can to read my bad writing correctly,' writes Smith, as 'the blunders made in printing your communications in the N. Y. Observer, have made me think particularly on this subject.'[88] Robinson's letters to the editors of the *New York Observer* that Smith notes in this correspondence were not located.[89]

Very shortly after his arrival in Berlin, while looking for appropriate lodging for his family, Robinson hastened to visit two scholars whom he had met about a decade earlier when studying in Berlin, the geographer Ritter and the theologian and historian Johann August Wilhelm Neander (1789–1850). Both offered 'spontaneously the use of their libraries in the kindest manner.' Lying on Ritter's table, Robinson found Smith's Armenia book, which Ritter had been using for his *Comparative Geography*.[90] These leading scholars, together with cartographer Heinrich Berghaus, encouraged Robinson continually, influencing considerably his decision to stay in Berlin and write the book. Robinson easily found his place there as one colleague later affirmed, 'I first saw him in Berlin, going in and out, like one of them, among the renowned Gelehrte [scholars] of that famous University.'[91] Even so, it is still quite certain that nobody expected the end result to be of such great scope. 'In

[87] Smith to Robinson, Leipzig, 28.1.1839, ESP, ABC 60 (105\10).

[88] Smith to Robinson, Leipzig, 28.1.1839, ESP, ABC 60 (105\10).

[89] Smith to editors of the *New York Observer*, private, Leipzig, 28.1.1839, ESP, ABC 60 (105/5. 1839). For further discussion see chap. 5.a.

[90] Robinson to Ritter, Berlin, 23.11.1838, ESP, ABC 60 (63). For studies by Neander see Weigel Williams (2009), 246–47, 252; Robinson to Norton, Halle, 24.4.1828, ERP, ER 1.17, where Robinson also mentions Neander's Jewish origin; Lazarus (1930), 4, 437; NDB Neander. For the phenomenon of leading German academics converting in the late 1830s, as understood by the LJS, see LJS (1839), 159.

[91] Prentiss (1889), 62.

general, our journey & observations seem likely to prove of more value & importance than I at least had anticipated,' writes Robinson about two months after reaching Berlin. In this letter to Smith, he repeats Berghaus' unprecedentedly enthusiastic words, that until that time, he had thought Burckhardt deserving of 'the highest praise on grounds of his meticulous attention to even the most unobtrusive phenomena and his precision in trigonometrical and time measuring when determining distances;' however, after meeting Robinson and seeing his and Smith's journals, 'I see myself under the necessity to considerably modify this opinion. (...) The observations made by the aforementioned travellers being so vast and comprehensive, their description of the shape and relief of the country so exact and beyond any doubt that they will enable the geographer to design a special map which will leave little to be desired.'[92]

Subsequently, Berghaus, who mentioned Robinson and Smith for the first time only in his letter of March 12, 1839,[93] played an important role in the decision to publish the book, his impact seen primarily in the additional cartography. This narrative, including that of the map of the Negev, the only map drawn by Berghaus for Robinson and Smith, has been published in detail.[94] Robinson's aforementioned letter to Smith, dated February 2, 1839, describes the way in which he was encouraged by Berghaus' enthusiastic reaction to the expedition notes and itinerary: 'Such a testimony, wholly disinterested, & from the person perhaps best qualified of all men living to form a correct judgment on the subject, was to me as gratifying as it was unexpected; & I send it to you, that you too may see & feel, that our labours are not likely to prove in vain, nor without acknowledgement.'[95] It is quite certain that Berghaus was a central figure behind the success of the *Biblical Researches* and Robinson's international reputation.

Robinson had no doubt that both Ritter and Alexander von Humboldt (1769–1859), who from 1827 lived and taught in Berlin, would help him achieve his goal of publishing the book. Ritter wasted no

[92] Berghaus to Washington, Potsdam, 30.4.1839, RGS-IBG Collections, Corr Block 1834–40, RGS/CB2. The German version: Ritter to Smith, Berlin, 11.2.1839, ESP, ABC 60 (63); Berghaus (1840), 531.

[93] Berghaus to Justus Perthes, Potsdam, 12.3.1839, FBG SPA ARCH MFV, MTV 19A/269–70. See also his cited letter in the JRGS: Berghaus (1839).

[94] Robinson to Smith, Berlin, 11.2.1839, ESP, ABC 60 (63). See in detail Goren (2017), 21–28.

[95] Robinson to Smith, Berlin, 22.1.1839, ESP, ABC 60 (63). See above, and Goren (2015); (2017), 21–28.

time in inviting Robinson to the next meeting of the Geographical Society that was held on December 8. Robinson, again writing to Smith, 'If possible, I hope to lay before them a general account of our journey, which can then be published perhaps with their sanction.' Ritter even postponed his own lecture in order to enable Robinson to present his 'brief report' as soon as possible, additional proof of the importance he attributed to Robinson's discoveries. Only three weeks after reaching Berlin, Robinson stood before its leading geographers, members of the highly respected *Geographische Gesellschaft zu Berlin*. According to his own impression, 'so far as I could judge, it was well received.'[96]

In accordance with standard scientific society protocol, Ritter forwarded his meeting reports for publication. Ritter also submitted Robinson's report, which he thought highly important, adding his own letters of recommendation to some of the target journals. Ritter read the first report of the expedition 'before the Crown Prince & several others of the royal family,' who showed much interest.[97] Throughout Robinson's stay in Berlin, Ritter continued to support Robinson, 'his kindness unremitted & unabated.' Robinson noted that Ritter also intended to read before the Berlin Geographical Society extracts from 'the translation of our route from 'Akabah to Jerus.'[98]

Robinson also updated Smith about the unexpected resonance of their expedition material: 'I hear from various (...) of the interest taken on this account, from the palace downward.' Robinson sent 'the best half' to his old acquaintance Emil Rödiger in Halle, an editor of the journal *Zeitschrift für die Kunde des Morgenlandes* (*ZKM*), for publication in its 1839 volume, adding that the rest would soon follow.[99] Ritter sent a shorter note for publication in the *Allgemeine Zeitung* and an English version was submitted to the *Biblical Repository* in New York.[100] In June 1839, Robinson sent 'an extract of the journal of travels' with a somewhat detailed version of the expedition

96 Robinson to Smith, Berlin, 23.11.1838, 10.1.1839, ESP, ABC 60 (63); Robinson (1839b).

97 Robinson to Smith, Berlin, 25.3.1839, ESP, ABC 60 (63). The Prince was Friedrich Wilhelm (1795–1861), from 1840 King Friedrich Wilhelm IV. For his various and multi-facetted connections to Jerusalem and the Holy Land see Schütz (1988).

98 Robinson to Smith, Berlin, 24.12.1839, ESP, ABC 60 (63).

99 Robinson (1839b). First volume of the periodical in 1837, the second in 1839, though Williams (1999), 354, gives 1838. Cf. Wokoeck (2009), 127–45.

100 Robinson (1839).

legs across the desert, to Jerusalem, and to Wady Musa, and a shorter note of its entirety to Secretary of the London Geographical Society John Washington, accompanied by a letter from Ritter. This extract was published in the *Journal* of 1839.[101] Robinson closed his letter to Smith with his expectation that the expedition results would soon become widely known.[102]

Following these initial papers, the 'Kurzer Bericht,' 'Extracts,' and 'A Brief Report,' Robinson began concentrating on the *Biblical Researches*. He continued submitting articles, but only to the *ABR* and other American forums, publishing three in 1840 and three the following year. In 1841, once in the US, Robinson also began sending many shorter comments and reports to newspapers, primarily the *New York Observer*.[103]

It is a bit more difficult to establish Humboldt's role in Robinson's advancing career. The young Robinson did attend Humboldt's lectures in Paris (1826) and Berlin (1828)[104] and, upon reaching Berlin in late 1838, he naturally looked for ways to meet the world-famous scientist. Humboldt wrote to Ritter that he would be happy to meet Robinson, noting that 'I already asked Berghaus twice to tell ER that I'll be happy to meet him.'[105] The letters do not reveal evidence of a meeting, though one might assume that they did meet, at least under social circumstances.

At this time, Robinson started to develop his methodology for studying the material collected, resolving questions, and writing a highly accurate text. He began by comparing both his and Smith's expedition journals, repeating his delightful surprise at their agreement, 'there is scarcely a difference in the relation of time, except as arising from the difference of our (?); & no discrepancy as to things seen, etc.'[106] This accord, a bit unexpected, reminded Smith of a prior experience when he compared his notes from his Armenia travels with those of his companion Dwight, 'it gave me great confidence in publishing the results of our inquiries.'[107] As he proceeded, Robinson repeated this impression in almost the same words. Naturally, Smith's journals were 'more full on

[101] Linke, 1984; Robinson to the RGS, 18.6.1939, RGS-IBG Collections, JMS/9/49; Robinson (1839a).

[102] Robinson to Smith, Berlin, 22.1.1839, ESP, ABC 60 (63).

[103] Bibliography by Williams (1999), 354–55. For example: Robinson (1832).

[104] Williams (1999), 119, 158; Weigel Williams (2009), 243–59.

[105] Päßler (2010), 69–70.

[106] Robinson to Smith, Berlin, 22.1.1839, ESP, ABC 60 (63).

[107] Smith to Robinson, Leipzig, 15.2.1839, ESP, ABC 60 (105\10).

many points of information derived from the Arabs, in the bearings etc.' and the complementary material was invaluable for producing a comprehensive study.[108]

There are numerous examples of Robinson and Smith tackling difficulties and questions through their 'ping pong' of letters between Berlin and Leipzig, two cities relatively close to one another. In an exceptionally long letter, even for Smith, sent on February 8, 1839, he writes, 'I told you in my last, that my next letter might entertain you with a new thing respecting the destruction of Sodom & Gomorrah. So here you have it.'[109] Robinson's answer (not found) triggered Smith to send another letter with corrections within a week. Robinson had included the destruction of the biblical cities, 'the Dead Sea and the Catastrophe of the Plain,' in the *Biblical Researches*, citing his letter exchange with the Berlin geologist Leopold von Buch (1774–1853).[110] There is also a long discussion in their letter exchange concerning the banks of the Jordan River and the Jordan Valley, where, trying to recollect the banks' nature, Smith relies on the descriptions of Burckhardt, Buckingham, and Maundrell. Smith questions whether to describe the phenomenon as 'the bank,' 'the shore,' or 'only a shallower part near the shore' adding, 'if by this additional explanation, our ideas can be brought together, please forward the letter, if not detain it till I hear from you again. I would not say anything which you may afterwards contradict in your book.' Smith opines that the book should contain 'as many places as we have materials for,' both in the text as well as in the cartography, and that the meaning of names should be explained when possible.[111]

Toponomy was another issue for decision making raised in their letter exchange. The etymological discussion of the name *Masada* and its affinity to the Arabic *Sabbeh* is one good example. 'I can discover no affinity,' writes Smith, probably answering Robinson's query, 'nor if we suppose them to come from Arabic roots, have these roots at all a similar meaning.'[112] An additional topic was the flora and fauna encountered and their scriptural references. The frequently mentioned

[108] Robinson to Smith, Berlin, 11.2.1839, ESP, ABC 60 (63).

[109] Smith to Robinson, Leipzig, 8.2.1839, ESP, ABC 60 (105/10).

[110] Robinson (1841), II, 601–08, 667–75; ADB Buch. For Buch's letter exchange with Robinson through American colleagues, see Robinson to Prof. Silliman in Yale, Berlin, 1.5.1839, ERP, ER 1.6.

[111] Smith to Robinson, Leipzig, 15.2.1839, ESP, ABC 60 (105/10).

[112] Smith to Robinson, Leipzig, 2.3.1839, ESP, ABC 60 (105/10).

terebinth was identified as the oak tree, though Burckhardt called it Buntam, which is actually the wild Pistacia.[113] Smith raises the question of plant names in general, asking Robinson how many names he would like in addition to those recorded in the expedition journals, as 'I can probably give you as many as you need.'[114] For additional sources, he mentions Peter Forsskål (1732–63), Carl Linné's student and the botanist of Carsten Niebuhr's 1761–67 expedition to Arabia Felix, whose book was edited *post mortum* by Niebuhr,[115] and 'Freytag in the lexicon,' both sources that he had access to in Leipzig. Georg Wilhelm Friedrich Freytag (1788–1861), philologist of Oriental languages, published his five-volume Arabic-Latin lexicon between 1830 and 1837. According to the letter, he was also responsible for Arabic types prepared in Paris by order of the Prussian Government.[116] Robinson expected to find additional material by the theologian Scholz, who published a book following his 1820–1 expedition to Egypt and the Holy Land[117] that he hoped to locate in Berlin. Robinson eventually sent Scholz' book to Smith, who writes to him deeply disappointed, that 'If I had known the nature of the book better, I should hardly have requested you to send it,' primarily due to its lack of value with regard to Arabic names.[118]

By late March 1839, Robinson had outlined the structure of the book. It would be divided into sections chronologically and, of course, geographically, each section also containing what he called 'remarks.' The first section was to be 'from Cairo to Suez, with remarks about Goshen & the Exodus at the close.' This would enable treating Smith's article about the religious sects of Palestine as a separate section and not an appendix. Smith fully agreed with this proposal[119] and soon thereafter Robinson started reporting on his progress.

113 Smith to Robinson, Leipzig, 11.3.1839, ESP, ABC 60 (105/10).

114 Smith to Robinson, Leipzig, 23.3.1839, ESP, ABC 60 (105/10).

115 Robinson to Smith, Berlin, 14.3.1839, ESP, ABC 60 (63); Forsskål (1775). Cf. Hansen (1964); Goren (2003), 30–36.

116 Robinson to Smith, Berlin, 14.3.1839, ESP, ABC 60 (63); ADB Freitag (written by Philipp Wolff!); Fück (1955), 166; Freytag (1830–37).

117 Scholz (1823). Cf. Goren (2003), 63–64.

118 Smith to Robinson, Leipzig, 8.4.1839, ESP, ABC 60 (105/11).

119 Robinson to Smith, Berlin, 29.3.1839, ESP, ABC 60 (63); Smith to Robinson, Leipzig, 8.4.1839, ESP, ABC 60 (105/11).

4.b.3. Attention to detail: Arabic place names and orthography

Robinson's December 13, 1838 letter to Smith opens their discussion concerning formatting the pre-existing lists of Arabic names. Smith wanted 'to rearrange' the lists and Robinson suggests 'to print them in columns; (...) to add also the Roman orthography of each name; & a translation of the title or heading before each province.' Robinson continues deliberating in subsequent letters, unsure, for example, whether the Arabic orthography should be included in the text or only in the relevant appendix.[120]

In January 1839, Robinson informs Smith that work on the maps was to commence with the highly recommended Heinrich Berghaus: 'Ritter says no person can do it so able as Berghaus, under my supervision.'[121] Smith responds with his hopes that Berghaus would, indeed, accept the task and that 'our list of words will help him much in regard to Orthography.' Later, satisfied with Berghaus' agreement, Smith writes, 'I shall be disappointed, if it [the map] is not better than any which has yet appeared,' and, already in Leipzig, requests Robinson's deadline for receiving the list as he was still very busy with the types.[122]

Smith started working with Burckhardt's book, which had been praised by Robinson, and soon found that his 'Arabic orthography differs so often from mine, or rather for that of the natives who wrote the greater part of my list (...). I think he had not a very accurate ear.'[123] It took Smith less than two months to share with Robinson his deep disappointment with Burckhardt's Arabic:

> I have examined all Burckhardt's names, & marked those whose orthography differs from mine. This examination has given me a much lower opinion than I had before of his Arabic. The mistakes were numerous & gross, occurring in words in daily use, & showing not only that his ear did not readily distinguish Arabic sounds, but that his knowledge of Arabic orthography was very imperfect. There is also evidence of carelessness, I find the same word spelled in different places, & there are four different ways.[124]

[120] Robinson to Smith, Berlin, 13.12.1838, 22.1.1839, ESP, ABC 60 (63).

[121] Robinson to Smith, Berlin, 10.1.1839, ESP, ABC 60 (63). Cf. Goren (2015); (2017); Weigel, P. (2014, some items).

[122] Smith to Robinson, Leipzig, 18.1.1839, 19.1.1839, ESP, ABC 60 (105/10).

[123] Smith to Robinson, Leipzig, 19.1.1839, ESP, ABC 60 (105/10).

[124] Smith to Robinson, Leipzig, 11.3.1839, ESP, ABC 60 (105/10).

Smith, still deliberating whether to include the Arabic names in the notes or put them in an index, writes 'perhaps (...) it would be best to insert in both places.' Eventually, deciding in favour of adding an index of the Arabic names, he reasons that 'to print them in the text would be awkward, surveyed and written by others, because Arabic letters occupy several spaces.'[125] However, reconsidering the issue, ten days later Smith writes that the Arabic names 'ought to be on the same page with the Latin, at least when it first occurs (...) either in the text or in the margin, as suites typographical convenience best.' In addition, there should be 'an index & vocabulary combined' noting the meaning of the names as well.[126] Robinson responds that 'I am still quite at a loss what to do with the Arabic names.'[127] In mid-March, Robinson writes that 'the manner in which he [Rüppell], & all the other recent travellers, as Lord Lindsay, Russegger [Jopseph von, 1802–63, Austrian Montanist], Laborde, etc., murder the Arabic names, made me very desirous to be very exact at this point.'[128] Smith took Robinson's expression so seriously that upon reaching London on his way to the states, he warns Robinson that Lord Lindsay 'is much respected here for his piety' and should be handled delicately.[129]

In late March, Robinson finally opted in favour of an index, still undecided as to whether it should be separate or part of the general index. Surely, Arabic letters within the text will 'confuse the unlearned reader,' and to put them as footnotes 'will often make a mass of short notes,' Robinson notes, given his habit of using and citing numerous references.[130] They continued collecting all potentially relevant orthographical material for the book, including missionary Isaac Bird's 'complete list of the villages of Mt Lebanon, made by a local Arab scholar.'[131] Whiting's list 'of the native names of places in & around Jerusalem,' reviewed and

[125] Smith to Robinson, Leipzig, 11.3.1839, ESP, ABC 60 (105/10); Smith to Robinson, Leipzig, 14.3.1839, ESP, ABC 60 (105/10).

[126] Smith to Robinson, Leipzig, 14.3.1839, ESP, ABC 60 (105/10).

[127] Robinson to Smith, Berlin, 25.3.1839, ESP, ABC 60 (63).

[128] Robinson to Smith, Berlin, 14.3.1839, ESP, ABC 60 (63). For additional detail on the aforementioned travelers in connection to the study of the region see: ADB Rüppell; Goren (2003), 57–60. ODNB Lindsay; Goren (2011), 192–93. ADB Russegger; BLKO Russegger; Goren (2003), 253–56. Laborde: Goren (2011), 146–47, and above.

[129] Smith to Robinson, Leipzig, 8.7.1839, ESP, ABC 60 (105/10).

[130] Robinson to Smith, Berlin, 29.3.1839, ESP, ABC 60 (63).

[131] Smith to Robinson, Leipzig, 23.3.1839, ESP, ABC 60 (105/10).

corrected by Smith, had been sent by the latter, now in England, to Robinson in September.[132] Shortly thereafter, Smith re-checked his earlier lists, adding the new information collected during the expedition. He organized the data according to administrative regions and added 'a few geographical remarks to each province.'[133] Robinson then asks whether it would be a good idea to add 'the signification of the Arabic names, where they have any definite meaning?'[134]

In March 1839, Smith was ready to forward his lists of orthography for 'the itinerary as far as Mt Sinai' after deciphering and comparing the Bedouin names to Burckhardt's list. He writes, 'when Burckhardt's orthography differs from mine, & I am certain he is wrong, I have noted it, where there was any room (?) doubt I have said nothing.' Smith opines that his orthography should serve as a model for subsequent writing, though 'If you wish me to make any alterations in the manner of writing the definitions in the rest of the Itinerary, please inform me.'[135]

According to Berghaus' instruction and his letter of January 22, 1839, Robinson spent 'the last fortnight' preparing '**a full itinerary**, of stations, courses, distances, bearings, with topographical notices' crucial for the planned map and also very useful for the book.[136] Ritter thought that the full itinerary should be added to the book as an appendix. Robinson, who at this stage was already conceptualizing the final publication format, accepted Ritter's suggestion, writing to Smith that the complete edition will include '4 appendixes, viz. 1. Religion Sects, etc., in Palestine. 2. Notes on the Arabic Language. 3. This Itinerary, etc. 4. Lists of Arabic Names, etc.'[137] Smith agreed that an appendix on regional religions was crucial, 'a distinct article upon the different sects in Syria,' offering to write the second appendix dealing with 'the spoken vulgar Arabic, especially its peculiarities in different provinces.'[138] In early February, Robinson writes that the 'diaring [*sic*] out from both the Journals a complete itinerary on a tabular form, containing the names of all the stations, the course, the time, the bearings, & all such topo-

132 Smith to Robinson, Yarnton, 17.9.1839, ESP, ABC 60 (105/11).

133 Smith to Robinson, Leipzig, 28.1.1839, ESP, ABC 60 (105/10).

134 Robinson to Smith, Berlin, 11.2.1839, ESP, ABC 60 (63).

135 Smith to Robinson, Leipzig, 1.3.1839, ESP, ABC 60 (105/10).

136 Robinson to Smith, Berlin, 22.1.1839, ESP, ABC 60 (63). On that day it was 'completed as far as Jerusalem, including all measurements in & around that city.'

137 Robinson to Smith, Berlin, 22.1.1839, ESP, ABC 60 (63). Also Robinson to Smith, Berlin, 11.2.1839, ESP, ABC 60 (63).

138 Smith to Robinson, Leipzig, 28.1.1839, ESP, ABC 60 (105/10).

graphical notices,' will be completed in a couple of days, indicating that it took him over a month to complete these lists.[139]

In some instances, such as with the inscriptions they duplicated from Sinai, Robinson and Smith needed the help of local German specialists. They consulted the Leipzig Orientalist, epigrapher, and paleographer Eduard Friedrich Ferdinand Beer (1805–41), who worked primarily on these Sinaitic inscriptions until his sudden death. He is first mentioned in Robinson's letter to Smith dated March 29, 1939, 'do you hear any thing further from Prof. Beer respecting the Sinaitic inscriptions? As I am now aiming upon them, I will thank you to ascertain whether he has been able to make any thing more of them, & what.'[140] Smith, who went to meet Beer, was unenthusiastic about Beer's scholarship, noting that 'he has kept himself for some time, very close in regard to his investigation, & I have had a good deal of doubt whether they would come to any satisfactory result.'[141] 'Prof. Beer's work does not account too much, after all,' answers Robinson, assuming that 'the inscriptions are doubtless the work of Christians of the early centuries. The mystery is the alphabet, and if from pilgrims, how all came to use that same unknown alphabet. This circumstance (of the alphabet) is explained by supposing this to come from the Nabbatean residents, but then how comes it, that they are all to the west of Sinai, & on the great pilgrims road?'[142]

Smith met with Beer repeatedly, who authorized use of his deciphering of the inscriptions in the *Biblical Researches*. Smith reassured Robinson that none of the material sent to Beer would reach other hands, adding, 'you seem to gone into the subject much more thoroughly than he. Of this he seems himself to be sensible.'[143] After nearly nine months of intensive activity, Robinson finally gave up, writing, 'I have not heard from Beer; & do not know whether he has made any further progress in his inscriptions. I presume not, for he recently referred Roediger to me for information respecting them.'[144] It was during this period of activity with Beer, while in London to check on the RGS transactions with regard to Robinson's report, that Smith decided against submitting two

139 Robinson to Smith, Berlin, 11.2.1839, ESP, ABC 60 (63).

140 Robinson to Smith, Berlin, 29.3.1839, ESP, ABC 60 (63). ADB Beer.

141 Smith to Robinson, Leipzig, 8.4.1839, ESP, ABC 60 (105/12).

142 Robinson to Smith, Berlin, 17.4.1839, ESP, ABC 60 (63). See also, in Robinson (1841), I, 552–56, 'Sinaitic Inscriptions. Deciphered by Professor Beer; his results and views. One at Petra.'

143 Smith to Robinson, Leipzig, 10.5.1839, 21.5.1839, ESP, ABC 60 (105/10).

144 Robinson to Smith, Berlin, 24.12.1839, ESP, ABC 60 (63).

papers to the RGS that he wrote in Leipzig. He decided to reserve 'on Arabic Pronunciation & the List pf places' for the *Biblical Researches*.[145]

What happened to Robinson's and Smith's original journals? They were not found in any of the relevant archives, where one would expect them to be housed for safekeeping, nor were they found at the UTS with Robinson's manuscript. A comment in one of Robinson's letters provides a clue. Smith, in Beirut, enquires as to their whereabouts, to which Robinson responds, 'I am sorry to lose your Ms Journals; & do not see of what exact use they are to be to you; but I shall send them as you wish to Dr Anderson, hoping you may still return them. I should like to deposit them & mine to place in some public library, say that of the A.B.C.F.M.'[146] To date, the journals have not been located.

4.b.4. The writing begins in earnest

In mid-April 1839, Robinson writes that 'I am just now ready to set off from Sinai.' In this section, he takes the reader on a journey, discussing at length the location of the Red Sea passage and the true Mount Sinai.[147] Later in the year, Robinson began adding dates to the manuscript, providing a parallel method to his correspondence for tracking his progress.[148] In July 1839, Robinson paused from his regular updates to express satisfaction from the widespread, positive reception of the initial published reports:

> I have seen the journal of the Geog. Soc. containing my report, as well as Count Bertou's; you have doubtless seen it also, the Lond. Athenaum for June 29, 1839, containing a letter from American correspondent, also speaks of our journey, & also of your letter in the N.Y. Obs.[erver], which he accounts to me, as 'operating on poor Mr Buckingham.' This shows at least that interest in the Mss. & producing the desired effect.[149]

[145] Smith to Robinson, London, 8.7.1839, ESP, ABC 60 (105/11). See: *Biblical Researches*. Printed texts by Smith from the 1856 volume, ESP, ABC 60 (119).

[146] Robinson to Smith, New York, 18.2.1843, ESP, ABC 60 (63).

[147] Robinson to Smith, Berlin, 17.4.1839, ESP, ABC 60 (63).

[148] ERP-UTS, Boxes 1–4; first date 22.11.1839, on p. 821.

[149] Robinson to Smith, Berlin, 10.7.1839, ESP, ABC 60 (63). For count Jules de Bertou, his studies and the almost illegal addition by Washington to Berghaus' map, see Goren (2011), esp. 211–17, 223–29; Goren (2015). For more comments on Bertou, his travels and reports: Smith to Robinson, Leipzig, 5.2.1839, ESP, ABC 60 (105/10) 1839 Jan.-Mar.; Robinson to Smith, Berlin, 15.4.1840, ESP, ABC 60 (63). The publication: Robinson (1841), II, 659–69.

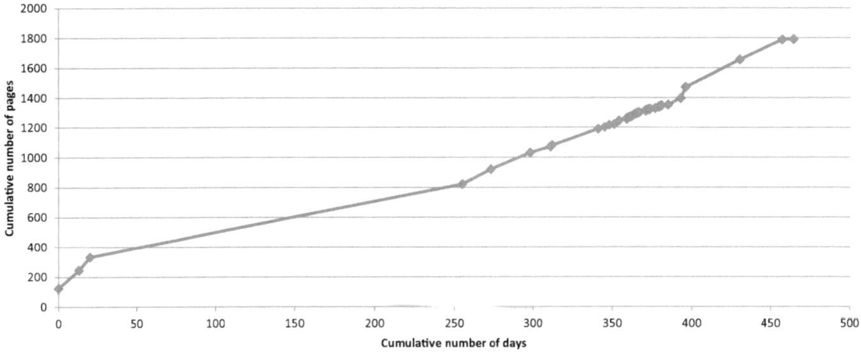

Ill. 18. Robinson's writing pace, March 16, 1839 – June 22, 1840.

Considering the substantial library research, one must admit that Robinson's writing pace was quite astonishing. In September 1839, he complained mildly that the section on Jerusalem, which he had hoped to complete by July, needed 'a wide field of investigation, & of a kind so interesting & important, that this part has become perhaps the most important of the whole book.' The writing was taking Robinson much longer than expected; yet, the fact that the library closed for four weeks of summer vacation provided him with days of forced relaxation.[150] Robinson did the bulk of his writing in the Royal Library (*Königliche Bibliothek*), today the State Library (*Staatsbibliothek Preußischer Kulturbesitz*, STABI), then and now the biggest and richest in Prussia and Germany.

Robinson feared that his writing was overly detailed, mainly due to the topographical issues, making it 'heavy & dry.' Seeking potential reviewers, he gave the manuscript ('as far as Jerusalem') to Henry Wheaton (1785–1848), who served between 1835–46 as US envoy to Berlin, and who was the most reliable transporter of mail between Robinson and his American and European colleagues. Wheaton's highly favorable reaction was very encouraging. He advised Robinson 'that the topographical notices cannot well be omitted, & that there is enough of personal narrative & interest attached to the subject, to overbalance any objections.' The section on Jerusalem was read by Gesenius, who 'expresses himself as being highly gratified.'[151]

[150] Robinson to Smith, Halle, 10.9.1839, ESP. ABC 60 (63).

[151] Robinson to Smith, Halle, 10.9.1839, ESP. ABC 60 (63). For Wheaton: DAB Wheaton; Wheaton (1889); Kellen (1902); Liu (2012). On transporting Robinson's letters to Ritter, see e.g. Robinson to Ritter, London, 2.11.1840, New York, 15.3.1840 [should be 1841], SBB-PK, Nachl. 133 (Carl Ritter), K 5, Q – SCH, Bl. 30.

On Christmas Eve, Robinson sent Smith 'a full account of my present structure.' He reiterates that 'the section upon the topography & antiquities of Jerusalem [came] out longer & more important than I anticipated.' The length resulted from his decision to include all accessible 'accounts of the ancients & those [of ne]wer travelers' and to attempt to resolve questions in dispute. Robinson also expresses his astonishment that he could not find a single written history of Jerusalem from the Moslem conquest to the crusades, even 'Neander, Ritter, & Wilken' could not help him: 'these I had to collect & arrange (...); all these investigations necessarily delayed me long.' Only after resolving these issues, was he able to proceed more quickly. He writes further that 'the Dead Sea & Jericho cause me some longer delay' as well, so 'I am now at Jericho; but hope to get back to Jerusalem before the new year.' Robinson quickly understood that he would not be able to complete the manuscript in three months as planned.[152]

Robinson's next report, in mid-February 1840, reflects his mood and acquiescence to the reclusive life of manuscript writing:

> I have little to add of our way of life; we live retired, & I avoid all new acquaintances, in order to devote myself & my whole time (?) to the work before me. Yet it is but the life I wish and need. To be so long shut up to one exclusive employment, & that retired, without active duties to call out & exercise the other faculties, is in itself dangerous in its influence on the hand & heart.[153]

Robinson was, however, involved during this period in various scientific and technical discourses that seemed unceasing. These matters included the publication of Berghaus' map in the *JRGS*, the various disagreements with Bertou, the letter exchange with Washington, then secretary of the RGS and editor of its journal, and the public quarrel with Buckingham.[154]

Expressing disappointment at his slow progress, Robinson adds in his February update that though 'I am sure that I have not lost a moment's time since September last,' in his writing he only 'just arrived at Hebron from Gaza.' He calculates what lay ahead and continues, 'I

[152] Robinson to Smith, Berlin, 24.12.1839, ESP, ABC 60 (63); cf. chap. 4.a. Friedrich Wilken (1777–1840, ADB Wilken) published from 1807–32 his seven volumes of *The History of the Crusades*, which had been used by Robinson as a textbook for that period (Wilken (1807–32)).

[153] Robinson to Smith, Berlin, 12.2.1840, ESP, ABC 60 (63).

[154] For Washington and his involvement in Palestine exploration and Bertou's connections, see Goren (2011). For the discourse with Buckingham see chap. 5.a. 3.

could hope to see the end before May.' This estimate gave Robinson reason to 'shudder,' for it seemed overly optimistic. He also admitted that what he termed 'the historical investigation' was responsible for significant delays, but certainly essential for the book. It was tremendously important for him to introduce his theories with caution, checking and rechecking them 'where there is not full proof; so that I have little fear that my positions will be after questioned.'[155] By mid-April, he had 'reached Ramleh,' and could now write that he expected to complete the manuscript in June, when he would go to London to explore printing options with the hope of returning home in September.[156]

It appears that during mid-1840, Robinson was in such haste to complete the manuscript that he did not allow himself to indulge in letter writing. His last letter from Berlin, according to Smith's files, is dated April 15, 1840. On October 3, he sent his next letter from London, writing, 'through the good providence of God, I was enabled to complete my work in Berlin about the 1st of Sept. & on the 6th of that month left by way of Halle & Hamburg to this place.' A 'very great amount of pictorial investigation & illustrations which came up along the routes' had caused significant delay, but now the maps were also completed, engraved by Heinrich Mahlmann (1812–48) in Berlin.[157]

Robinson had no illusions concerning the expected popularity of his book, as 'popular it cannot be in its present form,' although he thought that perhaps an abridged work would attract the public eye. He was not, however, seeking a popular audience, 'such a popularity as that of Stephens, I do not wish, except in the sale.'[158] John Lloyd Stephens visited Palestine in 1836 and his *Incidents of Travel* was published in the following year. It was broadly received and brought him a considerable sum of money, enabling him to embark on a second acclaimed voyage to the Yukatan Peninsula in Mexico, where he became the first in the new era to discover the Mayan culture. His biographer, Victor Wolfgang von Hagen (1908–85), wrote that on November 17, 1839, a new world was created, a new science – American archaeology.[159]

[155] Robinson to Smith, Berlin, 12.2.1840, ESP, ABC 60 (63).

[156] Robinson to Smith, Berlin, 15.4.1840, ESP, ABC 60 (63).

[157] Robinson to Smith, Berlin, 3.10.1840, ESP, ABC 60 (63). For Mahlmann and his role in producing the maps see Goren (2017), passim.

[158] Robinson to Smith, Berlin, 15.4.1840, ESP, ABC 60 (63).

[159] Stephens (1858), 1st pub. New York 1837, by 1851 10th ed., also in London 1838 and 1840 (Röhricht (1890), 384). Cf. Goren (2011), 146–57, mainly due to Stephens' part in revealing the story of the ill-fated Dead Sea explorer Christopher Costigan.

4.b.5. Publishing the manuscript

As early as January 1839, Robinson began seriously considering publishing house and language options, particularly the differences between publication in America and 'here' in Germany – and in German. The issue also troubled Smith, who, as early as January 1839, expressed his strong preference for an English edition. He reasoned that except for the interested German scholars, a German edition would be inaccessible to most people in England and America and, importantly, 'It will be a sealed book to all my brothers in Syria, & those parts, where it is of some importance that it should be known.'[160] In mid-April, Robinson was considering German publisher and philologist Gustav Parthey (1798–1872), who in 1823 visited Egypt and Palestine. Although Robinson introduced Parthey to Smith as a well-established publisher who went to the Leipzig book fair, he writes that Parthey 'may not be my publisher in German,' probably due to the expected costs.[161] Robinson, who naturally wrote in English, debated whether to hire a German-language translator, as it 'will nearly swallow up the honorar of the first edition; & a second is hardly to be looked for.'[162] On May 21, 1839, both Robinson and Smith wrote concerning the obstacles to translation and publication in German. Robinson had a translator about whom he writes, 'I am sometimes tempted to give up the idea of a German edition altogether. The translation turns out to be exceedingly flüchtig [volatile] und ungewissenhaft [unscrupulous], that I have come near a pulling-out with him.'[163] But at the end of 1839, he writes that although he was 'compelled to dismiss the first translator, who did not take the trouble to see whether he understood the original or made any sense in the German,' the translation was now progressing well as it was being done by a young theologian recommended by Neander.[164] Ritter, who closely accompanied Robinson's work, writes in his *Comparative Geography* that it 'was translated into German partly by the author himself, and wholly under his personal supervision.'[165] Following the various clues, Ritter's account appears reliable.

[160] Smith to Robinson, Leipzig, 19.1.1839, ESP, ABC 60 (105/10).

[161] Robinson to Smith, Berlin, 17.4.1839, ESP, ABC 60 (63). Cf., for Westphal's and Parthey's expedition, Goren (2003), 64; (2015a).

[162] Robinson to Smith, Berlin, 17.4.1839, ESP, ABC 60 (63).

[163] Robinson to Smith, Berlin, 21.5.1839, ESP, ABC 60 (63).

[164] Robinson to Smith, Berlin, 24.12.1839, ESP, ABC 60 (63).

[165] Ritter (1866), II, 70.

The more difficult problem, it seems, was finding the right publisher as attested to in Robinson and Smith's correspondence. Rödiger connected Robinson with a publishing house in Halle, 'which seemed very desirous to undertake the publication (…) [but] drew back when I ascended to their own terms.' Smith conveyed his and Rödiger's deep disappointment.[166] Both investigated new possibilities, including the *Halle Weisenhaus* publishing house, part of the *Franckesche Stiftungen* (Francke Foundation), a social and educational institute founded in 1695 by the pietist theologian August Hermann Francke (1663–1727).[167] Another potential publisher was Justus Perthes in Gotha, but Robinson did not know anybody there 'to whom I could entrust the version.' Robinson was referring to Wilhelm Perthes (1793–1853), son of Justus Perthes (1749–1816), who in 1785 established in Gotha the Geographical Institute publishing house named after him.[168] Other publishers mentioned were the Leipziger Barth ('Smith: I am not sufficiently acquainted'), Brockhaus (soon to be the central publisher of Talvj, 'I have no acquaintance at all'),[169] and Tauchnitz, Smith's host and producer of his Arabic prints, about whom Smith writes, 'I do not think it comes quite within the range of books, which he publishes.' Smith even offered to induce Perthes to print the work in Halle.[170]

Rödiger did exert his influence, and by September 1839, Robinson had in his hands an offer from the Orphanage Publishing House for a full German edition. Rödiger, who with Gesenius desired a complete German edition (with all indices, notes on Arabic, etc.), agreed to 'take charge of the correction' so that Robinson was now much more confident in the German translation. The latter did not mention it, but this might hint at the fact that he took an active part in the translation.[171] In mid-February 1840, Robinson was able to report substantial progress

[166] Robinson to Smith, Berlin, 24.12.1839, ESP, ABC 60 (63); Smith to Robinson, Leipzig, 21.5.1839, ESP, ABC 60 (105/11).

[167] ADB Francke.

[168] Robinson to Smith, Berlin, 21.5.1839, ESP, ABC 60 (63); Smith to Robinson, Leipzig, 21.5.1839, ESP, ABC 60 (105/11). For Perthes see, e.g., ADB Perthes; Brogiato (2008); for his connections to the study of the Holy Land: Goren (2008); (2017); Schelhaas (2017).

[169] Weigel Williams (2007), passim. Barth had been managed by Wilhelm Ambrosius B. (1790–1851, NDB Barth). The Brockhaus publishing house was then managed by Heinrich B. (1804–74, NDB Brockhaus, H.). Cf. chap. 5.2.

[170] See chap. 5.a. See also Talvj's letter to Kopitar, Berlin, 14.12.1838, Milović (1941), 99.

[171] Robinson to Smith, Halle, 10.9.1839, ESP, ABC 60 (63).

('already 250 pages, beyond Sinai') in the printing of the German edition under the supervision of Rödiger, who 'seems quite interested & satisfied with the results,' though the maps, which 'are to be engraved here in Berlin under my superintendence; which I regard as a great advantage,' were still not completed.[172]

Robinson had no difficulty choosing his American publisher. He turned directly to the leading Bostonian publishing house Crocker & Brewster (established in 1818 and operable until 1876). Uriel Crocker (1796–1887) was responsible for the printing, with whom Robinson had previously published a number of books.[173] In early July, when he began looking for a British publisher, the first two names that arose were Longman and Murray. The former, a renowned (since 1724) and well-established London publisher, was then managed by Thomas Norton Longman (1771–1842). The latter, known for its publications of travel and scientific literature, was founded in 1768 by John Murray I (1737–93), a native of Edinburgh and a retired marine lieutenant. The publishing house was now managed by John Murray III (1808–92). Smith consulted Colonel Thomas Aspinwall (1786–1876), US Consul in London (1816–54!) who lost an arm in the war of 1812. Aspinwall acted as literary agent and liaison between American authors and British publishers.[174] Smith, of the opinion that the 'Longmans are not very comfortable people to make a bargain with,' still preferred them to Murray, who lacked the resources. However, since Murray set aside some funds for the project, Smith noted that it might prove worthwhile contacting him. At the time, Smith was still hoping to circumvent a German edition and suggested including a section in German in the English edition. This idea was quickly rejected.[175]

As the year 1839 drew to a close, Robinson surveyed his situation. Through Rödiger's mediation and supervision, the *Weisenhaus* in Halle were the first to start printing. Robinson was unable to realize his plan to publish first in London, originally thinking that 'this would bring it out several months sooner.' Instead, he knew that he had a second option of bringing everything to print at home and only then to 'send out the sheets to London.' One of his biggest concerns, the printing of

[172] Robinson to Smith, Berlin, 12.2.1840, ESP, ABC 60 (63).

[173] Starting in 1832 with Robinson's American edition to Calmet's Dictionary of the Holy Bible (Wahl (1825)).

[174] Barnes (1984). For the Murrays as RGS members, see Herbert (1983), 86.

[175] Smith to Robinson, London, 8.7.1839, ESP, ABC 60 (105/11).

the Arabic letters, was now resolved, as 'Mr Trow in New York has now some of Tauchnitz Arabic type; & I have bespoken of Tauchnitz an additional quantity, in order to be able to print the lists & index.'[176] However, after expressing relative satisfaction with the progress of the printing, Robinson reveals that he was desperately disappointed by Berghaus: 'my maps alas! Get on less well.' It is here that he first mentions Kiepert as a possible replacement for Berghaus.[177]

As noted, in April 1840, Robinson estimated that he would reach London in July in order to check personally the possibility of printing the manuscript there. If that option did not materialize, then Robinson planned to take the manuscript with him across the Atlantic in September and have it printed only in the US. Supported by new recommendations from Ritter and Wheaton, he continued to look for a British publisher. At that time, Robinson had also finalized inclusion of a total of seven maps and names them.[178] In September 1840, after finishing the manuscript, he went to London, with finding a publisher his only objective. Robinson preferred Murray over Longman and others who would 'take the work & share the profits.'[179]

From London, Robinson returned to the US, anxious to involve Smith 'as much as he can' in the publication process, needing his expertise, of course, for editing the Arabic lists. But the latter was difficult to reach: 'I am called away to another part of the country in an agency for our society; & for the whole winter our time are taken up for travelling, & our attention diverted to literary pursuits.'[180] As 1840 came to a close, Robinson again wrote to Smith requesting his assistance with 'his' part, i.e. the 'Appendix.' When it became apparent that Smith would not reach New York in time to oversee the printing, and as 'the printers are not as driving as I could wish,' Robinson decided 'to let them go on with the whole appendix first, instead of merely with the Arabic part.' Robinson reasoned that the printers would reach the Arabic lists only in late January, hoping that this delay would enable Smith to personally 'aid in correcting the proofs.' Robinson notes at the time of writing that

[176] Robinson to Smith, Berlin, 24.12.1839, ESP, ABC 60 (63). Andover born John Fowler Trow (1810–86) was a printer and publisher in New York City. He moved from Andover to New York in 1833, so both Robinson and Smith were most likely acquainted with him.

[177] Cf. in detail Goren (2017), 26–28.

[178] Robinson to Smith, Berlin, 15.4.1840, ESP, ABC 60 (63).

[179] Robinson to Smith, London, 3.10.1840, ESP, ABC 60 (63).

[180] Smith to Washington, Boston, 20.1.1840, ESP, ABC 60 (105/12) 1840–41.

there were 'only about 216 pages done; which extent to the end of Sect. III, & the notices of Sinai.'[181]

During their long letter exchange, Smith repeatedly raises the issue of his contribution, speculating as to whether it was appropriate to name him co-author.[182] Robinson preempts him, writing that he was 'somewhat at a loss as to what will be the best title for our Book.' The simplest possible title, 'Travels in Palestine,' had already been 'captured' by the traveller George Robinson for what Edward Robinson coined 'his miserable Machwerk' [pathetic effort].[183] Therefore, Robinson suggests the title 'Biblical Researches in Palestine & the Regions of the South, in the year 1838. By E. Robinson & E. Smith,' and asks for Smith's opinion.[184] To this Smith answers that though he liked the phrase Biblical Researches, 'regions of the South strike as not sounding very well, though it exactly suits the nature of book.' Smith had nothing more to suggest in this matter, but he again expresses his doubts as to whether 'I shall have enough to do with the authorship of the book, to entitle my name to a place on the title page by the side of yours.' With regard both to the title and authorship, Smith requests that Robinson decide.[185] Two weeks later, Smith repeats his concerns at length, his difficulty with the word 'South,' as 'it will not strike the reader at first sight as a scripture term & will convey no very definite meaning.' He also expresses discomfort with Robinson's praise of his 'poor assistance (...) I feel constantly ashamed, that I made myself no more useful.' Smith also seemed to have his doubts concerning his exact role in writing the book. He regretted writing as if he were going to write the book himself, because that content might not be usable for Robinson. Now that his comments and hints were there, in the text, Smith adds, 'you must make the best of it if you can.' Of course, Smith also expressed his readiness to conduct any further research needed.[186] Their mid-April 1840 correspondence reveals that the name decided upon did include 'the countries on the South,' namely 'Palestine & the countries on the South. A journal of travels in the year 1838 by E. Robinson & E. Smith; undertaken in Reference to biblical Geography...'[187] In actuality, the final version of the title omitted the South.

[181] Robinson to Smith, New York, 28.12.1840, ESP, ABC 60 (63).

[182] See chap. 5.a.

[183] Robinson, G. (1837). Cf. Oliver (2014), 127–30; Goren (2011), 34–38, 48–51.

[184] Robinson, G. (1837a). Cf. Robinson to Smith, Berlin, 11.2.1839, ESP, ABC 60 (63).

[185] Smith to Robinson, Leipzig, 15.2.1839, ESP, ABC 60 (105/10).

[186] Smith to Robinson, Leipzig, 2.3.1839, ESP, ABC 60 (105/10).

[187] Robinson to Smith, Berlin, 15.4.1840, ESP, ABC 60 (63).

In April 1841, when Robinson received word that the next set of proofs was ready, he was already deeply engaged with his duties at UTS and Smith was still travelling in New England and about to embark on his return voyage to the East. The proofs for the second volume and a considerable part of the third were ready; 'this will be, I hope, the whole text & notes, & perhaps the whole of the Arabic, Index & all.' It meant that the book was nearly finished. Robinson hoped that Smith's vessel would be delayed so that he could review not only these proofs but even more material. From their correspondence, it appears that Smith was then in Boston, the central city of the ABCFM.[188] Robinson's next letter, written only three days later and after realizing that Smith's ship set sail on time, included a detailed list of the material he intended to send: 'Vol. II, done up; Vol. III, all the text; the few parts are printed not yet, Appendix, Signat. 18 to 25.' Consequently, what remained were the notes, preface, the greater part of the Arabic index, and the other two indices of ancient geography and passages of the scriptures.[189] 'The printing of the work is going on in England also,' Robinson reported to his friend and mentor Ritter, 'after all the American edition will be the Standard; & I shall forward to you a copy at the very earliest opportunity. Perhaps too I may send a copy for his majesty the King of Prussia.'[190]

4.b.6. Biblical Researches 1841 – distribution

The book was finished and, so it seems, printed, three thick volumes, including maps. Robinson, as noted, extremely busy both with teaching and private pursuits, occupied himself with the review and promotion of his creation as well. He was concerned about attaining publicity, public interest in an academic work, acceptance by the scholarly community, and whether enough private individuals (libraries and institutes excepted) had the money needed to buy the three volumes. Robinson was highly satisfied with the reception he and his work received in his homeland. His compatriots were 'rather proud of the idea, that an American should have adventured to make discoveries in regard to the Antiquities of the Old World, – a field which hitherto our friends of England & Germany, have had wholly in their own hands.'[191]

[188] Robinson to Smith, New York, 14.4.1841, ESP, ABC 60 (63).

[189] Robinson to Smith, New York, 17.4.1841, ESP, ABC 60 (63).

[190] Robinson to Ritter, New York, 15.3.1840, SBB-PK, Nachl. 133 (Carl Ritter), K 5, Q – SCH, Bl. 30. As mentioned above, this letter should have been dated 1841!

[191] Robinson to Ritter, New York, 15.3.1840 (should be 1841), SBB-PK, Nachl. 133 (Carl Ritter), K 5, Q – SCH, Bl. 30.

DR. ROBINSON'S RESEARCHES IN PALESTINE—in 3 vols. 8vo.

CROCKER AND BREWSTER, 47 WASHINGTON-ST., BOSTON,

Have in Press and will Publish June 1st,

BIBLICAL RESEARCHES IN PALESTINE, MOUNT SINAI, AND ARABIA PETRÆA.

Being a Journal of Travels in the year 1838, undertaken in reference to Biblical Geography; by the Rev. Dr. ROBINSON and the Rev. ELI SMITH. Drawn up from Notes made on the spot, with Historical Illustrations,

BY EDWARD ROBINSON, D. D.,

Professor of Biblical Literature in the New York Theological Seminary, and Author of "A Greek and English Lexicon of the New Testament."

With Maps and Plans of Jerusalem, Palestine, and Mount Sinai, from original Measurements and Observations, in five sheets.

OPINIONS OF THE WORK.

We look for the narrative of the Syrian Travels of Dr. Robinson and the Rev. Eli Smith, which will lay before us the results of a journey, made by keen observers and accomplished scholars.—*Address of the President of the Royal Geographical Society of London.*

The correction of many erroneous views as to the situation of places historically important, and accounts of the undoubted discovery of celebrated ancient monuments, long lost or overlooked, even in Jerusalem itself, give to the reading of Dr. Robinson's Report an unusual interest.—*Report of the Geographical Society of Berlin.*

The work on Palestine by the Rev. Dr. Robinson is very different from the hasty journals hitherto published by many tourists in the same region. It is the result of studies connected with long preparation, and of an important enterprise in behalf of a deeper scientific investigation of the geographical, historical, and antiquarian relations of the Holy Land. This enterprise has been happily carried out upon the spot under the most favorable circumstances; and the work thus resulting from it, with its historical illustrations, has been prepared with the aid of the richest literary means, and will certainly occupy one of the most important places among the sources for the study of the geography and antiquities of the Scriptures.—This work, by its numerous discoveries and new investigations, will be of lasting value for the knowledge of the East, and for the historical part of theological study. It contains in itself the materials for an entire transformation of the cartography of Palestine, hitherto so defective, as is proved by the very important maps already constructed, founded on the new observations afforded by this journey. These, as an accompaniment to the text, give the latter a peculiar value which can be regarded only as highly desirable.—CARL RITTER, *Professor in the University and Member of the Berlin Academy of Sciences, also President of the Geographical Society of Berlin.*

Of the high qualifications of Dr. Robinson as a traveller in oriental countries, particularly in the Holy Land, it is not necessary for us to speak. To sound judgment, comprehensive views, and habits of minute accuracy, is to be added a long and intimate acquaintance with biblical antiquities and with the languages of the original Scriptures. He has had, also, for some years, in contemplation, a work on biblical geography, and had thus ascertained the many deficiencies in the books of travellers and of geographical writers. The work has, likewise, been prepared for the press at the fountain head of knowledge on eastern subjects—Berlin in Prussia—in the vicinity of large libraries, and with all the encouragements and facilities afforded by the presence and countenance of the most distinguished oriental scholars in Europe. With such advantages in the preparation of the work, we might well anticipate what has been actually accomplished. The work is a noble present to all lovers of the Bible—a production, which, it is no exaggeration to say, will cast into the shade all the volumes on the Holy Land, which have gone before it, and which will certainly create a new era in biblical interpretation. It is not a journal which will be attractive to the unreflecting and superficial reader, who resorts to a book of travels for the same reason that he takes up a romance. Volumes on the Holy Land, like those of Chateaubriand and Lamartine, we have enough, and more than enough. A record of incidents and impressions, however correct and well drawn, we did not need. We have, in the work before us, that which will meet the real wants of the literary and religious community—the result of conscientious observations, and diligent and laborious inquiry, by men who felt that they were treading on sacred ground, and that they had no right to trifle with the credulity of the reader, or to furnish aliment for his imagination merely. The work must be read, or rather studied, in connection with the excellent maps and drawings, which accompany it. At the same time, it should be mentioned, that there are parts of it which will be interesting to the general reader, where he will not meet with the details of distances, bearings, measurements, and the like. There are many passages in the sheets which we have read, of fine description, by which the reader gets a vivid impression of that which the writer intended to communicate. There are no Arabic words in the work, and but very few Hebrew; and these are confined to the notes at the bottom of the page. A great number of Arabic names, and a large amount of other valuable matter will be inserted in various appendices to the three volumes.—It is matter for grateful recognition of the good Providence of God, that this important undertaking is brought so near to a conclusion. We cannot avoid looking at it in a national point of view. It will confer great honor on American enterprise and scholarship. But very few men in our mother country could have combined the qualifications necessary for the production of a journal like the one before us. We do not know of any Englishman who could have done it. And our German cousins, however learned and indefatigable, might have failed in making a book which should be acceptable to American Christians and scholars.—*Boston Recorder.*

March 15th, 1841.

Ill. 19. Pamphlet of the American book, a distribution letter from the publisher: Boston, March 15, 1841, Crocker & Brewster.

The three publishers, Murray in England, Crocker and Brewster in the USA, and the Halle Orphanage in Germany, were primarily responsible for the distribution and sale of the book, the American edition being of 2500 copies, the English one of 1500, and the German of 1000.[192] However, Robinson, together with Smith, Ritter, and Rödiger, the collaborators and friends most deeply involved in the project, followed developments closely and contributed where they could. The letters shed light on hurdles in the three distributing processes.

Murray faced difficulties due to the fact that the maps were being printed in the US. Robinson received an update about publication of the English edition from Robert Cooke (1816–91; John Murray's third cousin who joined the firm in 1837) in a letter sent August 3, 1841. Copies were 'issued to the public on the 7th of July (...) we are still waiting the arrival of the fifth map, as well as an additional supply of the other 4, as in the first instance we only received a partial supply.' Nonetheless, he was very optimistic about sales considering that, at the time, England was experiencing a 'very great depression of trade (...) the work was most favourably received by the booksellers & already 500 copies are sold.' He writes further, 'we shall not fail to give you timely notice when the present edition is exhausting itself.'[193] Cooke also notes that 'few reviews of the work have yet appeared, but those have been highly laudatory,' and the editor of the *Edinburgh Review* asked for a copy. Of course, Robinson remembered his friends, and asks Murray to send the three volumes to Ritter, 'because it is handsomer & you would get it quicker' than the German edition.[194]

In early 1843, Robinson updates Smith that the edition 'is nearly all sold (...) the London edition is exhausted & Murray desires to reprint it; the American one is not yet all sold.'[195] In 1849, Cooke writes to Robinson that 'we have been long since without copies of your "Biblical Researches" & have not written for more, for although the work is oc-

[192] Robinson to Ritter, New York, 24.12.1840, SBB-PK, Nachl. 133 (Carl Ritter), K 5, Q – SCH, Bl. 27–28.

[193] Robert Cooke (for his uncle John Murray) to Robinson, London, 3.8.1841, ERP, 1816 Robinson- 2. Murray chose to publish the maps in a separate volume (Robinson (1841b), Vol. Maps).

[194] Robinson to Ritter, New York, 1.10.1841, SBB-PK, Slg. Darm. Asien 1840; 31.12.1841, Nachl. 133 (Carl Ritter), K 5, Q – SCH, Bl. 34.

[195] Robinson to Smith, New York, 18.2.1843, ESP, ABC 60 (63); Robinson to Ritter, New York, 27.2.1843, SBB-PK, Nachl. 133 (Carl Ritter), K 5, Q – SCH, Bl. 36.

casionally enquired for, we have scarcely felt justified in expecting a large demand for the 3 vols.'[196] About a year later, Murray suggests that the American publisher send 50 copies, which 'we shall be happy to receive them on commission & account for the sales every six month.'[197]

The American edition was ready for distribution sometime in late 1841. On December 16, Smith's missionary brothers in Boston were able to write that 'a parcel marked at the outside "Five sets Robinson Travels for Rev. Eli Smith"' arrived at the missionary house a few weeks earlier and they were on the lookout for a vessel to ship the package.[198] The German edition is a bit of an enigma. As late as February 1843, Robinson updates Smith that 'the German edition was not all printed in November last!!' and that sections after the visit to the Galilee city of Safed were missing.[199] When in 1842, the German theologian and Orientalist Justus Olshausen (1800–82) wrote his highly favourable review based on the German edition, he only had access to the first and second volumes and the first part of the third.[200]

The newly published volumes generated great interest and in a surprisingly short time. This reception was due, in part, to their almost simultaneous release in 1841 in three important academic centres. Reviews began appearing before the year's end: 'few reviews of the work have yet appeared but those have been highly laudatory,' among them a detailed article by Schubert in 'his' periodical.[201] One of the best examples of the rapidly gaining popularity of the *Biblical Researches* is the direct testimony of travellers, who used the volumes as their guidebook.[202] An outstanding example is that of Samuel Wolcott (1813–86), who graduated from Andover Seminary in 1837, served as ABCFM missionary from 1839 to 1843, when he returned to the USA.[203] Wolcott, 'who spent the

[196] Cooke to Robinson, London, 29.8.1949, ERP, L.1.

[197] Cooke to Robinson, London, 16.8.1850, ERP, L.1.

[198] George E. Head to Smith, Missionary House, Boston, 16.12.1841, ESP, ABC 60 (43).

[199] Robinson to Smith, New York, 18.2.1843, ESP, ABC 60 (63).

[200] Olshausen (1842).

[201] Cooke to Robinson, London, 3.8.1841, ERP, L. Correspondence to E.R. 1840–60; Schubert, G. H. (1841). For a 'selected' list of 1841–42 reviews in US periodicals: Williams (1999), 359; and the long and detailed review in the *ABR*: Hall (1841).

[202] See, for example, Durbin (1845), iv, cited in chap. 6.

[203] Kawerau (1958), 258, 528; Henze (2011), 5, 531–32. Wolcott's application for the secretaries of the ABCFM, Forsyth, 28.9.1839, ABCFMA, 16.5 vol. 4: Near East 1823–60.

winter of 1841–2 at Jerusalem, and while there took the opportunity of carrying out several inquiries,' wrote that Robinson 'has traversed the country and explored Jerusalem with the eye of a keen and intelligent observer.'[204] With regard to the land of the Jordan, Ritter wrote after publishing his *Comparative Geography* that he used Wolcott's works extensively.[205] Wolcott's reports were published swiftly, first by Ritter and then by Robinson in the first volume of his *Bibliotheca sacra*, according to Wolcott's letters to Smith, dated January 10 and January 25, 1842.[206]

> I surely have none to impact, as my whole time has been hitherto occupied in following your tracks, with the 'Biblical Researches' as my guide-book (…) Through Mr Tipping I have most fortunately access to an English copy of the 'Researches,' & it has been a continual feast to me. From all other narratives I turn to the simplicity & fidelity of this, with high relish. I value it the more for being such a constant index to the Bible, which after all is the Book which one wishes to study in connection with the sacred localities.[207]

Wolcott's travel companion was the British painter William J. Tipping (1816–97), an amateur archaeologist, who from 1839 spent seven years in the East.[208] Wolcott explored Jerusalem, reached the Dead Sea, and was one of the first to climb the cliff of Masada (Sebbeh) 'descending the steep declivity by a zigzag path.'[209] In addition to Smith, he met British surveyor-officers who reached Syria and Palestine with the British fleet in late 1840, supported the Ottomans in reconquering lands from the Egyptians, and conducted important measuring and surveying work.[210] Wolcott and Smith received valuable resources from the surveyors, including the first measuring map of the sources of the Jordan, drawn by Major Frederick Holt Robe († 1852) and published in the aforementioned papers of Ritter and Robinson.[211] Wolcott also met Lieutenant

[204] Robinson (1843), 9–10, 17.

[205] Ritter to Robinson, Berlin, 18.3.1850, 27.1.1851, ERP, L.4 Ritter.

[206] Ritter (1842); Robinson (1843), 17; cf. Robinson to Smith, New York, 18.2.1843, ESP, ABC 60 (63).

[207] Wolcott to Smith, Jerusalem, 30.12.1841, ESP, ABC 60 (96).

[208] Ben-Arieh (1997), 193.

[209] Robinson (1843), 61–66; Wolcott (1846), 398–402.

[210] Jones (1973); Goren (2005).

[211] Detailed discussion: Goren (2000), 111–23; (2011), 102–11, 234–37.

John Frederick Anthony Symonds (1801–72), who 'had spent a fortnight between Yâfa & the Dead Sea, taking altitudes by measurements.'[212]

Not all the reviews were so overwhelmingly favourable like those of Schubert or Olshausen, who deemed the *Biblical Researches* 'a masterpiece.'[213] In 1843, Karl von Raumer added a sixty-page long 'contributions to Biblical Geography' to the second edition of his *Palästina*, where he argues with many of Robinson's theories and identifications. Naturally, Robinson could not ignore this attack. Using his periodical, *Bibliotheca Sacra*, he countered in brief, writing that Raumer provided 'an alphabetical list of eighty-seven places under which additions or corrections are made' and missed no opportunity for a polemic remark, 'while, on the contrary, there is nowhere the slightest word of acknowledgement in behalf of the Researches or its authors.' Robinson also responded to debates on three issues, 'Exodus of the Israelites,' 'Arimathea or Rama,' and the 'Depression of the Dead Sea.'[214]

4.c. Robinson's sources for the *Biblical Researches*

By the nineteenth century, the use of existing literature for the research and publication of Holy Land travelogues and studies of all sorts was common; two scholars of note, however, perfected this research method. One was Titus Tobler, the Swiss physician from German Switzerland who spent his post-retirement years living in Munich and working in the city library. Tobler and his bibliographic and monographic works are the subject of numerous studies, witness to his leading role in Palestine research.[215] The second was Edward Robinson.

[212] Wolcott to Smith, Jerusalem, 30.12.1841, ESP, ABC 60 (96). For Symonds and his triangulation see Goren (2011), 102–37, 231–47.

[213] Schubert, G. H. (1841); Olshausen (1842), 126.

[214] Raumer (1843); Robinson (1843a), citations 564–65.

[215] Heim (1879); Furrer (1878); Stern (1988); Goren (2003), 222–43. For Ritter's praising of Tobler in a letter to Robinson: Berlin, 18.3.1850, ERP, L.4 Ritter: 'He is about to publish his first scholarly work on Palestine based on fundamental research, which might seem too intellectual, but his striking effort is certainly recommendable and he has discovered many new things through his courage to penetrate into grottoes and caves.' Ritter used a double meaning, as Tobler actually did explore several caves in Palestine, such as the Chariton Cave near Tekoa (Judean Mountains), which he described in detail: Tobler (1847). Robinson does not mention this cave, and probably did not know of its existence (Robinson (1841), II, 181–84, visit of Tekoa and vicinity).

Tobler first visited Palestine as a tourist in 1835, when he was 29 years old; Robinson, already 44, came three years later. By 1865, Tobler had returned three more times, all for research purposes, whereas Robinson returned only once, in 1852. Robinson's book, published in 1841, reached Tobler, then living in Horn on Lake Constanz (*Bodensee*), only three years later. Having spent the last few years in libraries and monasteries collecting 'Holy Land material,' in early 1844, Tobler made up his mind to travel again, this time for research, and to focus on Jerusalem. Robinson's three volumes, which he probably received in the German edition,[216] gave Tobler a bit of a shock, had the American covered all that can be studied in the Holy Land? But upon deeper thought, Tobler reached the contrary conclusion: Robinson's research intensified his decision to concentrate on specific areas of the Holy Land, studying them as thoroughly as possible. Tobler realized that even with the multitude of new data and research, much remained to be written about the toponymy and history of the more important and well-known places, such as Jerusalem and Bethlehem. After copying all accessible, relevant sources for the study of Jerusalem and its sites, Tobler set out in late 1845, supplied with '696 folio pages full of abstracts of all possible books and manuscripts.'[217] Following his return in mid-1846, Tobler began publishing papers in the weekly *Das Ausland*. Three years later, he financed publication of his first monography, *Bethlehem in Palästina*, and between 1851 and 1854, he published four more monographies![218] As far as we know, Robinson did not meet Tobler, but Ritter did travel south to visit him in 1849, after receiving from Tobler 'some copies of his Bethlehem.'[219]

Tobler distinguished himself in his extraordinary and nearly unfathomable productivity, as well as in his remarkable control of historical sources in the field he defined as 'Palestine Literature.'[220] Though coming from different cultural and educational backgrounds and driven by different motivations, Tobler and Robinson shared these virtues so necessary for their scholarly achievements. Consequently, both were awarded the title 'father of Holy Land research / study.' These virtues stood behind Tobler's two primary innovations: the bibliography and

216 Robinson (1841a).

217 Tobler (1853–4), I, iii–v; Furrer (1878), 54.

218 Tobler (1849); (1851); (1852); (1853–4). Cf. Stern (1988a).

219 Ritter to Robinson, Berlin, 18.3.1850, ERP, L.4 Ritter; Tobler (1849).

220 Tobler (1867), III–IV.

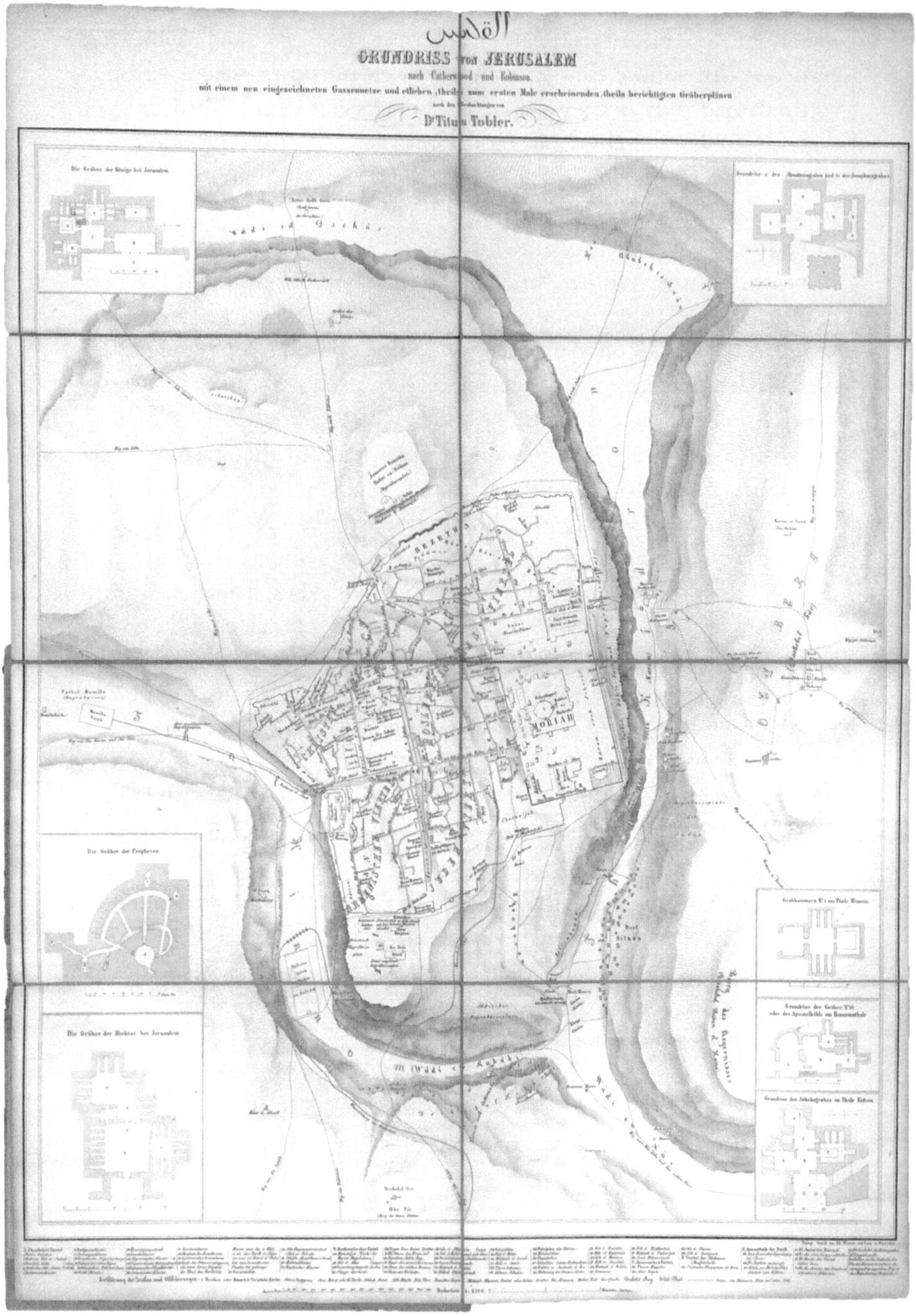

Ill. 20. Titus Tobler, Grundriss von Jerusalem nach Catherwood und Robinson, 1849.

the monography. He compiled and published the first comprehensive bibliography of Palestine literature and he was the first to adapt monography, a known literary and research genre, to the study of the Holy Land. Tobler also published critical editions of early traveller's literature. In contrast to Robinson, Tobler chose to focus on a single site, discussing various topics chronologically, while checking every detail and using every available source. As an example, his *Bethlehem* discusses the following topics: name, location, hills, valleys, climate, water systems, vegetation, city and its streets and alleys, houses, history, people, occupations, agriculture, arts and crafts, language, customs, etc. Naturally, he dedicated a detailed section to the Church of the Nativity.

Robinson's research method was different. He could not and did not travel to libraries and archives in order to collect material for preliminary research. Libraries in the US held almost no relevant material, so he had few extracts of different sources. He was much too busy and, unlike Tobler, had a family. Robinson and Smith did carry with them some books and materials on the expedition:[221] Bibles, in English and in their original language; Adriaan Reland's (1676–1718) *Palæstina* (Robinson: next to the Bible the most important book!; 'this yet remains the standard classic work on Palestine, as far down as the era of the Crusades'[222]), Karl Georg von Raumer's (1783–1865) 1835 *Palästina* ('The work is compiled with great diligence, and forms an excellent Manual'[223]), Johann Ludwig Burckhardt's *Travels in Syria and the Holy Land* ('As an oriental traveler, Burckhardt stands in the very highest rank'[224]), Léon de Laborde's *Voyage en Arabie Petrée* ('The chief value ... is in its splendid plates'[225]), and *Modern Traveller in Arabia, Palestine, and Syria* by Josiah Conder (1789–1855). The last book is neither mentioned in his bibliography nor in his footnotes, indicating that Robinson attributed no value to it.[226] After returning, Robinson noted that he would only have added *The History of the Crusades* ('compendious,' by William of Tyre (1130–86); 'the chief and most important historian of the crusades'[227]), the relevant volume of Ritter's *Erdkunde*, and for Jerusalem Josephus

221 Robinson (1841), I, 48.

222 Reland (1714); Robinson (1841), III, 1st Appendix, 27.

223 Raumer (1835); Robinson (1841), III, 1st Appendix, 27.

224 Burckhardt (1822); Robinson (1841), III, 1st Appendix, 23.

225 Laborde (1836) (English version); Robinson (1841), III, 1st Appendix, 25.

226 Conder, J. (1830).

227 Willermus Tyrensis (1560); Robinson (1841), III, 1st Appendix, 7–8.

and other travellers. As for maps, they carried along Laborde's map of Sinai and Berghaus' Map of Syria; both turned out to be of little use.[228]

When Robinson arrived in Berlin after the expedition, he frequented the STABI, its exceptional library, writing that it 'is rich in this department; and contains a large number of works on Palestine, from the sixteenth century and onwards, which I have not found quoted in any catalogue.'[229] He divided his bibliography of sources used for the *Biblical Researches* into 'Itineraries, Travels, etc.,' and 'Works on the Geography of Palestine.' 'I believe,' he adds, that the list is 'nearly if not quite complete, down to the time of Breydenbach and Felix Fabri in A.D. 1483.'[230]

Compiling a bibliography decades later for his *Bibliotheca geographica Palaestina*, the Berlin teacher and scholar Reinhold Röhricht (1842–1905) used Tobler's *Handexemplar* of his bibliography, to which Tobler added every source he found subsequent to its publication in 1867 until his death ten years later. Röhricht published his volume in 1890, but its bibliography contains publications dating only until 1878.[231] Both Tobler and Röhricht added 'list[s] of sources' from books published previously.[232] A careful comparison of bibliographies should support Robinson's claim as to the completeness of his sources until the late fifteenth century. Joseph Darling ('Bib') Ibbotson (1869–1952), Hamilton College and UTS alumni, made this comparison. Ibbotson studied in Berlin and Halle, and from 1895 served as assistant professor of English Literature, Hebrew instructor, and librarian at Hamilton's Burke Library (1911–36). In 1933, 'on the occasion of the exhibit of books from Robinson's Library,' he printed and read at a faculty club meeting a paper titled 'Edward Robinson and his Books: A Biographical and Descriptive Sketch':

> The Edward Robinson Library at Hamilton College affords unusual facilities for the study of the geography of Palestine and Syria. It is well equipped with the modern works which record the progress in the understanding of the lands of the Bible that the last century has accomplished; it is no less rich in the older & oldest literature of the subject. These books were for the most part carefully selected by Robinson himself.[233]

228 Laborde (1830); (1830a); Berghaus (1835).

229 Robinson (1841), III, 1st Appendix, 3.

230 Robinson (1841), III, 1st Appendix, 3, 12; Breydenbach (1486); Fabri (1557).

231 Röhricht (1890). Cf. Goren (1995).

232 Tobler (1867), 1–4; Röhricht (1890), VII–XVI.

233 Ibbotson (1933), ERP, AL COLL I2 E3 1933; ERP, AL COLL I2 E3 1933 Draft.

Ibbotson did, indeed, carefully compare the bibliographies with the library holdings. Röhricht, in his 'standard bibliography of Palestinian travel,' recorded 42 'itineraries and pilgrims books' originating before 1095, the year the first Crusade was declared. A careful check reveals Röhricht's mistake, as he actually included works of travellers and pilgrims who visited the land up until 1099, the year Jerusalem was conquered by the First Crusade forces.[234] Ibbotson notes, 'of these we lack only four. (...) We possess 62 editions and translations of the 37 earliest works.' As for the period of the Crusades, '1095 to 1291,' the Burke Library possessed 60 of the 104 printed works listed by Röhricht, as well as several others he did not mention. For 'the later Middle Ages,' 1291 to 1500, there were 'some 80 of Röhricht's 150 printed works,' and up until 1700, 175 of 325 works.[235] Robinson, of course, collected and purchased the great majority of these volumes. When and how he built this impressive collection of Palestine literature, and when and how it reached Hamilton's Burke Library is curious given that Robinson was never a faculty member at Hamilton. Surely, many of these books were from Robinson's personal library. Robinson's drive for purchasing libraries in Europe and sending them to the US and establishing relevant collections in all 'his' institutes, including Andover and the UTS, is described in section 4.e.

Returning to Berlin, it was not only at the STABI where Robinson found source material. As noted, his correspondence in late 1838 and early 1839 presents the extraordinary cooperation of Ritter and other German scholars. 'Ritter & Neander offered me spontaneously the use of their libraries in the kindest manner' he writes just after arriving in Berlin.[236] It was only a few weeks later when he could begin to evaluate Ritter's generosity:

> I consulted Neander & Ritter in the subject. The latter showed me in his ms. entries a long list of books, the sources of history in respect to former missions. He offered to copy them for me, & make further additions, which often I thankfully accepted. He has hardly as yet had time to complete it; when finished I hope to transmit it or a copy to the Rooms.[237]

234 Röhricht (1890), 1–21.

235 Ibbotson (1933), ERP, AL COLL I2 E3 1933; ERP, AL COLL I2 E3 1933 Draft.

236 Robinson to Smith, Berlin, 23.11.1838, ESP, ABC 60 (63).

237 Robinson to Smith, Berlin, 10.1.1839, ESP, ABC 60 (63). See also Ibbotson (1933). Does he mean the 'rooms' of the Stabi?

In the 21 volumes of his *Vergleichende Erdkunde*, Ritter dedicates 168 pages to Palestine and Syria without mentioning Sinai even once. Published in 1818, two years before Ritter moved to Berlin, it was one of the earlier volumes of this outstanding series. Here, Ritter primarily discusses the ancient periods, sacred places, and their related traditions without any treatment of the present and its connection to the past.[238] Referring to Ritter's series in the *Biblical Researches*, Robinson notes that 'a new edition, wholly re-written, may be hoped for soon,' hinting at Ritter's list of sources prepared for the revised publication.[239] It took seven years from the publication of Robinson's book, when from 1848 to 1855, Ritter published his four volumes of the *Vergleichende Erdkunde der Sinai-Halbinsel, von Palästina und Syrien*, comprising no fewer than 4505 pages, out of which 1141 deal with the Sinai Peninsula.[240] No other part of the world merited from Ritter such a detailed study.

In addition to the bibliography for his *Vergleichende Erdkunde*, Ritter gave Robinson additional lists, including a bibliography 'on the Hist. of Missions' that naturally was sent to Anderson in the ABCFM headquarters, and a copy of the Palestine Map, compiled and published in 1830 by Julius Ludwig Grimm.[241] Much has been written about Ritter's path to becoming a leading geographer, his teaching, the establishment and management of Berlin's geographical society (*Die Gesellschaft für Erdkunde*), his contacts with scientific and amateur travellers, and his leading role in establishing Berlin as a centre for geographical studies. One of the most important outcomes of Ritter's efforts was the building of Berlin's book and map collections.[242] These rich resources, located in the state library, the library of the geographical society, and in Ritter's private library, were available for Robinson's use. And use them he did!

Robinson also acknowledges Ernst Wilhelm Theodor Herrmann Hengstenberg (1802–69), a student of Neander and Tholuck, and strong neo-Lutheran theologian and rival of Gesenius. Robinson had translated and published Hengstenberg's *Christologie* in 1831. Robinson writes

[238] Ritter (1818), 299–467.

[239] Robinson (1841), III, 1st Appendix, 27. Ritter's long and detailed list of 'the Authorities on the geography of Palestine': Ritter (1866), II, 22–103.

[240] Ritter (1848–55); (1866). Cf. Goren (1999); (2003), 68–83.

[241] Robinson to Smith, Berlin, 11.2.1839, ESP, ABC 60 (63); Grimm (1830); cf. Goren (2017), 21–23.

[242] Engelmann (1977); Plewe (1978); Zögner (1979); Beck (1981); Richter (1983); Livingstone (1992), 137–41; Goren (1999); (2003), 67–83.

to Hengstenberg that he was looking for 'various suggestions & in the use of your very valuable library.'[243] 'He lives remote from me; but I intend calling on him soon, though he has never returned my former call,' writes Robinson to Smith in early 1839. He did succeed in establishing regular contact with Hengstenberg and used his collections.[244] Even more interesting is Robinson's request for special editions of Holy Land literature, for which 'I have watched all the auctions since I have been in Germany' in vain. He requests 'whether under these circumstances, you will perhaps consent to part with your copies, for a time, until you can replace them here, - for which you have the best opportunities?'[245] Rare and difficult to acquire volumes were currently not to be found in the New World, such as the second edition of the late sixteenth-century collection of traveller's reports,[246] of Franciscus Quaresmius (1583–1650), twice Franciscan Guardian in Jerusalem,[247] the collection of the Dutch traveller Johannes Cotovicus (?-1629),[248] and the *Gesta Dei per Francos*, of which there was 'only a single copy.'[249] Robinson presents Hengstenberg with this unusual request, 'Partly in behalf of myself; but yet more in behalf of science.' It is unclear whether Robinson, himself, received the volumes, but they exist today in his collection at the Hamilton College Burke Library.[250] Robinson extended his research to collections outside Germany, and was quite resourceful in locating in other libraries sources that did not exist in Berlin. For example, he found Irby and Mangles' *Travels in Egypt and Nubia, Syria, and Asia Minor*, published in London in 1822, in the library of the RGS.[251]

Some of the sources cited by Robinson, mainly in his preface, are highlighted in various chapters in this work. The list includes Richard

[243] Robinson (1831a) and other translations, see Williams (1999), 351; Robinson to Hengstenberg, Berlin, 24.8.1840, SBB-PK, NH. Hengstenberg (ADB Hengstenberg), a leading evangelical and emphatic protestor against every form of 'rationalism,' particularly with regard to Old Testament criticism, clearly opposed Robinson's approach.

[244] Robinson to Smith, Berlin, 10.1.1839, 22.1.1839, 3.4.1839, ESP, ABC 60 (63).

[245] Robinson to Hengstenberg, Berlin, 24.8.1840, SBB-PK, NH.

[246] Feyrabend (1609); cf. Simon (1998). In Robinson's list: Robinson (1841), III, 1st Appendix, 15.

[247] Quaresmio (1639). In Robinson's list: Robinson (1841), III, 1st Appendix, 16.

[248] Cotovicus (1619). In Robinson's list: Robinson (1841), III, 1st Appendix, 15.

[249] Bongars (1611). In Robinson's list: Robinson (1841), III, 1st Appendix, 9.

[250] Cf. Ibbotson's works and lists (Ibbotson (1933)), e.g. ERP, 1816 Robinson- 3: Edward Robinson Collection, U.

[251] Irby (1823); Robinson (1841), III, 1st Appendix, 24.

Pococke and Carsten Niebuhr as examples for 'the form of narrative to be adopted.' Robinson includes Pococke while strongly criticizing his maps and views,[252] and of Carsten Niebuhr he writes, 'The prince of Oriental travelers (...). His visit (...) was brief and hurried (...). His plan of Jerusalem is very imperfect.'[253] He noted Henry Maundrell and Burckhardt as examples for travelogues originating in 'a daily journal.'[254] Robinson relied upon and discusses extensively Burckhardt's book, published in 1822, praising his orthography, 'He is hitherto the only Frank traveler in Syria, who gave Arabic names also in Arabic letters;' as mentioned above, the book was later criticized by Smith. In this connection, Robinson also notes maps compiled and published in the French surveyor and cartographer Pierre Jacotin's atlas following the French campaign in Egypt and Palestine, and in Bonner Catholic priest Johann Martin Augustin Scholz's travelogue following his 1821–22 excursion, though they are 'so very incorrect.'[255] Robinson uses three books, Eusebius's *Onomasticon* for the fourth century, Brocardus (Burchard of Mt. Sion, 1200–85) for the crusader period, and Quaresmius for the beginning of the seventeenth century, as examples of 'three ages or periods (...) distinctly marked by documents, which show us, with tolerable completeness, its state and character at the time,' and demonstrate 'the process of forgetting (...) how the light of truth has gradually become dim.'[256]

In another instructive letter, Robinson mentions sources that he assumed Smith could find in Leipzig and use for his own research and writing: Schultens' publication of Baha' al-Din's *Vita Saladin*, *Flora Egyptiana* by Peter Forskål (1732–63), the botanist of Niebuhr's expedition to Arabia Felix, and Rüppell's first volume of his *Travels*, wherein he supplies a list of hypsometric measurements in Sinai.[257] Robinson also complains in the letter about 'the manner in which [they] murder the Arabic names,' accusing a list of contemporary travellers, Rüppell,

[252] Pococke (1745); Robinson (1841), III, 1st Appendix, 20–21; Goren (2003), 29; Medvekova (2014).

[253] Niebuhr, C. (1969); Niebuhr, B. G. (1836), his biography written by his son, translated by Robinson; Robinson (1841), III, 1st Appendix, 21; Hansen (1964); Goren (2003), 30–36.

[254] Maundrell (1703); Burckhardt (1822); Robinson (1841), III, Preface, vi.

[255] Jacotin (1810); Scholz (1823); Robinson (1841), III, ix.

[256] Robinson (1841), III, vii–viii.

[257] Robinson to Smith, Berlin, 14.3.1839, ESP, ABC 60 (63). Baha' al-Din (1755); Forsskål (1775), cf. Hansen (1964); Rüppell (1838), cf. Anon (1832).

Lord Lindsay, Russegger, Laborde, 'etc.'[258] Further astonished, he writes, 'do you recollect the name of the convent at Mt Sinai? Rüppell calls it the convent of the Annunciation & [I] do not remember to have heard the name at all.'[259] Criticizing the 'letters' of Lord Lindsay, Robinson prefers Ulrich Jasper Seetzen's letters and map, published in astronomer Franz Xaver von Zach's *Monatlische Correspondenz zur Beförderung der Erd- und Himmels-Kunde* (*MCEHK*).[260] 'Seetzen was judicious, enterprising and indefatigable,' writes Robinson, giving a rather long list of Seetzen's letters, which 'had been communicated to me in manuscript by Prof. Ritter.'[261]

In his preface, Robinson also lists the most important maps used by Kiepert when compiling cartographic depictions of areas not visited during the expedition. These include East India Company officer Robert Moresby's (1794–1863) map of the Red Sea, Laborde's map of Sinai, Burckhardt's, Rüppell's, Jacotin's, Seetzen's, Irby's and Mangles's maps of the Dead Sea, the German mikrobiologist Christian Gottfried Ehrenberg's (1795–1876) maps of the Red Sea and Sinai, and fellow ABCFM missionary Isaac Bird's map of Lebanon. This is only a partial list, as Kiepert used many additional maps when compiling his maps published in the *Biblical Researches*, described in detail in his *Memoir*.[262]

This was not the first time that Robinson used some of the sources noted. In his revised and enlarged 1832 American edition of *Calmet's Dictionary of the Bible*, he added a map of the Exodus 'after Burckhardt, corrected after Rüppell and Ehrenberg,' an 'Exodus' plan of the Extremity of the Red Sea near Suez' taken from Niebuhr, as well as a map of Jerusalem and its environs.[263] In 1836, Robinson translated into English

[258] Rüppell (1838); Russegger (1841–3), first volume published only in 1841, hence it is difficult to determine which material Robinson had in his possession; the book is not mentioned in his bibliography; Lindsay (1838); Laborde (1836).

[259] Robinson to Smith, Berlin, 14.3.1839, ESP, ABC 60 (63).

[260] Robinson to Smith, Berlin, 14.3.1839, ESP, ABC 60 (63); Lindsay (1838); (1838a); Seetzen (1805–6); (1810); Lindenau (1810). Cf. Goren (2003), 36–48; (2011), 139–43.

[261] Robinson (1841), III, 1st Appendix, 22–23; Robinson erroneously calls him 'Ulrich Jakob Seetzen.' Ritter received only the manuscripts and published letters, his repeated call for a collection and publication of Seetzen's letters was only fulfilled in 1854 (Kruse (1854)).

[262] Robinson (1841), III, Preface, xi–xii. Cf. In detail Kiepert (1841); Goren (2017).

[263] Taylor, C. (1832), title, illustrations between 400–01 and 408–04. A copy in ERP, Hamilton 5, Cellar. Cf. Robinson (1832).

Niebuhr's biography, written by his more famous son, the historian Barthold Georg N. (1776–1831).[264]

The *Biblical Researches* contains two chapters dedicated to sources, one written by Robinson, 'Chronological List of Works on Palestine and Mount Sinai,' the other Kiepert's 'Memoir on the Maps Accompanying this Work.' Kiepert enters into great detail, listing the maps, itineraries, and manuscripts he used for each part of the country.[265] Whereas the American and British editions are quite identical, there are some significant differences from the German edition, changes that were probably editorial decisions, most likely made with Rödiger's consent. The all-important bibliography and Kiepert's Memoir in the American and British editions appear in the third volume; in the German edition they are placed at the beginning of the first volume before the table of contents.[266] This placement created a difference in the chapters included in Volumes I and II, though both end with the same chapter, XI. In the German edition, the third volume is divided into two parts and the second volume is missing the sections that were moved to the first volume. The German editors also decided to divide the Second Appendix into two in order to have a *zweiter Anhang* with Smith's 'Essay on the Pronunciation of the Arabic,' and a *dritter Anhang* with the lists of Arabic names. Another significant difference exists in the index. Whereas the English editions have two indices, the first for 'Arabic Names and Words' and the second for 'Ancient Geography, Antiquites, etc.,' the Germans preferred one long and detailed '*Namen- und Sach-Register.*'

Robinson's list of 'Itineraries, Travels, etc.' begins from *c.* AD 330–400 with Eusebii et Hieronymi *Onomasticon*[267] and ends with the multilingual British Diplomat John Bowring's (1792–1872) 1838 report, which Robinson describes as containing 'a large and valuable body of facts.'[268] Robinson met Bowring in Berlin, where he was stationed by his

[264] Niebuhr, B. G. (1836).

[265] Goren (2017), 37–38, list prepared according to Kiepert (1841).

[266] Robinson (1841), III, 1st Appendix, 3–55; Robinson (1841a), I, xvi–lxxiv.

[267] Robinson used some editions of this 'important work (…), the production of the successive labours of Eusebius and Jerome'; Robinson (1841), I, 375, III, 1st Appendix, 4.

[268] Bowring (1840). Bowring, British statesman and prominent linguist, 'travelled in Egypt and Syria in 1838, as the accredited agent of the British Government,' collecting information for Palmerston. His report, submitted to the House of Parliament, was published in 1840; Robinson (1841), III, 1st Appendix, 26; Williams, J. B. (1972), 294–301; Goren (2003), 87–88.

government in 1839.[269] The 'Works on the Geography of Palestine,' dating from Christian van Adrichom (1533–85) 1590 to Raumer 1835, contain geographical descriptions by authors who did not travel or reside in Palestine;[270] three additional works are under 'Jerusalem.'

Naturally, this impressive list of sources raises numerous questions deserving of answers, even if partial. The questions include: how many sources did Robinson actually use? How were they divided chronologically? Did he also use non-European/American sources? How did Robinson use itineraries and what he calls 'geographical books,' and is there any difference between the two? What led Robinson to define various sources as 'the more important works,' marking them with a star in his bibliography?[271] How did he use 'historical books'? Which sources did he use for the different chapters in the book, divided according to different regions? Finally, which sources did Robinson use most frequently?

The attached table, naming over 230 sources used by Robinson for the *Biblical Researches*, was compiled through a detailed and meticulous reading of all footnotes that appear in the text. Robinson's system of references did not make this task any easier, as he failed to write them systematically and, in some cases, he used two different abbreviations for the same source. Sometimes the author is mentioned in the footnote; however, in many instances, he notes it solely in the text itself. To give one example for repeat citing of the same source, the note on p. 61 in the first volume states 'Travels in Syria and the Holy Land, p. 462,' whereas two pages later the text states, 'Burckhardt's Travels in Syria, etc., p. 464.'[272] In many places, Robinson includes a long list of sources, frequently arguing with them and pointing at what he sees as inaccuracies, mistakes, and even fantasies, sometimes in the text (and then adding references) and sometimes in very long notes.

The table included here is not fully accurate, as the counting was complicated. The results, however, represent careful study and are helpful in answering the above questions. Both Tobler's and Röhricht's bibliographies were consulted in building the table, together with Robinson's list of sources. All three scholars present the 'Holy Land literature' chronologically, generally according to the year/s of the visit to the East. Following Robinson for 'works on the geography of Palestine' and his-

[269] Talvj to Kopitar, Berlin, 12.8.1839; Milović (1941), 100–01.

[270] Robinson (1841), III, 1st Appendix, 4; Adrichom (1590).

[271] Robinson (1841), III, 1st Appendix, 4.

[272] Robinson (1841), I, 61, 63.

Ill. 21: Sources for the Biblical Researches: List of Sections

Section 2: Cairo and Suez
Section 3: Suez to Mount Sinai
Section 4: Mount Sinai to 'Akabah
Section 5: 'Akabah to Jerusalem
Section 6: Jerusalem – Incidents and First Impressions
Section 7: Jerusalem – Topography and Antiquities
Section 8: Jerusalem – History, Statistics
Section 9: Jerusalem to Bethel
Section 10: 'Ain Jidy, the Dead Sea, the Jordan
Section 11: Jerusalem to Gaza and Hebron
Section 12: Hebron to Wady Musa and Back
Section 13: Hebron to Ramleh and Jerusalem
Section 14: Jerusalem to Nazareth and Mount Tabor
Section 15: Mount Tabor by the Lake of Tiberias to Safed
Section 16: Safed by Tyre and Sidon to Beirut
Section 17: Religious Sects in Syria and Palestine

Year	Source	s2	3	4	5	6	s7	8	9	10	11	12	13	14	15	16	17	Total
410 BC	Socrates							3										3
60 BC	Diodorus									1		6				3		10
24	Strabo	3					1			1	1	7		2		9		24
66	Josephus	1	1	1	2		71	9	7	32	26	18	17	72	40	17		314
100	Pliny									1		4		4	3	7		19
100	Dio Chrysosotom					1	7	7					2					17
150	Ptolemy		1		3					1	4	1		4	1	3		18
150	Justin Martyr							3		1								4
180	Galen									5								5
235	Dio Cassius							3										3
330	Eusebius		1		5	4	10	31	11	20	32		19	44	11	5		193
333	Bordeaux				3	3	11	6		2	1		2	5				33
350	Cyril					3	3	6						1				13
384	St Jerome													45	18	6		69
400	'Eusebius'		1	1	9	3	14	27	16	27	42	28	24	14				206
400	Peutinger Map		1		5					1	5	5		3				20
420	Nilus of Sinai		2		1													3
440	Sozomen				1			9			2	1						13
555	Cyrillus							5						3				8
560	Procopius		2	1				4	1	2			1	8	1			20
575	Gregory of Tur							5		1								6
600	Antoninas		2		1	1	3	3		3	9		1	7	2	1		33
670	Arculf				2	3	9	7	1	5	3		2	9	3			44
728	Willibaldi					2		2		7	2		2	3	3			21
750	Epiphanius							1		1			1	3	8			14
810	Theophanes						2	7										9

Year	Source	s2	3	4	5	6	s7	8	9	10	11	12	13	14	15	16	17	Total
870	Bernhardi						4	4		1			2					11
933	Eutychius		3				4	5			1		1	1	2			17
1040	Eugesippus								2	1								3
1095	Guibert							3										3
1096	Fulcher						4				1	8	1	2		2		18
1102	Saewulf					1	1	1		1	2			7	1			14
1125	Albertus						6				4			2		5		17
1154	al-Idrisi	3	1	3	2	2	12	1		5	3	4	1		1	5		43
1165	William / Tyre		1		1	3	30	15	2	8	19	17	12		15	20		143
1170	Benjamin / Tudela				1	2	8		2		5		1	17	7	3		46
1185	Phocas						2			1			1	10	1	1		16
1210	Jacques / Vitry					2				6	4	5	6	9	6	7		45
1211	Willebrand									3			2					5
1226	Jakût													4				4
1235	Abu'l-faraj							6										6
1300	Abu'l-Fida	2	1	3	1		3	5	1	5	6	18	7	18	15	7	1	93
1310	Sanudo					3	9		1	1	5	2	7	21	9	11		69
1320	Ibn al-Wardi				1		2	1				2			1			7
1320	Nikephoros					3		1							1			5
1332	Brocard					2	17		10	17	5	1	8	33	13	7		113
1332	William / Baldensel													5	6			11
1336	Maundeville		1		1	1	6			3	3			8	4			27
1336	Suchem	1	5		1	1	8	1		4	44		7	13		2		87
1400	al-Maqrizi	1	1	1	1									2	19	1		26
1449	Gumppenberg						2				4		1		1			8
1450	edh-Dhâhiry												1	2	4	4		11

Year	Source	s2	3	4	5	6	s7	8	9	10	11	12	13	14	15	16	17	Total
1479	Tucher	1	4		1	1	3			3	4		3	0				20
1483	Bernhard / Breidenbach	1					7		5	7	9		3	5	1	1		39
1483	Felix Fabri	1	4		1		11	1		2	14		3					37
1495	Mudjir ad-din						9	1			9		8	7				34
1508	Anselm													1	2			3
1522	Barthelemy / Salgnac						5			3			1					9
1547	Belon, Pierre	2	3			6	1			2	4		1	1	1			21
1552	Bonifacio									1	1		2	4	2			10
1565	Helffrich	1	1				1				1							4
1566	Haimendorf													1	9			10
1575	Rauwolf						2			2								4
1583	Radzivill														3			3
1584	Adrichom					6	4		1	3	2	1	1	3	5			26
1586	Zuallart						8		1	2	2		3	3	1			20
1589	Villamont						2									1		3
1596	Cotovicus						19	2	3	3	4		5	18	8	2		64
1610	Sandys						6		1		1		1	1	1	8		19
1616	Della Valle													12				12
1616	Quaresmius					2	38	2	5	16	11		4	27	17	5		127
1618	Lightfoot						8	1						2	14	1		26
1620	Hottinger														3	1		4
1636	Neitzschnitz		6											2		1		9
1644	Surius						1				1		1	3				6
1650	Buxtorf														6			6

Year	Source	s2	3	4	5	6	s7	8	9	10	11	12	13	14	15	16	17	Total
1650	Beer ha-Gola		4															4
1652	Doubdan						9		2	3	3		1	7	1			26
1666	Troilo		6				2		3	4	6		2	4	2			29
1674	Nau								2	1		1	1	1	7	11		24
1679	Arvieux													4	3	25		32
1681	Bruijn						7			3								10
1683	Huntington													1		20		21
1697	Maundrell		2			1	12	1	3	15	2			20	2		1	59
1697	Morison		1							3	1		2	10	1			18
1700	Egmond & Heymann		4				3				2		3	1	9			22
1708	Reland		1	1	6	2	7	3	7	17	32	32	19	32	14	10		183
1720	Bohaeddin						1				3	4	6	10	11	5		40
1722	Shaw		2	1				1						3				7
1738	Korte						4	1					1	3	2			11
1738	Pococke	2	11		1		16	1	4	18	4		4	8	29	20	1	119
1740	Le Quien, M.												3	7	6	6	2	24
1750	Baumgarten		2			1	3				2				1	1		10
1750	Hasselquist									18	1	1	1	1	6	8		36
1754	Schultz, Stephan					1									6			7
1757	Gignes							8										8
1758	Bachiene						1			2	1							4
1763	Forsskål	1	5		1					4	1			1				13
1766	Büsching		1								3	2			3	2		11
1766	Niebuhr	10	7	2	1		6			1	1	1			3	5	5	42
1767	Mariti									3	2			5	3			13

Year	Source	s2	3	4	5	6	s7	8	9	10	11	12	13	14	15	16	17	Total
1783	Volney										7	1	1		4	9	5	27
1795	Rossi														3			3
1798	Browne	1	1											4	1			7
1799	Le Père	4																4
1799	Jacotin	2									3		4	8	1			18
1800	Sicard	1	2															3
1801	Clark						10	2	1				2	7	7			29
1805	Seetzen	1	1	1	5		3			21	1	19			24	2	1	79
1806	el Abbassi						7			1	5				1			14
1807	Wilken												8	25	26	39		98
1810	Burckhardt	6	53	19	11		2	1	2	29	2	56		15	46	9	6	257
1810	de Sacy							1						26			1	28
1815	Turner						3	3		6			1	5	8	6		32
1815	Richter						1							8		1		10
1815	Jowett							1							2			3
1816	Irby & Mangles,		1				4		1	37	3	36	4	8	14	1		109
1817	Belmore, Earl						3				2				1			6
1818	Rosenmuller,	1					3			2	1	4	1					12
1818	Richardson						10	1	1	1	7		1	5	6			32
1818	Ritter, Carl	1								1		9						11
1818	Legh									4	2	7						13
1820	Henniker		1		2													3
1821	Berggren							1		4				10	9	1		25
1821	Scholz						5	7	1	2			5	7	2			29
1822	Leake											1			2			3
1823	Ehrenberg	1	4															5

Year	Source	s2	3	4	5	6	s7	8	9	10	11	12	13	14	15	16	17	Total
1823	Fisk									1	2				9			12
1823	Gesenius	1	3						1		1	12		8	5	1	4	36
1825	Buckingham						1	2	1	8		1		8	16	2		39
1826	Lord Prudhoe		1		3							4						8
1828	Laborde		9	8	1		1					29						48
1829	Reinaud														20	15		35
1829	Poujoulat										4							4
1829	Prokesch-Osten					2	7	1	1	1	2		7	3	2			26
1829	Hengstenberg		1											3		2		6
1830	Neander							7										7
1831	Linant de Bellefonds			1	2													3
1831	Rennell	3			2							1						6
1832	Cornelis															3		3
1832	Callier				3							4						7
1832	Spencer														4			4
1833	Catherwood					1	18											19
1833	Munro						3			6	2		1	8	9	4		33
1833	Olshausen						5											5
1834	Eli Smith								1	5	4	4				1		15
1834	Thomson							1						2	6	10	1	20
1835	Raumer						4	1		4	6	3	2	10	2	1		33
1835	Rüppell	3	13	11	2							1						30
1836	W. Rose									1		1						2
1836	Lane		1		1					4		1				5		12
1836	Schubert		4	2	3		7	3	3	6	4	16		30	10	1		89
1836	Stephens		1			1				1	3	2						8

Year	Source	s2	3	4	5	6	s7	8	9	10	11	12	13	14	15	16	17	Total
1836	Lindsay		3						1	1		6						11
1836	Paxton										1		1	3		1		6
1837	Salzbacher						1	5			2		2					10
1837	Moore & Beek						1			2	1	1						5
1838	Bowring							3										3
1838	Russegger		5		5		1				1			1				13
1838	de Bertou									1		14		2	10			27
1839	Kinnear	1										2						3
1839	Letronne											5						5
1839	Berghaus	2		1	2		1				4		1		1			12
	Total	55	192	57	94	64	545	240	105	450	429	396	249	793	629	361	28	4.687

Ill. 21. Table of Edward Robinson's sources for the Biblical Researches, number of references according to book sections. Seventy-two sources, mentioned only once or twice, were omitted from the list. However, they are included in the total numbers.

torical works, 'the date prefixed refers to the time of first publication.'[273] In the best scientific style, Robinson added, where needed, the edition he consulted. Comparison of the lists yielded some 'disagreements' between the sources concerning the exact date of expeditions in question.

Following Ibbotson's division, Robinson used 45 publications dating before 1099, 18 for the crusader period prior to 1291, 17 prior to 1500, 52 prior to 1700, 32 for the eighteenth century, and the remaining 73 for the first four decades of the nineteenth century. A better classification would follow accepted Holy Land periodization. Eighteen consulted publications predate the Byzantine Empire, originating earlier than Eusebius. The best example here is, of course, Josephus, whose works are 'the chief source next to the Bible for the History and Antiquities of Palestine' for which Robinson used the 1726 Amsterdam edition.[274] Palestine travellers' reports date after the Byzantine occupation, and 18 of Robinson's sources were dated within that period, only nine from the early Moslem period, 18 from the crusader period, 20 from the Ayyubid and Mamluk periods, and the remaining 156 from the Ottoman period and up until 1840.

Robinson's sources can be divided according to the extensiveness of their use. Josephus Flavius and 'the important work of the *Onomasticon*, the production of the successive labours of Eusebius and Jerome,' are by far the most often cited, though Robinson considered the latter only 'in a historical respect, only as a record of traditions current in their day.'[275] Hadrian Reland's 1714 treatise was considered by Robinson, as noted, as 'the standard classic work on Palestine' until the crusades.[276] As for travel literature, Burckhardt is by far cited most frequently (258 times!), followed by Richard Pococke, 'a classical scholar, but not a good biblical one, and had but a slight knowledge of the Arabic' (119), and then the two adventurous British officers Charles Leonard Irby (1789–1845) and James Mangles (1786–1867), 'well written and full of accurate information' (110).[277] The monk Burchardus of Mt. Sion and the Franciscan Quaresmius are also cited more than 100 times each.[278] The most

[273] Robinson (1841), III, 1st Appendix, 26.

[274] Robinson (1841), III, 1st Appendix, 3; Flavius (1895).

[275] Robinson (1841), I, 375.

[276] Robinson (1841), III, 1st Appendix, 27.

[277] Burckhardt (1822); Pococke (1745); Irby (1823); Robinson (1841), III, 1st Appendix, 20–21, 24. Cf. ODNB Irby.

[278] Burchardus de Monte Sion (1519); Quaresmio (1639).

frequently used non-Christian source, the Kurdish Ayyubid historian Abulfeda (Abu al-Fida, 1273–1331), Emîr of Hamah, 'describes the country as an eye-witness' in the early fourteenth century.[279] Also extensively cited are the early travellers Ludolph von Suchem, who travelled from 1336–41, Johannes Cotovicus, and Marino Sanudo.[280] To this mostly frequently relied upon and cited source list belong the later travellers Maundrell, Seetzen, and von Schubert.[281]

The division of sources according to the geographical areas of Robinson's book 'sections,' is a second window into the state of Holy Land scholarship predating Robinson. There are fifteen relevant sections for this study, tracking the voyage itinerary and chronology beginning in Cairo and ending in Beirut, omitting the first ('Introduction. Greece and Egypt') and the seventeenth ('Religious sects in Syria and Palestine'). The detailed table indicates significant variance in sources used for different regions. Earlier travellers' reports are rarely mentioned in the three Sinai sections, II. From Cairo to Suez, III. From Suez to Mount Sinai, and IV. From Mount Sinai to 'Akabah, except by those who visited the St Catherine Monastery (Ludolph von Suchem, Bernhard von Breidenbach, Felix Fabri, Johannes Tucher[282]). Early scientific travellers of the eighteenth century, Richard Pococke and Carsten Niebuhr, are followed naturally by Burckhardt (78 times), Laborde (17), and Eduard Rüppell (27). When Robinson wrote about their 'Excursion to 'Ain Jidy, the Dead Sea, the Jordan, etc.' (Section X), he consulted different sources, starting with extensive use of Flavius and the *Onomasticon*, William of Tyre, Bishop of Acre Jacques de Vitry (1160/70–1240), and Burchard of Mount Sion for the crusader period, followed by Quaresmius, Maundrell, and Reland, and then later travellers, Pococke, Linées student Fredrik Hasslequist (1722–52), Seetzen, Burckhardt, and Irby and Mangles.

The picture differs significantly in Section VIII on the history and statistics of Jerusalem. Naturally, Eusebius and Jerome are a leading source in the various discussions, as well as William of Tyre. There are very few references to other sources. Sections XIV. From Jerusalem to Nazareth and Mount Tabor, and XV. From Mount Tabor by the Lake of Tiberias to Safed, contain the largest numbers of references by far,

279 Abulfeda (1776); Robinson (1841), III, 1st Appendix, 10.

280 Suchem (1477); Cotovicus (1619); Sanudo (1972).

281 Maundrell (1703); Seetzen (1805–6); Schubert, G. H. (1839).

282 Suchem (1477); Breydenbach (1486); Fabri (1557); Tucher (1561).

801 and 645 respectively. As to be expected for the Galilee, Josephus and the *Onomasticon* are each cited over 100 times total in both sections. In addition, there are numerous references from the Jewish traveller Benjamin of Tudela (1130–73), Quaresmius, Maundrell, and the late Middle Ages sources. The same applies for Pococke and Seetzen. Descriptions of and following Napoleon's failed expedition are also cited relatively frequently, and there are many references to late eighteenth- and early nineteenth-century travellers.

Robinson used thirteen Arabic sources, and three Hebrew ones, even though for the itinerary of Benjamin of Tudela he could have used translations, including the 1840 Berlin and London editions edited and commented on by the German publisher and bibliographer Adolph Asher (1800–53).[283] Robinson could probably also have used a later edition of *Beer Hagolah* by Rabbi Judah Loew, the Maharal of Prague, first published in 1598.[284] This interesting source is cited only four times, and only in Section III. From Suez to Mount Sinai. On the other hand, Benjamin of Tudela's itinerary is cited 46 times, about a third in Section XI. From Jerusalem to Gaza and Hebron. But Robinson lost faith quite early in this source, writing to Smith:

> On Benjamin of Tudela one cannot place much reliance. His best translators & commentators are very much disposed to doubt, whether he ever visited the places he describes. The same opinion has recently been expressed to me here by Dr Lebrecht, a young Jewish scholar, one of the most distinguished orientalists of Germany (…). He had read Benjamin in the original, & has no faith in him.[285]

Additional examples of Robinson's pedantic scouring of all available sources for relevant data are in his letters of January 22 and March 14, 1839, written at the beginning of his work in Berlin. In the January letter discussing Mount Tabor (see below), Robinson notes that he already consulted Buckingham, Burckhardt, Maundrell, Korte ('a very sober, plain (?) traveller'), Richter ('an accurate observer'), Hasselquist, Rosenmüller, Nicolayson, Schubert, Moore, Fleischer, Wilken, Berghaus, and

[283] Benjamin of Tudela (1840–41); Robinson (1841), III, 1st Appendix, 7. Cf. Starkey (2019).

[284] Robinson could, for example, use a Hebrew 1838 Warsaw edition (Loew (1838)).

[285] Robinson to Smith, Berlin, 11.2.1839, ESP, ABC 60 (63). For Fürchtegott Lebrecht (1800–76), who made the interesting way from an Orthodox Jewish *Jĕahiwa* in Preßburg to Gesenius in Halle: ADB Lebrecht; Rosenfeld (1929).

Hengstenberg.[286] Researchers such as missionary Nicolayson and the Irish Dead Sea explorer George Henry Moore (1810–70)[287] did not publish books, rather Robinson had access to their papers and letters. The March letter also contains a long list of sources, including a new version of the life of Saladin written by his secretary, Baha' al-Din ibn Shaddad, published in 1755, with the Arabic text translated into Latin by the Dutch philologist and orientalist Albert Schultens (1686–1750).[288] Robinson made special mention of its geographical index where Smith could find 'many names known to us.'[289] Forsskål's *Flora Egyptiana* was noted above. The German traveller and natural sciences scholar Eduard Wilhelm Peter Simon Rüppell (1794–1884) had just published the first volume of his travels in Ethiopia, 'beginning with a very accurate & impartial account of the improved state of Egypt under Mahammed Ali,' and including heights measured of important mountains and other observations during his third visit to Sinai.[290] But Robinson also criticizes Rüppell, first for mistakenly calling St Catherine's Monastery the 'convent of the Annunciation' and, more importantly, for his and other recent travellers' scandalous treatment of Arabic names.[291]

There are numerous examples for Robinson's extensive use of sources when trying to answer questions of geography. For example, with regard to Mount Tabor, Robinson sought to know what exactly can one observe from its summit, how far can one see and, primarily, to what extent can the Mediterranean be seen from it, as Buckingham remarked that 'a small ship is seen in the N.W.'[292] 'The case,' Robinson writes, 'affords a curious illustration of the teaching of memory & of the importance of writing down everything on the spot.' Checking the testimonies of Maundrell, Korte, Richter, and Hasselquist, Robinson concludes that 'the testimony is pretty

[286] Robinson to Smith, Berlin, 22.1.1839, ESP, ABC 60 (63). Cited and used for his work: Maundrell (1703); Korte (1743); Hasselquist (1766); Wilken (1807–32); Buckingham (1821); Burckhardt (1822); Berghaus (1835a); Schubert, G. H. (1841).

[287] Moore (1837); Robinson to Smith, Berlin, 22.1.1839, ESP, ABC 60 (63); cf. in detail Goren (2011).

[288] Baha' al-Din (1755). For Schultens: Fück (1955), 105–07.

[289] Robinson to Smith, Berlin, 14.3.1839, ESP, ABC 60 (63); cf. Robinson (1841), III, 1st Appendix, 8.

[290] Rüppell (1838). Cf. Martens (1949); for his hypsometric work in Sinai: Anon (1832); Goren (2003), 57–60; (2011), 136–37.

[291] Robinson to Smith, Berlin, 14.3.1839, ESP, ABC 60 (63).

[292] Robinson to Smith, Berlin, 22.1.1839, ESP, ABC 60 (63); Robinson (1841), III, 215–17. For the controversy between Smith and Buckingham, and subsequently Robinson, see chap. 5.a. 4.

strong, that the Mediterranean is visible from Tabor; & according to my impressions, a much larger extent of it than Buckingham makes out.'

Additional examples of Robinson's scientific and highly critical approach are worth noting. A letter to Smith written in spring 1840, discusses two French contemporaries. The first is the famous novelist Alexandre Dumas Snr. (1802–70): 'I am happy at being able to satisfy you respecting Monsieur Dumas & his friends.' After consulting with Bowring, a personal acquaintance of Dumas, Robinson was able to elaborate:

> Mr Dumas, you know, is a famous French writer of the romantic school. In order to make a new book, he formally applied to the French government for a sum of money to travel in the East. He caused it to be announced in the journals, that on face a day Mr Dumas has set off from Paris on Marseilles on his oriental journey. This being done, he set himself quietly down in his study & wrote his book![293]

The book *Quinze jours au Sinai* was first published in 1839, under the names of Dumas and the French romantic painter Adrien Dauzats (1804–68), 'who did visit our area twice before.'[294]

Count Jules de Bertou had a central role in the process of establishing the exact level of the Dead Sea surface.[295] Robinson, who met de Bertou in Jerusalem, had many reservations regarding the accuracy of his results and their acceptance by the scientific community:

> In connexion with the 'Arabah, I have had to point out the errors & inconsistencies of Bertou's account; & send off today my remarks on him to Capt. Washington, probably they will be (printed?) in his journal [the *JRGS*]. I have been exceedingly cautious as to the tone & remarks; Ritter has read & approves them; & will read the Brit. Geog. Socie. & publish them in their monthly report. In this way I trust some of his observations may not be counteracted.[296]

It is not difficult to imagine the tall American entering the Berlin library every morning except Sundays. The 'Old Library,' housed in the

[293] Robinson to Smith, Berlin, 15.4.1840, ESP, ABC 60 (63).

[294] Dumas (1841). Ben-Arieh (1997), 145–47, 163, claims that it was Dauzats who printed the book with the help of Dumas. Dawson, W. R. (1995), 117, and Taymanova (1998), argue that Dauzats' diary 'was utilized' by Dumas.

[295] Goren (2011), for Bertou 211–17.

[296] Robinson to Smith, Berlin, 15.4.1840, ESP, ABC 60 (63). Cf. Goren (2011), 221–22.

majestic Baroque building called *die Kommode*, erected from 1775 to 1784 under Friedrich the Great, is home today to the Law Faculty of Humboldt University. Entering the reading room, he surely walked straight to what would have been his permanent seat, where a high pile of books awaited him. He might have, of course, walked directly to the handwritten catalogue, diving into the search for more and more sources. When he was not at the central library, Robinson went from his Berlin apartment, 'Tauben Strasse Nr. 16, frühe Hotel de l'Europe,'[297] to Carl Ritter's new house on the corner of the *französischen und der Markgrafenstraße*, the city center and very close to the *Gendarmenmarkt*.[298] A third possibility was to find him in the library of Berlin's geographical society, located from 1830 to 1843 in the headquarters of the old maps publisher and seller *Schropp'sche Kartenhandlung*, in *Jägerstrasse* 24.[299]

Robinson never wasted a minute, and his tempo of writing, as can be seen from the table in ill. 18, was astonishing. Most of his time, however, was probably not spent in writing but in reading his collected sources. As his writing advanced geographically, according to the chapters of the *Biblical Researches*, he must have developed his own system to organize his sources. When consulting a certain source while writing about Jerusalem, for example, he needed a system for recalling the many details he found on other regions, such as the Galilee, that he would need for future chapters. The same held true for the opposite situation, while writing about the Galilee, he needed a system for cataloguing an unused source for Sinai.

The manuscript of the *Biblical Researches* bears evidence of his solution for the second situation, of cataloguing new sources found during research. Robinson used booklets, writing on only one side of the page so that he could add many comments and changes on the back of the opposite page.[300] The solution to the first question can be given only as hypothesis. Robinson must have had used additional booklets for entering all comments for future chapters. Unfortunately, like Robinson's and Smith's diaries, no such booklets have been found.

297 Robinson to Smith, Berlin, 23.11.1838, ESP, ABC 60 (63); Robinson to Barth, Berlin, 3.11.1838, ERP, 1816 Robinson- 1.

298 Zögner (1979), 52 nr. 113.

299 Lenz (2003), 33.

300 This can be observed in all his manuscripts: ERP-UTS, Edward Robinson Papers, 1836–38. Series 1: Writings, boxes 1–4.

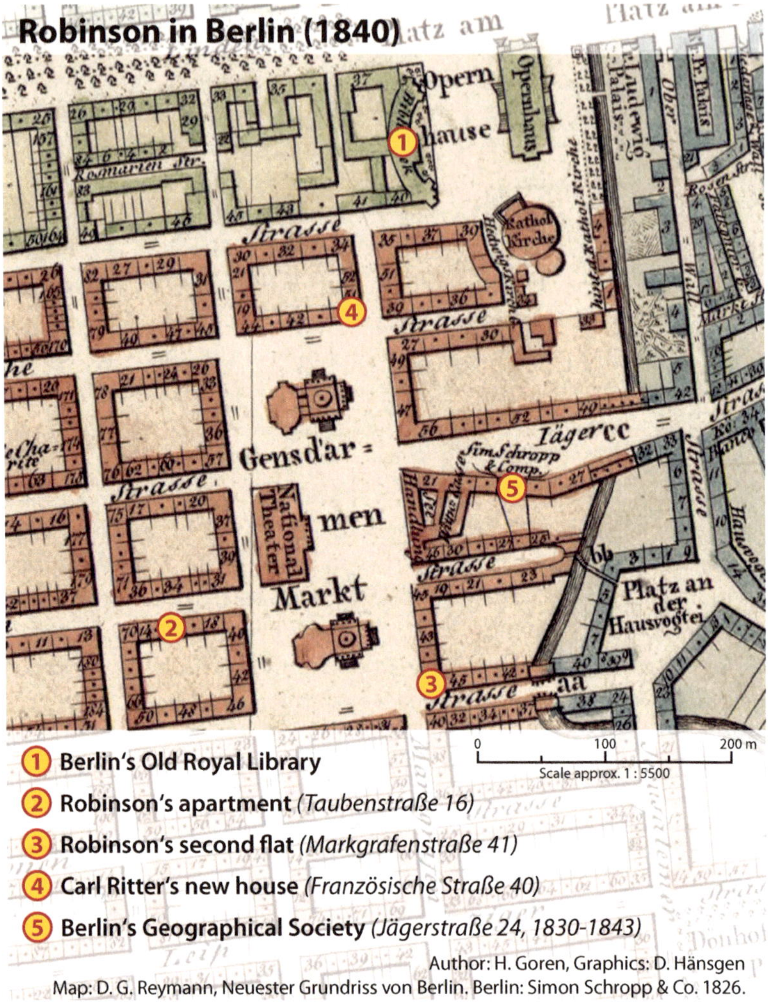

Ill. 22. 'Robinson's Berlin'.

4.d. Scientific instruments for Smith's further research

Quite early on in the expedition, Robinson and Smith clearly understood the ramifications of not having the necessary instruments for producing geodetic data.[301] In his introduction to the *Biblical Researches*

[301] For detail on topics such as geodesy development, methods, instruments, and manufacturers, see the following reference works for eighteenth- to nineteenth-century measuring devices: Brown, L. A. (1977); Deumlich (1982); Bennett (1987); Smith, J. R. (1997); Turner (1998); Withers (2013).

(see Chapter 2.d.), Robinson admits that they carried few instruments on the expedition, 'an ordinary surveyor's and two pocket compasses, a thermometer,[302] telescopes, and measuring tapes.' There is no hint as to the value and reliability of these simple and basic instruments. In a letter to Smith asking him to arrange the necessary equipment for the expedition, Robinson mentions the need for a 'pocket sextant, or any other instrument, (not too enormously expensive) by which it may be possible to ascertain the latitude of places N. of Akaba.'[303] It is difficult to guess his intentions or whether he hoped to achieve more than simply ascertaining latitudes with this minimal equipment.

Clearly, Robinson met the contemporary definition of a 'philosophical traveller' used frequently by Charles W. J. Withers in his study of 'scientific instruments and questions of method in nineteenth-century British geography.'[304] The question of 'observation in the field,' namely, of what and how to observe, was primary to British travellers of various backgrounds. The *Manual of Scientific Enquiry*, published for the first time in 1849 by authority of the Lords Commissioners of the Admiralty, was the standard guidebook for any British traveller – explorer – collector, and mainly, of course, for officers. Edited originally by John Frederick William Herschel (1792–1871), the leading astronomer, it includes contributions of 'some of our most eminent men of science' in various fields.[305] In addition to technical innovations, the proper outfit for travellers, and the use of a notebook, the instructions in geography, written originally by William Richard Hamilton, traveller, antiquarian, and acquaintance of Robinson, include gathering of geographic data of the relevant country, such as mountain ranges, rivers, springs, lakes, coastlines, oceans, population, languages, government, buildings, agriculture, trade and commerce. The first *Hints to Travellers* of the RGS, a manual for field observation, was published some years later, in 1854, and was discussed thoroughly by Felix Driver in 2001.[306]

[302] Thermometers were also used for measuring latitudes: Sykes (1838). The first to publish the fact that the Dead Sea 'is considerably lower than the Mediterranean,' George Henry Moore and William G. Beek, established the fact by measuring the temperature of boiling water: Moore (1837); cf. Goren (2011), 158–206.

[303] Robinson (1841), I, 47; Robinson to Smith, Alexandria, Egypt, 3.1.1838, ESP, ABC 60 (63).

[304] Withers (2013).

[305] Herschel (1851), iv. Cf. ODNB Herschel.

[306] Herschel (1851), iv; Freeman (1980); Driver (2001), 49–67. To Hamilton see ODNB, Hamilton; Mill (1930), 38–39; Hallett (1965), 358; Fagan (1975), 80–90; Dawson, W. R. (1995), 188; Manley (2001), passim.

Naturally, the Near East was a sought-after destination, with fascinating artefacts scattered throughout, many of great scientific significance, and of value to Western cultural heritage. Consequently, their discovery, study, and purchase, legal or illegal, became central to Western scientific activity in the Levant, filling European collections and museums.[307] Comparing the *Biblical Researches*, based on Robinson's and Smith's diaries, with developing British geographical science as detailed in Withers' study, it appears that Robinson and Smith understood and employed many of the accepted principals of documentation and publication, even though they were not educated geographers fully versed in contemporary methods of the geographical traveller.[308]

After returning to Berlin and beginning to work on the *Biblical Researches*, it became clear that the lack of reliable measuring instruments had been a major deficiency. Robinson admits that 'I became aware of the value of the materials thus collected, in a geographical point of view, from the judgment passed on them by eminent geographers.' It was certainly a period when 'instruments were identified as a key part of how geographers should undertake procedure and conduct themselves methodically,' in the words of Withers in his discussion of Julian Jackson's *What to Observe*, published first in 1841.[309] It seems that Robinson was unaware of 'the Traveller Manual' published in the *JRGS* as early as 1832, 'containing a clear and concise enumeration of the objects to which a Geographer's attention should be especially directed.'[310] Robinson was not a geographer, nor did he claim to be one. As a student, he rarely went to listen to the early geographers, but he understood the need for an accurate and reliable report of his observations, recorded as much as possible in real time (though he was unable to record 'on the spot'). Memory cannot always be trusted. Pointing to the encouraging reactions of Ritter and Berghaus, Robinson could 'look back with painful regret on the circumstances, which prevented me from taking along more perfect instruments,' necessary for 'obtaining a more exact knowl-

[307] Fagan (1975) is a good example of a modern approach to this phenomenon.

[308] As example: Herschel (1851). See Withers (2013) discussion, mainly 167–69; Collier (2002); Goren (2011), 26–30.

[309] Withers (2013), 171(citation)-3. In the coming year, Jackson (1790–1853), as secretary of the RGS (Mill (1930), 52–57), would write to Robinson informing him that he was chosen recipient of the RGS gold medal: Jackson to Robinson, RGS 29.5.1842, ERP, I 2; cf. chap. 4.b.1.

[310] *JRGS* 2 (1832–33), vii, cited by Withers (2013), 172–75.

edge of the observations necessary for the trigonometrical construction of a map.'[311]

It did not take Robinson long to realize that 'in constructing the net of triangles, Prof. Berghaus starts from the base I measured at Jerusalem & [on] the Mount of Olives,' but he 'does not take any note of our routes & courses.' The leading cartographer had his reasons; the 'uncertainty of the hour with horse & mules' in and of itself made calculating reliable distances difficult, and together with variance in the type of 'ground travelled over,' made it nearly impossible.[312] The fact that 'there are around Jerusalem as yet no points astronomically fixed, at least not sufficiently so,' also made it impossible 'to determine the length of the hour with horses.'[313] Later in Berlin, while writing the book, Robinson's self-assessment changed and he began considering himself an 'ancient geographer.' Writing to Smith from New York, Robinson discusses two of the Prussian scientific consuls to the eastern Mediterranean countries, Anton Albert Heinrich Louis von Wildenbruch (1803–74) and Ernst Gustav Schultz (1811–51), both of whom 'I have not the pleasure of knowing,' and mused, 'I almost wonder sometimes, that the King of Prussia does not send a scientific expedition to Palestine,' to which 'I should be almost tempted to offer myself as ancient geographer.'[314]

The need for more advanced instruments was not raised in letters written en route. Only when in Berlin after the expedition did Robinson begin considering the issue. The first letter in which he discusses at length the need for and possibility of procuring more reliable and advanced instruments was written in February 1839, less than three months after arriving in the Prussian capital. Smith, who knew that he would be returning to the east and might conduct further studies, had raised the issue previously. However, it was a coincidental meeting in Berlin that led Robinson, who was nearly *tabula rasa* in this field, to make 'a discovery, which was to me in part pleasing; & in part, for the moment, exquisitely painful.'[315] In Berlin, Robinson met 'our brother-in-law, Major Schlüsser [who] is here attached to

311 Robinson (1841), I, 48.

312 Robinson instructed his readers about 'The Measure of Distance': Robinson (1841), I, For the Reader,' xv–xvi.

313 Robinson to Smith, Berlin, 14.3.1839, ESP, ABC 60 (63).

314 Robinson to Smith, New York, 18.2.1843, ESP, ABC 60 (63). For the Prussian scientists who served as consuls cf. Goren (2003), 181–85, 194–201.

315 Robinson to Smith, Berlin, 11.2.1839, ESP, ABC 60 (63). All following information and citations are from this letter.

the General Staff, (Aid of Prince Karl[316]).' In 1831, Adolf Friedrich Emil Schlüsser (1793–1863) married Auguste Luise Emilie von Jakob (1794-?), an elder sister of Talvj's whom Robinson met on his frequent visits to the von Jakob's while in Halle in the late 1820s. Having been educated in Berlin's military academy, the officer turned out to be the right man for Robinson, and he began showing him various scientific instruments for future expedition use. The fact that Robinson added prices noting that 'it was pleasing to me to discern that such instruments may be had, for some 30$,' is the best proof of Robinson's future plans.

In his letter to Smith, Robinson notes two specific instruments that Schlüssel had in his possession. He initially asked for a simple protractor for measuring angles; instead, Schlüssel showed him: '(...) a beautiful little English instrument, Douglas's Patent reflecting Semicircle, usually called in German a patent reflector. It is a small semicircle with light & reflector, for taking angles, precisely in the plan of the Sextant, with a small scale, & answering also all the purposes of a protractor.'[317] Then Robinson visited Schlüssel's home and was shown a 'Smalkalders Patent Compass, a pocket instrument having every (?) that we desired, an eye-sight, an object-sight, & on the former a scale reflection, which enables one to read the number of the degree at the same moment he sees the object,' and a *Höhenvermesser*, an instrument for measuring heights produced by the same leading London manufacturer of Württemberg origin.[318] For Robinson, it was as if his dreams had come true, 'these are just the things for you, & instrument for all purposes,' he writes happily to Smith. In his letter, Robinson laments the fact that when in Berlin on his way to the East he surely met Schlüsser, but he had no idea 'that Schlüsser had to do with such things, it so happened that the subject was never spoken of before here.' Robinson exclaims that 'it would have been a pleasure to him to have put them into my hands; they have been being the whole time idle in his drawer!' Consequently, he recommends that Smith bring these instruments upon his return to the East, as 'it

[316] Friedrich Karl Alexander (1801–83), Prince of Prussia, third son of King Friedrich Wilhelm III. and brother of Emperor Wilhelm I.

[317] For an early description of the instrument see Cary (1811); Brown, L. A. (1977), 180–207 ('The Latitude'). William Cary (1759–1825) was an English scientific-instrument maker.

[318] Charles Augustus Schmalcalder (1781–1843), inventor and maker of optical and mathematical instruments, arrived in London from Stuttgart around 1800, starting the English family line. See Smith, J. (1993).

seems to me, that exclusive of astronomical observations, these are just the things for you.'[319]

In the end, it seems that Robinson never purchased the instruments from Schlüsser. It is not clear whether they were too expensive or whether there was another impediment. March 1839, however, turned out to be a tremendously active month for acquiring them. Smith thanks Robinson on March 11th, 'for your generous & unexpected offer to contribute half the price of Kater's Repeating Circle[320] & Smalkalder's compass' and promises to 'furnish the other half.' Together, Smith adds, 'they will furnish me just the means & so far as I know, the only means, I shall need to make such observations as I now contemplate in Palestine.' But Smith was not satisfied, writing that 'I had had in mind, all along, something like a theodolite.' This letter testifies that Robinson had been considering the two instruments, as Smith adds that 'you can take them at once, or wait till I come as you choose,' but cautions Robinson that he must ensure that they are not sold to other customers and so 'slip out of our hands.'[321] Smith repeats the warning five days later, adding that 'it is desirable they should be sent [to Beirut] with my types.'[322]

Shortly thereafter, Robinson paid several visits to a Berlin workshop for optical engineering. Established in 1813 by Carl Philipp Heinrich Pistor (1778–1847), the workshop was famous for inventing the most precise astronomical instruments, primarily for European observatories. In 1836, the young Carl Otto Albrecht Martins (1816–71) joined the workshop and it was renamed 'Pistor & Martins.'[323] On March 25th, Robinson, who in his scholarly way seemed to excel rather quickly in geodetic instruments and practice, met Pistor and presented him with an order for 'a Kater's circle; & Schmalkalder compass,' for which 'Pistor fixed the time for the 15th of June; but doubtless in a case of importance they might be made by the 1st of June.' Robinson writes that Pistor promised that 'he will make the circles somewhat larger, so that

[319] Robinson to Smith, Berlin, 11.2.1839, ESP, ABC 60 (63). For the long-time rivalry between astronomical observation-based and 'terrestrial' instruments, see Sobel (2014). See also Jackson's list of 'absolutely essential' instruments for the geographical traveller: Withers (2013), 171.

[320] For observations of magnetic declination with the help of a theodolite, Henry Kater (1777–1835). See Turner (1983), 50; Smith, J. R. (1997), 172.

[321] Smith to Robinson, Leipzig, 11.3.1839, ESP, ABC 60 (105/10) 1839 Jan.-Mar.

[322] Smith to Robinson, Leipzig, 16.3.1839, ESP, ABC 60 (105/10) 1839 Jan.-Mar.

[323] Robinson to Smith, Berlin, 25.3.1839, ESP, ABC 60 (63). See Zaun (2000).

the numbers may be read off more easily, but this will not increase the size nor weight of the instrument itself by a grain. He strongly urges that you should also have a pocket chronometer,' a precision timekeeper, which is a highly accurate clock generally used to measure longitudes on sailing ships.[324] After consulting Ritter, Robinson sent various suggestions. Ritter was, perhaps, uncertain about the price and use of chronometers for land measurements, and had recommended that Smith, while in Leipzig, study 'the mode of taking observations of longitude by the moon & stars' at the city observatory. Robinson asks Smith to determine the longitudes of Jaffa 'or perhaps of Jerusalem itself' by observation, and then notes that since 'the rules for calculation are given in Bowdich's navigator (...), all the other longitudes could follow from the chronometer.'[325]

Smith next wrote in early April, expressing hesitation concerning his ability to perform such measurements. His doubts primarily concerned 'observation for ascertaining the longitude of a place' using a maritime instrument. He repeats that reliance on Bowditch's navigation instrument and calculations 'would give me all the assistance I should need.'[326] Smith's discussion of the way he understood the performing of these celestial latitude and altitude observations and calculations is a unique glimpse into the geodetic difficulties and unclarity in those early years:

> If I recollect the principle correctly, it requires where the latitude is not exactly known than points to be mentioned (?). The time at the place of observation the altitude of one of the heavenly bodies, & the distance of the two bodies from each other. This requires two observers & instruments for measuring angles, unless the observer has acquired great tact of the business. But when the latitude is known, & the time ascertained, I shall think the altitude could be calculated. Perhaps the Circle is so constructed that both the angles can be taken by it at the same time.

[324] Robinson to Smith, Berlin, 29.3.1839, ESP, ABC 60 (63). For the invention of the chronometer and the ascertaining of longitudes, see Brown, L. A. (1977), 208–40 ('The Longitude'); Sobel (2014).

[325] Brown, L. A. (1977), 208–40, 'for Bowditch used to train his sailors, & his black cook, to find the longitude by lunar observations.' The American mathematician Nathaniel Bowditch (1773–1838) is remembered for his work on ocean navigation and often credited as the founder of modern maritime navigation, see Thornton (2016). For the observatory in Leipzig: Ilgauds (1995).

[326] Smith to Robinson, Leipzig, 8.4.1839, ESP, ABC 60 (105/11) 1839 Apr.-Dec.

Smith continues, discussing the exact instrument he thought right for obtaining the most precise results:

> A chronometer brings with it so much core that I shall hardly think of purchasing another. And besides for the short distances of Palestine, I have an idea that a good patent lever Match, with a chronometer balance wheel, which can be had in England for $50, will be all that is needed. (...) I suppose that the longitude of Beyroot & several other places on the coast has already been accurately ascertained.[327]

Smith, who did not have a direct relationship with the Berlin cartographer, most likely did not know for which places in Palestine the longitude had been 'accurately ascertained.' Actually, Kiepert, in his *Memoir*, used Berghaus' data and discusses various latitudes, for example of Tyrus, Saida, and Beirut.[328] The above-cited letter clarifies Robinson's and Smith's expectations that Smith would be returning to the East, and that supplied with the right instruments, he could conduct the needed measurements. At this point, they do not detail the nature and goals of Smith's research nor exactly what kind of geodetic data he should obtain.

Smith's next letter, sent from England on his way to the US, is discussed in the section on Berghaus' cartography and Smith and Robinson's plans with John Washington, then secretary of the RGS and editor of its periodical.[329] On his way from Leipzig to Boston, Smith met Washington in London, probably several times, and raised the issue of the instruments. Washington suggested that he procure for them a Cater's circle at a cheaper price, but that 'for astronomical purposes it is far inferior to a sextant,' which Washington gladly proposed to obtain in addition to an artificial horizon.[330]

Robinson doubted Washington's proposals and Smith did not accept them. In mid-July, Robinson sent the instruments purchased in Berlin to Tauchnitz in Leipzig, in order to be transported to Beirut together with

[327] Smith to Robinson, Leipzig, 8.4.1839, ESP, ABC 60 (105/11) 1839 Apr.-Dec. For a detailed discussion of these geodetic problems, and the development of measuring methods and instruments for longitude, see Sobel (2014).

[328] Berghaus (1835a), 22, 26–28; Kiepert (1841), 52–53.

[329] Goren (2015); (2017), 21–28.

[330] Smith to Robinson, London, 8.7.1839, ESP, ABC 60 (105/11) 1839 Apr.-Dec.; Washington to Smith, London, 20.10.1839, ESP, ABC 60 (101) ('I regret extremely to learn that you have been suffering from intermittent pain, the result I fear after so much labour in the good cause to which you are devoted').

Smith's types. This the publisher did in early September. The sextant is 'better for celestial observations on ship board (...), but I understood you to wish principally a terrestrial instrument; & in this respect there can be no question as to the superiority of the circle.'[331] In this letter, Robinson also hints at creeping doubts considering Smith's real intentions, 'our letter made me for sometime doubtful, whether I ought not to decline to buy the instruments,' but he decided to send them 'as proposed.'[332] Washington did not give up, and in a much later letter he offered a new chronometer, manufactured by the London firm Arnold & Dent, at a reasonable price: 'I cannot hear that a second hand watch can be had for less.' With this, he adds, they could 'settle the vexata quaestio of the level of the Dead Sea,'[333] a question in dispute since the publication of Moore and Beek's letter in the *JRGS*.[334]

The *Biblical Researches* reveals that Smith did return to Beirut with the instruments in his possession, hoping to use them on future expeditions. Robinson, who at this early stage could not see himself returning to the Holy Land, offers: '(...) it was stated, that the Rev. Eli Smith was about to return to his station at Beīrût, taking with him instruments of the best kind in order to verify our former observations, and prosecute further researches in parts of the country not visited by us; and that I hoped to be the medium of communicating his subsequent observations to the public.'[335] Although Robinson had no sign of Smith engaging in research, he did not give up. In April 1841, the book already on the market, he wrote again to Smith encouraging him to travel and suggesting 'to place 100$ at your disposal.' Robinson repeats his suggestion to serve as the 'medium (...), that in this way, as we have begun, so we may continue to labour together for the illustration of the Word of God. May this hope be fulfilled!'[336] Robinson sent Smith an additional donation of the same sum to aid him in his educational work.[337]

331 Robinson to Smith, Halle, 10.9.1839, ESP, ABC 60 (63); Tauchnitz to Smith, Leipzig, 4.9.1839, ESP, ABC 60 (80).

332 Robinson to Smith, Halle, 10.9.1839, ESP, ABC 60 (63).

333 Washington to Smith, London, 20.10.1839, ESP, ABC 60 (101). John Roger Arnold (1769–1863), son of watchmaker and inventor John Arnold (1736–99), continued the firm, in 1830 taking watchmaker Edward John Dent (1750–1893) into partnership. See Sobel (2014), chap. 14 and 15.

334 Moore (1837); see in detail Goren (2011), passim.

335 Robinson (1841), I, xiii, (1843), 9. See also Robinson to Smith, New York, 18.2.1843, ESP, ABC 60 (63).

336 Robinson to Smith, New York, 14.4.1841 and 17.4.1841, ESP, ABC 60 (63).

337 Robinson to Smith, New York, 18.2.1843, ESP, ABC 60 (63).

Whether Smith resumed field research is a question answerable primarily through publications and letters, and garnered from silence rather than information supplied. The best proof for Smith's near total neglect of scientific research can be deduced from the first issue of Robinson's *Bibliotheca Sacra*, published in 1843. It includes 'several communications from Palestine' by Samuel Wolcott, none by Smith! In his letters, Smith writes clearly that 'I have been full of business since I came back, & very much distracted by the state of our mission.'[338]

The political climate upon Smith's return in April 1841 was prohibitive; he had neither the time nor the possibility of conducting any research. The security situation was difficult and dangerous after the late 1840 battles and reversion of the countries from Egyptian to Ottoman rule, 'the state of confusion and anarchy and war (...) by which the mission has of course being greatly affected,' prevented freedom of movement. In the meantime, British officers with easier access conducted surveying projects in the Holy Land. Robinson mentions by name 'Majors [Frederick Holt] Robe, [Charles Rochfot] Scott and [Richard] Wilbraham,' as well as Lieutenant Symonds.[339] 'I have read several interesting communications from my friend Mr Smith. One is a map of the country around the sources of the Jordan by Maj. Robe, the Officer of Engineers who explored that region, making many corrections necessary in our map,' writes Robinson to Ritter.[340] Robinson praises 'the intercourse subsisting between the English officers and the American missionaries,' which led the former to 'often communicate to Mr Smith so much of their observations as was compatible with their confidential duty to their own government.' In this way, both Robinson and Ritter were able to publish considerable new material based on the British geodetic work, such as Robe's pioneering map.[341] Robinson sufficed himself with being the 'medium' for publication of Wolcott's letters that he received from Smith (see also Chapter 5.a.).

Smith's first report of a travel study, which reached Robinson on December 6, 1842, was a description of a short excursion to a temple

[338] Smith to brother Benjamin, Beirut, 7.4.1842, ESFP-YD, Box 1 Folder 4. See also there, Box 2, Folder 11: Smith's 'Memorandum of Letters,' April 1842 to August 1845.

[339] Smith's 'Memorandum of Letters,' April 1842 to August 1845, ESFP-YD, Box 1 Folder 4, 9–10. Cf. Jones (1973); Goren (2005).

[340] Robinson to Ritter, SBB-PK, Nachl. 133 (Carl Ritter), NL Carl Ritter, K 5, Q – SCH, Bl. 35.

[341] Robinson (1843), 10. Cf. Goren (2000); (2011), 102–15, 230–38; (2017), passim.

in Lebanon, conducted without the use of measuring instruments. He was accompanied by Wildenbruch, Prussian consul in Beirut between 1843–48.[342] Further papers of his were printed in the same periodical, the *Bibliotheca sacra*, but only from 1846 and are similarly lacking geodetic work.[343] No further information concerning the instruments and their whereabouts was located. The instruments are not mentioned in Robinson's next letters to Smith of record dated February 18, 1843, and December 11 and 22, 1845 (though they discuss journals of Smith's travels), and January 11, 1846.[344] It is unknown whether the instruments were used in subsequent research by Smith, other scholars, or other ABCFM missionaries. Robinson returned to the East more than ten years later to conduct 'further researches;' yet, no instruments or geodetic work are mentioned in his letters from that period.

4.e. Robinson as library emissary: purchasing European books and collections for American libraries

At the turn of the nineteenth century, US libraries had sparse offerings of scientific works in the fields of theology, history, historical-geography, Oriental studies, and philology. This lack preoccupied theologians, churchmen, and scientists alike, including, of course, Edward Robinson. From his first stay in Germany, Robinson viewed purchasing and collecting, for the Andover Divinity School Library and later the UTS library, as many works of 'classical' and historical literature as he could locate and finance a central part of his mission. These included works of 'Palestine literature' in the broadest sense, primarily history, geography, and travelogues.

Halle, a scientific center with a relatively large library, and Leipzig, with its many publishers, booksellers, and leading annual book fair, were ideal for book purchasing, as the young Robinson discovered during his early years in Prussia.[345] Robinson wrote extensively about his purchasing activities not only in his letters to Stuart, but also in letters to his sister Elisabeth, letters that were collected, printed, and edited by

[342] Smith (1843); Wildenbruch (1846); Goren (2003), 181–85.

[343] Smith (1843); (1846); (1846a).

[344] Robinson to Smith, New York, 18.2.1843, 11. and 22.12.1845, 11.1.1846, ESP, ABC 60 (63).

[345] Williams (1999), 146–47.

Hermine Weigel Williams.[346] His socializing with von Jakob and Gesenius was very effective for making the right acquaintances to achieve his mission. In February 1827, some four months after reaching Halle, writing to Stuart, Robinson notes that he dined with 'many booksellers (…) I think in 2 or 3 weeks to send a box for you to Hamburg, with books for the library…' The next letter, dated June 18, contains the list of books in the parcel that had already been sent.[347] His subsequent letters recount 'technical' problems in finding the quickest, cheapest, and most convenient method for transporting the packages (through Hamburg or Bremen?).[348] Robinson visited Vienna later that year in October 1827, supplied with a recommendation, probably from Talvj or her father, for the Slovene Slavicist Kopitar, who had been living in Vienna and working in the Austrian Imperial Court Library since 1808.[349] In this letter, Robinson mentions various books that he was seeking. In 1826 and 1827, he spent 'between 400$ & 500$ in the purchase of books for Andover library.'[350]

Subsequently, while in Bonn on his honeymoon, Robinson received a letter from Samuel Farrar (1773–1864), co-founder and treasurer of Andover Seminary, who had been involved with the book purchasing from the start, 'informing he is about to send ER 1000$ for books for Andover library.' Robinson asks to be 'more explicitly informed, whether among these are to be reckoned the books which I have already bought.'[351] In late 1828, Robinson writes to Stuart that 'before leaving Halle I sent off three large boxes of books principally for the Seminary library.- the rest for the lib. of the Rhetorical Society.' A month later, from Paris, he acknowledges receipt of the $1000, announces the shipment of four boxes of books, and adds that 'I have forwarded to Germany enough to pay for all the books yet bought.'[352] While in Paris in the winter of 1828–29, he bought additional books and sent them directly to Boston,

[346] Weigel Williams (2003); (2009).

[347] Robinson to Stuart, Halle, 16.2.1827, 18.6.1827, Weigel Williams (2003), 53, 58–59.

[348] Robinson to Stuart, Halle, 18.12.1827, Weigel Williams (2003), 85–86.

[349] Edward to Elisabeth Robinson, Vienna, 1–13.10.1827, Weigel Williams (2009), 232–33. For Kopitar and his relations with Talvj see chap. 5.b.; NDB Kopitar.

[350] Robinson to Elisabeth Robinson, n.p., 1.6.1828, Weigel Williams (2009), 265.

[351] Robinson to Stuart, Halle, 18.6.1827, 20.5.1828, 5.8.1828, Weigel Williams (2003), 59, 97, 100; Robinson to Elisabeth Robinson, Bonn, 18.7.1828, Weigel Williams (2009), 290.

[352] Robinson to Stuart, Paris, 13.12.1828, Weigel Williams (2003), 111.

'which are just gone off.' Robinson adds his hope that the $1000 was in addition to the $200 he received previously.[353] He also notes that he made an arrangement 'with Prof. Kieffer & Mr H. Edwards, which will enable the library at Andover to obtain readily any books which may be wanted from Paris.'[354] Back in Germany, he again sent some 'promised boxes of books.'[355]

Robinson probably wrote that this sum might be insufficient for his needs, as in early 1839, Farrar writes, reassuring him, that 'if you have the opportunity of purchasing valuable books, if it falls short of the 1000$, or exceeds it,' exploit the opportunity for 'purchasing good books & cheap.'[356] The letter contains many other details, including information on the royalties received from sales of Robinson's 'translation of Christian Abraham Wahl's *Clavis Philologia*,' published in 1825 by Andover's publishers Flagg and Gould. The book 'was to become one of Robinson's most successful contributions to Biblical scholarship.'[357] Robinson's archive also includes a certificate stating that Robinson, 'assistant in the Dept of sacred literature, Theological Seminary, Andover, deposited a book consisting of 850 pages, entitled *A Greek and English Lexicon of the New Testament of Christ*. Abr. Wahl, late senior Pastor of Schneeberg, now superintendent of Oschaz, Saxony' is listed as the copyright holder of this publication.[358]

Robinson continued to purchase books when in Germany before the 1838 expedition and upon his return while living in Berlin until late 1840. Robinson had been appointed librarian at the UTS; hence 'the

[353] Robinson to Stuart, Paris, 30.1.1829, Weigel Williams (2003), 114–15; Edward to Elisabeth Robinson, Paris, 30.1.1829, Weigel Williams (2009), 338.

[354] Robinson to Stuart, Paris, 13.2.1829, Weigel Williams (2003), 116. Unfortunately, Jean-Daniel Kieffer (1767–1833; Fück (1955), 149) died a few years later, on 29.1.1833. Kieffer was professor of the Turkish language at the Collège de France and esteemed teacher and close friend of Robinson (Kieffer (1833)). He was born in Strassburg in 1767; hence German was also his native language.

[355] Robinson to Stuart, Halle, 10.8.1829, Weigel Williams (2003), 126.

[356] Farrar to Robinson, Andover, 14.2.1829, ERP, Weigel Williams (1994), I, F.

[357] Wahl (1825), ERP, HBA, ERP, R65Z5; AL COLL R65 G7 (two copies). See Williams (1999), 108–09 (citations).

[358] ERP, AA-4. Wahl (1773–1855), German Protestant theologian (ADB Wahl). The certificate reads: 'Under the provisions of "An act for the encouragement of Learning, by securing the copies of maps, charts and books, to the Authors and Proprietors of such copies...," passed on the 31st day of March, A.D. 1790, and of "An act supplementary...", passed on the 29th of April, A.D. 1802,' signed by the secretary of the relevant US government department. See, e.g., John Quincy Adams' certificate for Robinson's 1822 translation of *Homeri Iliadis Libri*, ERP, Weigel Williams (1994), I, 3 K 3.

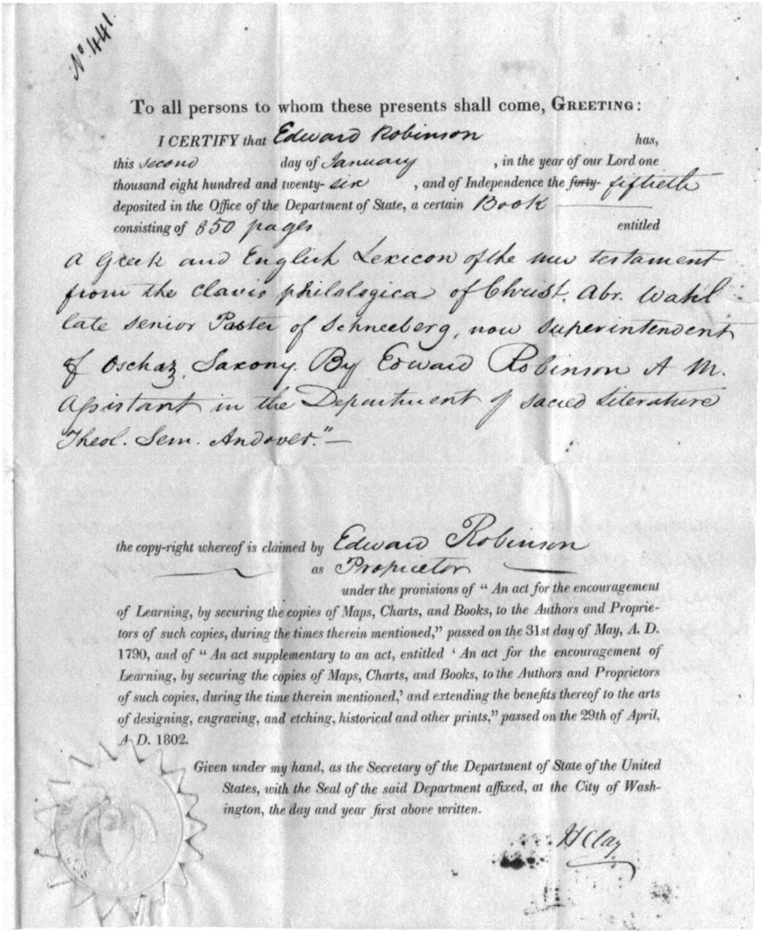

No. 441.

To all persons to whom these presents shall come, GREETING:

I CERTIFY that Edward Robinson *has, this* second *day of* January *, in the year of our Lord one thousand eight hundred and twenty-* six *, and of Independence the* ~~*forty-*~~ fiftieth *deposited in the Office of the Department of State, a certain* Book *consisting of* 850 pages *entitled*

"A Greek and English Lexicon of the new testament from the Clavis philologica of Christ. Abr. Wahl: late senior Pastor of Schneeberg, now Superintendent of Oschaz, Saxony. By Edward Robinson A. M. Assistant in the Department of Sacred Literature Theol. Sem. Andover." —

the copy-right whereof is claimed by Edward Robinson *as* Proprietor

under the provisions of "An act for the encouragement of Learning, by securing the copies of Maps, Charts, and Books, to the Authors and Proprietors of such copies, during the times therein mentioned," passed on the 31st day of May, A. D. 1790, and of "An act supplementary to an act, entitled 'An act for the encouragement of Learning, by securing the copies of Maps, Charts, and Books, to the Authors and Proprietors of such copies, during the time therein mentioned,' and extending the benefits thereof to the arts of designing, engraving, and etching, historical and other prints," passed on the 29th of April, A. D. 1802.

Given under my hand, as the Secretary of the Department of State of the United States, with the Seal of the said Department affixed, at the City of Washington, the day and year first above written.

H Clay

Ill. 23. Certificate stating that Edward Robinson, assistant in the Dept. of Sacred Literature, Theological Seminary, Andover, deposited a book consisting of 850 pages, entitled *A Greek and English Lexicon of the New Testament of Christ.* Abr. Wahl, late senior Pastor of Schneeberg, now superintendent of Oschaz, Saxony, the copyright claimed by E. R. as proprietor. Signed and sealed by H[enry] Clay. This is E. R.'s English translation of Wahl's book, 1822, 3 April.

purchase of books for our library' was one of his central missions. 'We have information of a theological library in Germany of 13000 volumes, which is for sale, & we hope to obtain it,' he writes to Norton before embarking for Europe.[359] Various letters in his archive reveal the long

[359] Robinson to Norton, New York, 1.7.1837, ERP, ER 1.12.

negotiations that began much earlier, in 1837, for the purchase of the Rosenmüller library for the UTS. Both father and son, Johann Georg (1736–1815) and Ernst Friedrich Karl (1768–1835), were Orientalists and theologians who lived and taught in Leipzig. The father, Protestant German Bible exegete whose Bible commentaries and Arab lexical studies became significant scholarly achievements, was also editor of the catalogue for the renowned traveller Ulrich Jasper Seetzen's Oriental manuscripts.[360]

Robinson tried to acquire the library through the mediation of the Leipzig publishing house J. A. Barth, named for its founder Johann Ambrosius Barth (1760–1813) and managed by his son Wilhelm Ambrosius (1790–1851).[361] The negotiations began after Robinson's return to Berlin from the East. 'You were so good as to say, you would give me the price of the Library of Dr Rosenmüller, before I left Leipzig; – I did not however receive it,' he writes to Barth in November 1838, again announcing his interest in purchasing it.[362] Barth's answer, which we do not have, greatly disappointed Robinson. 'When you had the goodness to offer me the library of Rosenmüller, while in Leipzig, I certainly understood you to give me the absolute refusal of it until about the 1st of March next,' Robinson protests. Barth's response left insufficient time for Robinson to receive the decision of his New York contacts. Moreover, Robinson apparently understood from Barth's letter that a different party was currently negotiating for purchase of the library, as Robinson expresses his understanding of the situation this way, 'that should the university or the government wish in the meantime to purchase the Library, you would feel yourself under obligation, to sell it to them;' hence, continuing the negotiations was currently impossible. However, he adds, should the library 'remain unsold, I do not say that I may not wish to purchase it at a future time.'[363] Smith, who was in Leipzig, checked in March 1839, and found that 'Rosenmüller's library is not yet sold.'[364]

360 Rosenmüller: ADB Rosenmüller; Seetzen (1810a). Cf. for Seetzen, his journeys, collections and publications: Kruse (1854), I, III–XXXIX; Ben-Arieh (1972); Schienerl (2000); Goren (2011), 139–43; Henze (2011), 5, 124–29; Weigel (2014), 34–39.

361 Meiner (1930); ADB Barth.

362 Robinson to Barth, Berlin, 23.11.1838, ERP, ER 1.6. Someone, probably Barth, added in the bottom righthand corner of the letter: 'Reisender Palästina.'

363 Robinson to Barth, Berlin, 13.12.1838, ERP, ER 1.11.

364 Smith to Robinson, Leipzig, 11.3.1839, ESP, ABC 60 (105/10) 1839 Jan.-Mar.

This was not the end of the story. A year later, in March 1840 and still in Berlin, Robinson sent another letter to J. A. Barth, writing to the effect that you now have the library 'in your hands.' The rumor that it had been sold to the University of Leipzig was apparently incorrect, as Robinson adds, 'when I had the pleasure of seeing my countryman Mr Appleton in September, I was informed that the library had been offered to him, & that he had looked at it with a view of purchasing.' Robinson asks directly, as was his style, 'whether the library is still in your hands & for sale? & whether, if I should be disposed to take it, I could have it at once, without having to wait the pleasure of a third party?' Robinson was keen for an answer, 'an early reply will never oblige.'[365] The purchase never transpired and 2,500 volumes from Rosenmüller's library reached the University of Leipzig library, probably those dealing with theology and Oriental sciences.[366] Smith's archive includes a handwritten document, 'Titles of Arabic Books printed since 1810. Found in E.F.C. Rosenmüller's Library,' which he could have prepared while in Leipzig.[367] 'An abstract of the more important geographical illustrations of Messrs. Robinson and Smith' was added to an 1841 Edinburgh edition of Rosenmüller's *Biblical Geography of Asia Minor, Phoenicia, and Arabia*, translated into English.[368]

In contrast with this unsuccessful effort, Williams describes Robinson's role in the fruitful purchase of 'several thousand books' from the collection of the Catholic theologian Leander (Johann Heinrich) van Eß (1772–1847), whom he visited in Darmstadt in September 1837.[369] 'Van Ess,' writes Williams, 'was, to say the least, an extraordinary character,' a monk who took upon himself the translation and printing of Bibles in the vernacular German for lay people. He became disillusioned and disappointed in the face of strong opposition and 'eventually moved to Darmstadt, a sick and somewhat bitter man.' Over the years, he built a vast collection from libraries of closing monasteries in Protes-

[365] Robinson to Barth, 6.3.1840, ERP, ER 1.10. Daniel Appleton (1785–1849) began importing and selling books in the early 1830s, establishing the New York City firm D. Appleton & Co., specializing in travel literature. His eldest son, William Henry Appleton (1814–99), joined his father as a partner in 1838; see DAB Appleton, D.; DAB, Appleton, W. H.; Wheaton (1889). It is not clear from the sources whether father or son went to Germany.

[366] https://www.ub.uni-leipzig.de/ueber-uns/geschichte/zweite-periode-1833-1932.

[367] ESP, ABC 60 (118), Lists of Books; see also Seetzen (1810a).

[368] Rosenmüller (1841).

[369] ADB Eß; Williams (1999), 130, 209–11; Prentiss (1889), 74–75.

tant Germany, and then in 1829, decided to sell everything. However, it took nine more years before the books were shipped to the UTS in New York, creating 'the first genuine research library for theology in the United States.'[370] Robinson's role in the acquisition, that took place under the harsh conditions of the 1837 monetary crisis and national panic, is described by George Lewis Prentiss in his history of the UTS: 'Dr Robinson, on his departure for Europe and the Holy Land, in July, was instructed to obtain the refusal of the collection. After a careful examination of the books by [*sic*] Mr Phillip Wolff, of Erlangen University, (...) the purchase was effected, in April 1838.'[371]

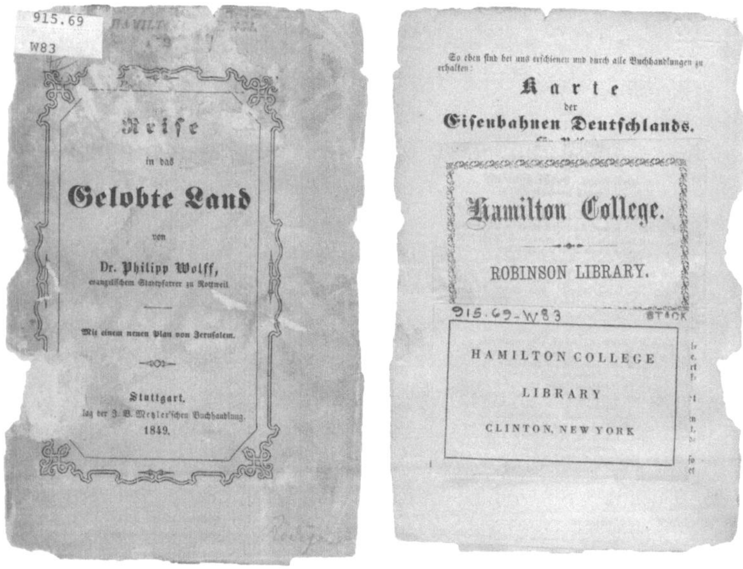

Ill. 24. Philipp Wolff, *Reise in das Gelobte Land*, Stuttgart: J. B. Wetzler, 1849 [cover, Robinson library ex libris; Hamilton library].

Robinson went to Darmstadt to examine the collection.[372] He was travelling with his family on their way to Halle, leaving little time for a thorough inspection. The prudent Americans sent, therefore, another expert. They chose Philipp Wolff (1810–94), student of Gesenius and Rödiger in Halle and Göttingen, who later became a leading German

[370] Williams (1999), 209.

[371] Prentiss (1889), 75 (citing an article in *The Evening Post*, 8.5.1888).

[372] For Robinson's role in the acquisition of this library see also Gatch (1996), 9–10.

Holy Land scholar.[373] The young Wolff, whose sister lived in New York, had been a *Privatdozent* at the University of Tübingen, not Erlangen as the Americans wrote, and had just left his position due to financial difficulties to become a village priest in Rottweil, Württemberg. It seems that Wolff recommended and the UTS did purchase the 16,000-volume library:[374]

> Contains complete editions of the Fathers – extensive and important works on Church History – Ecclesiastical Law – the Acts of the Councils of the Church – Didactic and Polemic Theology, &c. (...) A most valuable collection of Bibles in more than thirty languages. Also Commentaries, Concordances, Lexicons, Polyglotts and Harmonies, as well as a large variety of other works needful to the student of the Scriptures, and interesting to the general scholar.[375]

In the collection, contemporary scholars could find all requisite topics for a serious study of Church history and the Scriptures.

Fellow theologians encouraged Robinson in his efforts to enrich America's theological and biblical literature libraries. Luther Halsey (1794–1880), professor in the fields of natural philosophy, natural history, theology, and Church History and Polity (Princeton 1824–29, Western Theological Seminary in Allegheny, PA 1829–37, Theological Seminary in Auburn, NY 1837–44, and Union Theological Seminary), discusses the issue in a letter to Smith:

> In this country, our Theological Seminary are the only places where biblical literature can be cherished, we feel therefore peculiarly anxious to avail ourselves of the best method of securing a predicious & extensive collection of Biblical apparatus. Moreover as they are the Missionary schools, it is important that we should have such books in oriental languages, as this Library is likely to supply.[376]

Though extremely busy with the last stages of writing the *Biblical Researches* and dealing with the maps and publishers, Robinson did not neglect his book purchasing efforts. As noted above, Robinson requested

373 Hecklinger (1981); Goren (2003), 243–53.

374 Gatch (1996).

375 Gatch (1996), 15.

376 Halsey to Smith, Auburn, 11.1.1840, ESP, ABC 60 (43).

specific special editions of Holy Land literature from Hengstenberg.[377] This letter reveals Robinson's ingenuity in locating and purchasing desired volumes. 'I have watched all the auctions since I have been in Germany,' he writes in August 1840, realizing that he would be leaving Germany in a few short weeks. Robinson was persistent until the day of his departure. In September 1840, on his way to the US (Robinson completed his duties in Berlin on September 1, 1840, and five days later the family left for Hamburg through Halle), Robinson sent a letter from Hamburg to Johann Friedrich Lippert (1795–1878), bookseller and und auctioneer in Halle, who, it appears, used to purchase books at auction for Robinson, this time requesting books from the collection of the deceased Orientalist Peter von Bohlen (1796–1840) from Königsberg.[378]

Upon returning from Germany to New York in late 1840, Robinson 'brought home three or four thousand more [books].'[379] These marked a significant contribution to the existing UTS library, then holding 14,000 volumes, including the books purchased from the library of van Eß. It seems that Robinson kept many of them in his house, as it was only after his death that the collection reached the Burke Library at Hamilton College.[380]

It is not clear why Robinson's volumes were not donated to the UTS, but rather 'were acquired by his alma mater, Hamilton College.' Other documents clearly record Robinson's 'original gift in 1863' to Hamilton College, probably by his widow, which 'consisted of 1420 volumes and 100 maps'[381] scattered among different departments of the Burke Library, according to regular cataloguing rules. Only about seventy years later, 'on the occasion of the [1933] exhibit of books from Robinson's Library,' Joseph Darling ('Bib') Ibbotson, Hamilton College librarian from 1911–36, noted that 'our entire Palestine collection numbers some 1200 volumes,' and that 'a few had been added since the collection came to us.' An effort was made to relocate the volumes originally belonging

377 Robinson to Hengstenberg, Berlin, 24.8.1840, SBB-PK, NH.

378 Robinson to Lippert, Hamburg, 18.9.1840, ERP, 1816 Robinson- 1, ER 1.11; ADB Bohlen.

379 Robinson to Ritter, New York, 21.12.1840, SBB-PK, Nachl. 133, K 5, p. 32; Robinson to Lippert, New York, 15.7.1841, SBB-PK, Slg. Darm. Asien, 5; Robinson to Lippert, 18.10.1840, ERP, 1816 Robinson- 1, ER 1.10.

380 Ibbotson (1933), also: 13 leaves, Typewritten, 1933, ERP, AL COLL I2 E3 1933; idem, 'Edward Robinson and his Books,' print, signed 12.7.1952, ERP.

381 Weigel Williams (2007), 214; J. Malkolm Allen to current librarian, Burke Library, Hamilton College, Stirling, Scotland, 21.9.1985, ERP, 1816 Robinson- 1. Allan (born 1935) moved from Hamilton to Scotland.

to this collection that had been scattered in different areas of the library: 'we have located items from the collection as follows: Rare Book Room – approx. 700 volumes at 915.69 Compact Storage.'[382]

After locating the majority of Robinson's library, nothing more was done for the next three and a half decades. The next step came in 1969, when one of the reference librarians sent a letter to Walter Pilkington (?-1983, college librarian 1952–76), describing the current situation of Robinson's collection. She tried to use the library copy of Röhricht's *Bibliotheca* containing Ibbotson's handwritten comments to determine 'whether it can be taken as a complete guide to the collection.' The negative answer caused them to renew the search, locating Robinson's books primarily in the Rare Books Collection, but also in other departments. 'The space situation in the Rare Books Room is becoming desperate,' so it was suggested to transfer Robinson's books, with their right slips, to the compact storage.[383]

Subsequently, in 1985, J. Malcolm Allan, Hamilton Library reference librarian from 1962 to 1966, wrote to the current librarian from Scotland concerning Robinson's library. Allan was writing 'the history of Hamilton College Library 1793–1963,' consulting 'Bib' Ibbotson as well as his acquaintance, Jordan archaeologist and historian Rev. John Bartlett of Trinity College, Dublin. Allan asked for a copy of Ibbotson's paper and for 'any current information about the Robinson collection.'[384] The librarian answered quite quickly, recalling the history and the current situation in which 'the Robinson collection does not form a unit but rather had been scattered among the stack books many years ago.' Consequently, books that lost their book-plates 'and others [that] had been withdrawn' were difficult to identify. The books that could be identified were transferred – again – to the Rare Book Room 'but no doubt many are now gone.' An interesting thread of information included in the letter is that 'Jay Williams had for a time been interested in Robinson,' but 'could not get a "handle"' on him, as 'there are apparently no personal papers of consequence to be found.'[385] It took nearly another decade for

[382] Ibbotson (1933); HBA, ERP, AL COLL I2 E3 1933 (draft with handwritten corrections and additions); cf. for Ibbotson chap. 4.d.

[383] VH [Vivien M. Hart, A. L. A. (Britain), Reference Librarian] to Pilkington, Hamilton College, 18.1.1969, ERP, 1816 Robinson- 1. The copy in the library: RB Z 3476.R71 1890.

[384] J. Malkolm Allen to current librarian, Burke Library, Hamilton College, Stirling, Scotland, 21.9.1985, ERP, 1816 Robinson- 1.

[385] Frank Lorenz to Allan, Hamilton, 2.10.1985, ERP, 1816 Robinson- 1.

Robinson's archive to find its way to the college and into the custody of the Williams.' Williams' list of 'a partial bibliography' of 'books about Palestinian travel and geography drawn largely from Robinson's own library,' includes 73 items.[386] Generally, it appears that during the second decennia of the twenty-first century the books were still scattered throughout the library, and reference librarians continued the efforts to locate as many books as possible from the original collection.

Robinson had a second purchasing goal beyond that of books for 'his' institutional and private libraries. He needed types of letters and symbols in various Oriental languages that were difficult to find in the US. His first order, probably only for Arabic letters, was placed in late 1827 with Breitkopf & Härtel, the world's oldest music publishing house (Leipzig, founded 1719). The order noted that 'the former is just dead.'[387] The package was paid for in late December 1827 and ready for delivery and included 'the fonts of small Hebrew.'[388] Before returning from Europe to Andover, Robinson ordered types of letters for languages including Hebrew, Coptic, Ethiopian, Samaritan, and Armenian. Delivery was not without its problems. Initially, the company wrote that 'they had made a sad mistake in respect of the small Hebrew type' and it would cost him twice what he expected.[389] The types were sent, and the rather unlucky Robinson, who was still in Germany, wrote to Breitkopf & Härtel that he heard from home that a box of his ordered types, after reaching New York safely, 'was totally lost on the passage between N. York & Boston,' probably on its way to Andover. This was the second box, as the first, including 'the smallest Hebrew [types],' did reach its destination. Robinson requested replacements, 'to make them all according to my former order,' and since 'the money will be ready for you probably when the type is delivered,' to inform him of the delivery date by a poste-restante letter.[390] Needless to say, Robinson made much use of the types in his books published in Andover and in later years as well.

386 Williams (1999), Appendix, 341–48.

387 Robinson to Stuart, Halle, ?.?.1827, Weigel Williams (2003), 67. Gottfried Christoph Härtel, born 1763, died on 25.07.1827: ADB Breitkopf und Härtel.

388 Robinson to Stuart, Leipzig, 19.12.1827, 18.6.1827, Weigel Williams (2003), 86.

389 Robinson to Stuart, Berlin, 5.1.1828, Weigel Williams (2003), 88.

390 Robinson to Messrs Breitkopf, Härtel and Co., Leipzig, Dresden, 15.6.1828, ERP, ER 1.6; Weigel Williams (2009), 242.

CHAPTER 5
CO-TRAVELLERS AND COMPANIONS

5.a. Eli Smith, the underacknowledged partner

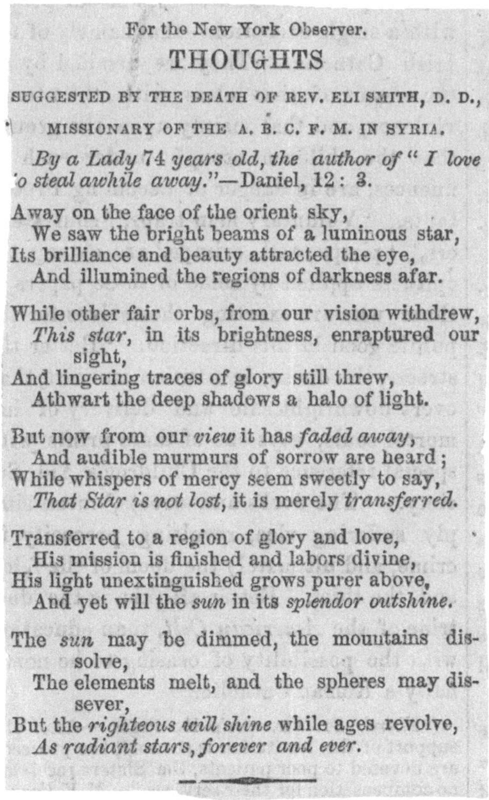

For the New York Observer.

THOUGHTS

SUGGESTED BY THE DEATH OF REV. ELI SMITH, D. D.,

MISSIONARY OF THE A. B. C. F. M. IN SYRIA.

By a Lady 74 years old, the author of "I love 'o steal awhile away."—Daniel, 12 : 3.

Away on the face of the orient sky,
We saw the bright beams of a luminous star,
Its brilliance and beauty attracted the eye,
And illumined the regions of darkness afar.

While other fair orbs, from our vision withdrew,
This star, in its brightness, enraptured our sight,
And lingering traces of glory still threw,
Athwart the deep shadows a halo of light.

But now from our *view* it has *faded away*,
And audible murmurs of sorrow are heard;
While whispers of mercy seem sweetly to say,
That Star is not lost, it is merely *transferred*.

Transferred to a region of glory and love,
His mission is finished and labors divine,
His light unextinguished grows purer above,
And yet will the *sun* in its *splendor outshine*.

The *sun* may be dimmed, the mountains dissolve,
The elements melt, and the spheres may dissever,
But the *righteous will shine* while ages revolve,
As radiant stars, forever and ever.

Ill. 25. 'Thoughts suggested by the death of Rev. Eli Smith, D.D.,' for the New York Observer, 1857. ESP, ABC 60 (139).

Transferred to a region of glory and love,
His mission is finished and labors divine,
His light unextinguished grows purer above,
And yet will the *sun* in its *splendor outshine*.

'Thoughts suggested by the death of Rev. Eli Smith, D.D.,' for the *New York Observer*, ERP, ABC 60 (169).[1]

> It is interesting that throughout these volumes which chronicle his [Robinson's] adventure, very little is said about his companion and co-author Eli Smith. He says nothing about the fact that Smith was recently bereaved, having lost his wife of only a few years to death. Neither does he spend much space describing what Smith himself may have discovered or opined. In fact, he devotes more space describing the bedouin guides than Smith's contribution to the expedition. For the most part, his eyes are on the landscape, not upon their very important and productive relationship.[2]

With these words, Jay Gomer Williams accurately summarizes a pronounced deficiency in Robinson's publications. Though the *Biblical Researches* bears both their names, it is obvious that Smith's contribution is, to a certain degree, undervalued and even obscured. Robinson's impersonal approach, as demonstrated elsewhere in this book ('James Adger, the anonymous fellow traveler,' 'Therese Albertine Luise von Jakob-Robinson, "Mediator of the Balkan Slavs," and Holy-Land Studies'), probably reached its peak with the relative neglect of Smith, his background, and his role in the expedition and in writing the *Biblical Researches.*

This assumption, bordering on accusation, merits closer scrutiny. A careful examination of the preface written by Robinson reveals that it does hint at Smith's role. When discussing his approach to presenting the expedition findings, Robinson writes that 'I have therefore everywhere interwoven the personal narrative; (...) the manner in which the Promised Land exhibited itself to <u>our</u> eyes, and the process by which <u>we</u> [emphasis added] were led to the conclusion...'[3] Throughout the book, he describes experiences, ideas, and findings using the plural. In what

[1] The poem was written by Phoebe Hinsdale Brown (1783–1861) in New Haven, Connecticut in 1818. Inspired by Daniel 12:3 and originally titled 'An Apology for my Twilight Rambles, Addressed to a Lady,' the poem became a Christian hymn. Brown, also the author of 'I love to steal awhile away,' was 74 in 1857, the year of Smith's death.

[2] Williams (1999), 221.

[3] Robinson (1841), I, iii.

is, perhaps, a degree of underestimation, Robinson refers to 'my companion, Mr Smith,' who 'had already made some preparation for our journey in obtaining the names of places in many of the provinces and districts, written by educated natives,' without noting exactly who these 'educated natives' were.[4] Robinson concludes by citing the contribution of 'my friend' who took 'pains to exhibit (...) the principles which govern the pronunciation of the spoken Arabic at the present day,' and the fact that Smith 'looked through' and corrected the entire manuscript.[5]

In the beginning of the *Biblical Researches*, Robinson recounts the fact that the idea for the expedition arose and was agreed upon in their 1832 meeting in Andover. Most importantly, he devotes nearly half a page to his appreciation of Smith:

> I count myself fortunate in having been thus early assured of the company of one, who, by his familiar and accurate knowledge of the Arabic language, by his acquaintance with the people of Syria, and by the experience gained in former extensive journeys, was so well qualified to alleviate the difficulties and overcome the obstacles which usually accompany oriental travel. Indeed, to these qualifications of my companion, combined with his taste for geographical and historical researches, and his tact in eliciting and sifting the information to be obtained from an Arab population, are mainly to be ascribed for more important and interesting results of our journey.[6]

Here, Robinson emphasizes Smith's knowledge and experience over 'an ordinary uneducated interpreter,' with whom he would have undertaken much less than what they actually accomplished. He acknowledges that 'many points of interest would have been overlooked, and many inquiries would have remained without satisfactory answers.'[7]

From these passages, it may appear that Robinson did dedicate a respectable amount of text to Smith, his unique virtues, and irreplaceable contribution. However, Williams was actually rather delicate in choosing his words. Smith's contribution to the expedition and the publication should be presented in its entirety, illuminating background information necessary to fully understand the expedition's success. Robinson acknowledges this point clearly as cited above: Without Smith, he might

4 Robinson (1841), I, ix.

5 Robinson (1841), I, x–xi, xiv.

6 Robinson (1841), I, 2; Robinson (1839), 400. See also Bliss (1907), 197.

7 Robinson (1841), I, 2; Robinson (1839), 400.

have embarked on the expedition, but its scope would have been vastly limited. Robinson might also have published a travelogue, but would never have been able to produce such a highly important scientific treatise, bringing him international stature. Smith's expertise and scholarship were crucial to the planning, preparation, and execution of the expedition. His expertise included exceptional language fluency,[8] in European languages as well as in Arabic and its different dialects, and acquaintance with local leaders, authorities, and manners and customs in the different regions they travelled. Smith contributed significantly to the tremendous success of the expedition and its outstanding results and, consequently, to Robinson's growing reputation.

5.a.1. From influential missionary to America's first Arabist

Smith's reputation derived from being one of the most influential and key ABCFM missionaries in the Near East. 'Eli Smith became America's first Arabist' is only one of the titles bestowed upon him.[9] Smith is featured in every study devoted to ABCFM activity in the East, from the establishment of the Beirut printing house to the translation and distribution of the Bible into Arabic, the idea for which originated in a missionary conference in Beirut in 1834.[10] In the early years, Smith dedicated himself to learning Arabic, and was known in some cases to contravene ABCFM board plans and instructions or simply not to report to his superiors.[11] His influence is also evident in the hundreds of letters kept in his well-organized archive, current to his death in 1857, at the age of 56. His extensive correspondence with perhaps all ABCFM missionaries in the Near East and with ABCFM secretary Rufus Anderson, located in the Boston headquarters, deals with missionary issues, financial and other obstacles, campaigns and, of course, accomplishments.[12]

His employers quickly detected his unique virtues and abilities. It was Smith who accompanied Anderson in 1829 on his first tour of the ABCFM stations in the Mediterranean and the Near East. Responsi-

8 For a list of Smith's languages: Leavy (1993), 7.

9 Kaplan (1995), 22–27.

10 e.g. DAB Smith (The entries for Smith, Robinson, and Moses Stuart were written by leading Holy Land archaeologist William Foxwell Albright (1891–1971)); Smith (1900); Phillips (1939); Kawerau (1958); Roper (1998); Makdisi (2008); Auji (2016); Ayalon (2016).

11 Leavy (1993), 7–9, also cites Smith's letter from the ABCFM Archive.

12 ESP. Fortunately, Smith kept almost every letter he received, as well as a copy of the letters he sent.

ble for ABCFM foreign activity, Anderson was sent to resolve the crisis sparked by the 1828 abandoning of the station in Beirut.[13] Smith's archive contains a thick booklet, probably never published, titled 'Notes on Greece taken during a journey in that country in 1829,' dedicated to a careful and detailed examination of the potential for missionary activity in different localities in Greece.[14] In 1830–1, he travelled with fellow missionary Dwight to Armenia, resulting in a publication of their missionary travels, *Including a Journey through Asia Minor and into Georgia and Persia*.[15]

Amidst intensive activity, Smith found the time to write, publishing papers in the periodical *Bibliotheca Sacra and Theological Review* (*BSTR*), as well as a summary of the history of the translation of the Bible.[16] Smith's superior, even adventurous biography and his missionary achievements are the focus of contemporary as well as later studies.[17] His publications following the Holy Land expedition include papers, perhaps better termed 'letters,' in which he discusses various aspects of the Holy Land. The letters were published in *The Missionary Herald*, the organ of the ABCFM, between 1838 and 1839.[18] One particularly outstanding letter describes the Bedouin tribes in the Sinai and Jordan Valley, and discusses the potential for establishing missionary stations to facilitate their conversion to Christianity. Smith wrote the text in Vienna and Leipzig, probably as a report to Anderson in Boston.[19] In 1840, he published his personal missionary 'creed,' on a missionary's duty and life, originally prepared 'for the Society of Inquiry at the Theological Seminary at Andover.'[20]

[13] Phillips (1939), 142–43; Kawerau (1958), 211–14; Anderson (1963), 352–54;.

[14] Eli Smith, Notes on Greece taken during a journey in that country in 1829, ESP, ABC 17.3, vol. 1; cf. Anderson (1830).

[15] Smith and Dwight (1833). Cf. for the outstanding value of the book, Dwight to Smith, Pera, 12.7.1841, ESP, ABC 60 (19): 'I know of no other work in the English language that can in any measure supply the place of your book.'

[16] Smith (1846); (1846a); (1900, several chapters).

[17] 'Dr Goodell's Memoir,' Constantinople(?), n.d., ESP, ABC 60 (139); Antonius (1965), 36–54; Leavy (1992); (1993); Glass (1997). For Goodell as leading Armenian missionary: Leavy (1993), 9–11.

[18] See the list in Williams (1999), 373.

[19] Smith (1839); Smith to Anderson, Syra, 27.8.1838, ESP, ABC 60 (105/9) 1838 Jul.-Dec. In ESP the text is recorded as a letter to brother missionary Henry Augustus Homes (1812–87), who served in Constantinople from 1835 to 1850: ESP, ABC 60 (105/5), 1838.

[20] Smith (1840), cover.

Smith was constantly 'possessed by the mission,' never missing even the slightest opportunity to carry out his central obligation, as attested to by his many reports and letters.[21] Immediately upon arriving in Egypt, Smith sought out his fellow missionaries, and sent Anderson a long report on the activities of the CMS missionaries in Cairo. The report focuses on missionary schools for girls and their relations with the Pasha, Muhammed 'Ali, who wanted to replicate their success.[22] When travelling with Robinson, 'he hoped to be able to perform (...) a good deal of the labor appropriate to the traveling missionary.'[23] The meeting of the ABCFM missionaries they attended in Jerusalem during Easter 1838, was clearly a highlight of the expedition about which Smith expresses his delight to Anderson, 'I had the pleasure of meeting all my brothers.'[24] As noted, Smith continually discussed the possibility of missionary work among the Bedouin tribes, fully aware of the challenge presented by their Muslim faith.

Smith also never neglected his central (though not formal) position among the Near Eastern missionaries, a fact that emerges strongly from many, if not most, of his letters. For example, while in Leipzig, he tried to convince Isaac Bird, who came to Beirut in 1823 but returned to the US in 1836, to return and 'join the forces' in Lebanon. Some months later, he wrote to missionary Whiting that he regretted pressing Bird.[25] Smith's musings demonstrate that even during his travels to Leipzig for the types, he prioritized looking for 'the missionary character [which] is not in very high,' even describing his host, Tauchnitz, as 'a devotedly pious man, & a firm patron of mission.'[26]

From the early 1830s, Smith was recognized as a leading expert on the Near East, a reputation that grew over the years. His advice on local populations, officials, language, history and more was sought by European and American officials planning to settle in the region, by re-

[21] e.g. Smith to Anderson, Cairo, 10.3.1838, ESP, ABC 60 (105/5) 1838; Thomson to Smith, Beirut, 5.8.1838, ESP, ABC 60 (83); Smith to Anderson, Smyrna, 9.8.1838, ESP., ABC 60, (105/9) 1838 Jul.-Dec.

[22] Smith to Anderson, Cairo, 10.3.1838, ESP, ABC 60 (105/8) 1838 Jan.-June.

[23] Recent Intelligence (1838).

[24] Smith to Anderson, Smyrna. 9.8.1838, ESP, ABC 60 (105/9).

[25] Smith to Bird, Leipzig, 9.12.1838, ESP, ABC 60 (105/5) 1838; Smith to Whiting, Leipzig, 17.4.1839, ESP, ABC 60 (105/5) 1839.

[26] Smith to Hallock, Leipzig, 12.4.1839, ESP, ABC 60 (105/11) 1839 Apr.-Dec.; Smith to Hallock, 16(?).12.1838, ESP, ABC 60 (105/9) 1838 Jul.-Dec. For Tauchnitz see chap. 3.c.

nowned travellers, and by noted scholars alike. One of his early tasks was integrating John William Perry Farren (?-1864) into his new position as British consul general for Syria, whose seat was to be in Damascus. Farren's nomination had already been reported by the French consul on December 15, 1830, but his arrival in Beirut was postponed for six months, coinciding with the beginning of the first Egyptian-Ottoman crisis. Strong local opposition in Damascus, followed by threats to assassinate him if he arrived in the Syrian capital, forced Farren to remain temporarily in Beirut, settling in Damascus only in January 1834.[27] When Farren sought out an expert, Anderson was only too happy to recommend Smith, who answered in detail the consul's 'inquiries respecting the regions north and south of Damascus.'[28]

A fine example from travellers seeking Smith's advice is a letter received in spring 1840 from John Gadsby (1809–93), writing that he planned:

> to set out the coming autumn on a tour to Egypt & Syria, and should doubtless derive much advantage in view of such an undertaking from a conference with one so familiar as yourself with the state of things in Syria; and hoped to have seen you at this time [Smith in England on his way from Leipzig to the US]. But as this expectation has been disappointed, and I may not meet you anywhere in person, I should like to propose a few questions, and should ask an answer in writing, in case your multiple engagements will permit. The time allotted in my plan for travel in Egypt, to Mt. Sinai and Syria, is three, four or five months. Two things are to be consulted – time and economy.[29]

Gadsby, a London printer, publisher, traveller and collector, travelled frequently in the East, the first time in 1846–47, and published his two-volume travelogue in 1855.[30]

In late 1840, Smith received a curious request from John Maclean Jr. (1800–86), D. D., Professor of Ancient Languages at the College of New Jersey (later Princeton University), seeking information about wine growing in Palestine and Lebanon, specifically 'whether the wines (...)

[27] Whiting to Bird, Beirut, 6.12.1830, IBP-YSM, Box 1; Tresse (1936). Cf. Williams, J. B. (1972), 424–27; Goren (2011), 86.

[28] Farren to Anderson, Beyroot, 9.5.1834, ESP, ABC 60 (105) 1834; Smith to Farren, Beyroot, 9.5.1834, ESP, ABC 60 (105) 1834.

[29] Gadsby to Smith, Chartentown, 10.4.1840, ESP, ABD 60 (31).

[30] Dawson, W. R. (1995), 59; Gadsby (1855).

are unfermented and incapable of producing intoxication?' Maclean needed the answer for his rather long review of two essays, titled *Bacchus* and *Anti-Bacchus*, published in the *Princeton Review* of 1841. Without entering into a detailed explanation, Maclean adds, 'It is stated by the author of Anti Bacchus on the authority of Mr Buckingham, that the wines of Lebanon are usually boiled.'[31] A copy of Smith's response is not among his correspondence but Maclean quoted it in full in his review. In 1845, Smith wrote a paper titled 'The Wines of Mount Lebanon,' published in the *BSTR* the following year.[32] These are but a few examples of an impressive number of letters among Smith's correspondence.

Scholars and travellers alike admired Smith's control of the local languages, 'to the study of the Arabic, he applied himself with all the earnestness of his nature, and his flexible organs soon made the hardest gutturals seem so soft and sweet, that it was a pleasure to listen to him.' In his letter eulogizing Smith, Constantinople ABCFM missionary William Goodell (1792–1867) writes:

> The Arabic, too, with all its idioms must have been to him like his own vernacular tongue; and this itself being not so much a cognate as an effect of the Hebrew – in fact its very offspring. No man ever enjoyed greater advantages than he possessed, for expressing in a translation, the full force and meaning of the Hebrew text – such an exact expression as no other translation can reasonably claim.[33]

The 'List of Arabic Names' is perhaps Smith's central academic contribution to the expedition and publication. Smith, whose fluency in Arabic was praised as unprecedented by his fellow missionaries,[34] took on the standardization of Arabic pronunciation and transliteration as a personal mission, working continuously from his early missionary period at the Malta printing house through his post in Beirut. As noted in Chapter 4.b., Robinson's focus on Holy Land topography illuminates Smith's major contribution to the research. The orthography of Arabic names, 'both in Arabic and Roman letters,' a 'point which we gave particular

[31] John Maclean to Smith, 6.11.1840, ESP, ABC 60 (55). Cf. Maclean (1841); Hirrel (1998), 126–31, 209 n. 36.

[32] Smith (1846a).

[33] 'Dr Goodell's Memoir,' Constantinople(?), n.d., ESP, ABC 60 (139).

[34] By way of example, the leading LJS missionary in Jerusalem, John Nicolayson, wrote: 'You are so much better a judge of the Arabic...;' Nicolayson to Smith, Jerusalem, 3.1.1851, ESP, ABC 60 (15).

attention,' is interwoven throughout the study. Most likely an outcome of their 1832 meeting in Andover or at least discussed there, in 1834, Smith started preparing lists of Arabic place names according to Ottoman administrative regions. As mentioned above, he obtained some of the names during his own travels while others were 'written by educated natives (…) according to the established rules of the language.'[35] Uniform orthography for the ABCFM was established by adopting a system proposed in 1818 by John Pickering 'for the Indian languages of North America,' whose principles Smith adopted for his Arabic orthography.[36] Smith's orthography remained the standard for many years. In 1860, when the leading Scottish geographer and cartographer Alexander Keith Johnston (1804–71) published a map of 'Palestine or the Holy land' for his Royal Atlas, he added the following note on the spelling of Arabic names, 'the Orthography followed generally is in accordance with that adopted by Dr E. Smith in "Robinson's Biblical Researches" which is the best in use for the spelling of Arabic names as spoken in Syria.'[37]

Development of Arabic calligraphy and production of the Arabic types, discussed above, are further significant achievements of Smith's career. Without Smith's scholarly abilities and devotion, the types probably would not have been produced. Following his expedition to Armenia, Smith went first to the US, then to Malta where he stayed until December 1833, returning to Beirut (after almost six years) through Egypt in late January 1834. The transferring of the printing house from Malta to Beirut in May 1834 had been one of the most important and influential moves of the ABCFM and its work in the Near East. It facilitated the printing of manifests of nascent Arab national movements written by two brilliant Arab scholars raised by the American mission, Butrus al-Bustānī (1819–83) and Nāṣīf al-Jāziğī (1800–71).[38] Smith's

[35] Robinson's and Smith's archives hold vast material on this subject, see, for example: Lists of the Arabic names of places in Palestine and the adjacent regions, ESP, ABC 60 (119) B; Rules of the orthography of Arabic and Turkish names and words, in Roman letters, adopted by the Syrian Mission of the ABCFM, April 1838, ESP, ABC 60 (116); Smith (1841a): ESP, ABC 60 (119); and as a separate publication in German: Smith (1842). For Smith's handwritten manuscripts: ESP, ABC 60 (120).

[36] See discussion in Chap. 4.b. Robinson, Preface, I, x; Smith (1841); (1841a); (1842). Cf. Phillips (1939), 306 (citation); Pickering (1820); Stevens (1956).

[37] Keith Johnston (1861); ERP, Palestine from Keith Johnston's Royal Atlas; Alexander Keith Johnston to Carl Ritter, 3.6.1848, K 3, J – L, Bl. 8, SBB-PK, Nachl. 133 (Carl Ritter). Cf. Goren (2017), 46–47; Faehndrich (2017), 86; Schelhaas (2017), 110–29. No wonder that the map can be found in Robinson's archive.

[38] Cf. e.g. Fück (1955); Kawerau (1958), 260–66; Antonius (1965), 45–51; Leavy (1993).

archive holds extensive correspondence concerning the move, activation of the printing press and its use. The earliest correspondence dates to Hallock's letters in April 1834 about the 'home project' in Smyrna, which are very long and discuss mostly technical issues and finances.[39] The letters continue until Smith returned to Beirut equipped with types manufactured by Karl Christian Philipp Tauchnitz, the son.[40] 'I have now the pleasure to acquaint you,' writes Tauchnitz, 'that last week your types, and all the other objects, which I had received for you, have been sent to (...) Vienna, who are instructed immediately to forward them on board of a steamer by way of Constantinople to (...) Smyrna.'[41] 'This type still excites much interest,' writes Tauchnitz in a late 1840 letter to Smith in which he wonders about his whereabouts. 'The head Librarian of the royal Library in Dresden, Dr Falkenstein [Konstantin Karl F. (1801–55)], has also very honourably mentioned your type and given the (whole?) history of it in a splendid work which he edited at the Jubilee of the invention of printing.'[42]

Smith's reputation continued to grow and his correspondence contains numerous 'ask the expert' letters dated up until his death. The following notable examples are brought here. Rödiger, who in the late 1840s started dealing 'enthusiastically with Phoenician inscriptions and coins,' wrote to Smith asking whether 'in the area, where you are currently residing, should still some Phoenician monuments to be found.'[43] Schultz, first Prussian envoy to Jerusalem, sought his advice writing, 'allow me to send you with this, short extracts from Wilken's history of the Crusades vol. I & II, concerning geographical & topographical points and questions, which you will perhaps be able to identify or to answer

39 Hallock to Smith, Smyrna, 20.4.1834, 28.4.1834, ESP, ABC 60 (32). See also Leavy (1993), where the letters are used extensively.

40 Prominent examples, in chronological order: Hallock to Smith, Smyrna, 20.4.1834, ESP, ABC 60 (32); Smith to the Brothers of the Syrian Mission, Smyrna, 16.12.1837, ESP, ABC 60 (105/5) 1837; Smith to Anderson, Syra, 31.1.1838, ESP, ABC 60 (105/5) 1838; Smith to Anderson, Smyrna, 9.8.1838, ESP, ABC 60 (105/9) 1838 Jul.-Dec.; Smith to Anderson, Leipzig, 1.11.1838, ESP, ABC 60 (105/9) 1838 Jul.-Dec.; Smith to Hallock, Leipzig, 16.12.1838, ESP, ABC 60 (105/9) 1838 Jul.-Dec.; Smith to Anderson, Leipzig, 9.1.1839, ESP, ABC 60 (105/5) 1839; Smith to Thomson, Leipzig, 20.3.1839, ESP, ABC 60 (105/10) 1839 Jan.-Mar.

41 Tauchnitz to Smith, Leipzig, 4.9.1839, ESP, ABC 60 (80).

42 Tauchnitz to Smith, Leipzig, 16.11.1840, ESP, ABC 60 (81); ADB Falkenstein; Falkenstein (1840).

43 Rödiger to Smith, Halle, 12.10.1848, ESP, ABC 60 (65).

when travelling in the neighbourhood of Antioch, Aleppo and in the Nossairee Mountains.'[44]

The final two archival examples cited here are recommendations made at the highest governmental levels. 'Mr Smith, as you are aware, stands in the first rank of Arabic scholary at the present day, not less in the scientific knowledge of the language & literature, thence in the practical & daily use of it,' writes Robinson in 1846, when introducing Smith to George Marsh, most likely George Perkins Marsh (1801–82), a member of Congress who was preparing for his 1849 appointment as minister resident in the Ottoman Empire.[45] Another interesting recommendation was written in late 1840, when Smith was still in the US, by Chancellor of New York Reuben Hyde Walworth (1788–1867), addressed to 'His Exc. The President of the United States,' Martin Van Buren (1782–1862, president 1837–41):

> You will find him a very intelligent gentleman who has travelled extensively in that country and who will be able to give to the government correct information upon the various questions whatever to that country which are connected with the interest of the United States. (...) you can place the fullest reliance upon any information he communicates either in relation to the situation of that country, & as to individuals with whom he has become acquainted during his residence & travels there.[46]

5.a.2. Smith and the German Orientalists

Smith was surprised, so it seems, by the scholarly recognition he received from leading German scholars, mainly Orientalists, whom he met or at least contacted while in Leipzig following the expedition. Some of these connections lasted for years. 'Many things have made me fear quite like a child, when standing by the side of these giant scholars,' he wrote in spring 1839.[47] Smith probably had not intended, upon coming to Leipzig, to be-

[44] Schultz to Smith, Beirut, 19.4.1848, ESP, ABC 60 (71); Wilken (1807–32) cited often by Robinson for the *Biblical Researches*, cf. chap. 4.c. For Schultz: Goren (2003), 194–201. The mountain range in northwest Syria, named after its inhabitants, the Nusayris, a former name of the Alawites, a Moslem ethnic group.

[45] Robinson to Marsh, New York, 16.2.1846, ERP, ER 2.25; Findling (1989), 339–40; Oren (2007), 169–71 (with the fantastic idea of establishing an American Camel Corps).

[46] Walworth to the President, Albany, 12.11.1840, ESP, ABC 60 (112).

[47] Smith to Hallock, Leipzig, 12.4.1839, ESP, ABC 60 (105/11) 1839 Apr.-Dec.

come so involved with the flourishing German Orientalism centred in this Saxonian city and its neighbouring Halle on the river Saale, of Sachsen-Anhalt.[48] The groundwork had been probably been laid in 1835, when Heinrich Leberecht Fleischer (1801–88) received a teaching position in Leipzig and, ten years later, was able to establish the *Deutsche morgenländische Gesellschaft* (German Oriental Society, DMG) in the city.[49] During Robinson's early stay in Halle and visits to Leipzig between 1826 and 1829, he met some of the local scholars studying Oriental languages, including Emil Rödiger, Wilhelm Gesenius (who was also his host) and August Tholuck. Robinson maintained these connections after he returned to the US, and met with some of these colleagues in late 1837 on his way to the East. Now, Robinson and Smith's return to Germany afforded the ideal opportunity to develop personal and scientific relationships.

Though he came to Leipzig because he believed that Tauchnitz's production plant would be the preferable and most professional for producing his Arabic types, Smith could not have found a better place to enter the community of Oriental scholars. Smith's interest in Arabic printing and letters led him, on his way to Leipzig, to a thorough exploration of the libraries in Vienna, where he could examine Persian and Armenian as well as Arabic types being produced there. This endeavour occupied Smith for the weeks he stayed in the Habsburg capital nursing Robinson before Talvj's arrival, upon which he continued on his way to Leipzig.[50] He writes to fellow missionaries in Beirut:

> I propose to give you as concisely as I can, in a few letters, a summary of the most important information, bearing upon Arabic typography, + Arabic literature, which has, or may yet, come into my possession during my present visit to Germany. I will begin with Arabic typography in Austria. The first attempt to make Arabic types in that country, occurred in 1554.[51]

This long and detailed letter, like so many others written by Smith, presents a unique, learned typographical history of Arabic and other eastern

[48] Wokoeck (2009), mainly 86–145, and there Fleischer's role; Mangold (2004), passim; ADB Fleischer.

[49] ADB Fliescher; Fück (1955), 170–73 and passim; Goren (2003), 174–75; Mangold (2004); Wokoeck (2009), 117–45.

[50] See chap. 3.c.

[51] Smith to Thomson and Hebard, Leipzig, 10.12.1838, ESP, ABC 60 (105/9) 1838 Jul.-Dec.; Smith to Anderson, Leipzig, 1.11.1838, ESP, ABC 60 (105/9) 1838 Jul.-Dec.

languages that surely required in-depth study of rare treasures housed in the Vienna central library. The types produced to date, the 'old works,' were truly deficient, and the famous Austrian Orientalist, historian and diplomat, the Baron Joseph von Hammer-Purgstall (1774–1856), had initiated production of new types. Smith was able to examine Hammer's types, 'began to be made in 1829, it is not yet complete' and provide a detailed description of them.[52]

On his way to Berlin, Robinson went through Halle and wrote to Smith that 'the orientalists here, Gesenius, Roediger, & Tholuck, received me with open arms, & seem truly to rejoice at the results of our journey in Palestine. The two former have shown the most interest.'[53] Smith reached Leipzig in late October and immediately started writing letters, first about missionary issues, but soon after, he began his intensive correspondence with local scholars of the Orient. Though the scholars were clearly familiar with the *Researches in Armenia*, they did not know Smith, and it was probably Robinson who made the initial connections and he did so quite quickly! Rödiger was the first to correspond directly with Smith, as early as November 11, 1838. Rödiger wrote a recension of the *Researches in Armenia*, adding that he hoped to make Smith's acquaintance in Halle and to thank him for the new-Syrian manuscripts delivered through Robinson sometime earlier, which he – Rödiger – already published in the *ZKM*.[54] Robinson showed him both his and Smith's travel journals, and Rödiger stated that they hold 'much new and important' material. He also demonstrated great interest in Smith's initiative to cut 'a new Arabic script' with Tauchnitz. Rödiger expressed his wish that both a larger type for the texts and a smaller one for the notes would be produced, 'which seems to be highly necessary for a more comfortable organization of our print,' most likely referring to the printing of the periodical *ZKM*, which needed types for letters and other printing signs of the various Oriental languages.[55]

Robinson continued to pave Smith's way into the scholarly community, first in Berlin and later in Leipzig and Halle. The Berlin He-

[52] Smith to Anderson, Leipzig, 1.11.1838, ESP, ABC 60 (105/9) 1838 Jul.-Dec. Cf. ADB Hammer-Purgstall; BKLO Hammer-Purgstall; Wokoeck (2009), 84–95.

[53] Robinson to Smith, Halle, 14.11.1838, ESP, ABC 60 (63).

[54] Rödiger (1839); (1839a). The *Zeitschrift für die Kunde des Morgenlandes*, published in Göttingen from 1837 (to 1850), Rödiger was one of seven editors, see Ewald (1837).

[55] Rödiger to Smith, Halle, 11.11.1838, ESP, ABC 60 (65); Smith to Anderson, Leipzig, 9.1.1839, ESP, ABC 60 (105) 1839.

braist Fürchtegott Lebrecht had been shown a specimen of Smith's types and 'he was much pleased with it.'[56] Fleischer was also introduced through Tauchnitz to Smith's work while in Leipzig, though it is not clear whether they actually met.[57] It took some years until he and Smith began corresponding, with Fleischer always writing in French. Though relatively young in 1839, Fleischer already had a leading position among the German Orientalists.[58]

Taking advantage of their network in Germany, Robinson also tried to advance translating Smith and Dwight's *Armenia* into German. He naturally spoke with Ritter first, who favored the translation, estimating 'it could be valuable & saleable.' Ritter opined that Leipzig would be the best place for finding a translator, noting that 'it is a work which would come directly within the class of books which Mr Tauchnitz likes to publish, a missionary work.' In a letter to Smith, Robinson wrote that he believed that Gesenius and Rödiger would also be very positive, adding, 'I already spoke about it with Tauchnitz, wrote to Anderson and consulted Ritter and Neander in the subject.'[59] Upon receiving Robinson's letter, Smith wrote to Rödiger on the very same day. Rödiger responded exactly two weeks later to a letter that Smith sent him on November 11, 1838, which was not traceable. Rödiger's letter contains much information concerning Smith, his work, and growing reputation. The former sent Smith's *Armenia* book to Neander in Berlin. 'Your initiative will be very advantageous for the Oriental studies,' writes Rödiger, adding that he would gladly do a small favour for Smith and Tauchnitz by covering it in the Oriental periodical.[60] His last sentence is quite interesting. Rödiger believed that American missionaries were the most suited to translate the Bible into Oriental languages, as 'the Evangelium should be first brought back to the recognition of the Christians of the Orient, before they can be offered with success to the Moslems.'[61]

Smith answered in a rather long letter, beginning with his gratitude for Rödiger's offer 'to give notice of my endeavors to make a new font of Arabic types.' Smith believed, however, that the step was premature; it

[56] Robinson to Smith, Berlin, 22.1.1839, ESP, ABC 60 (63); ADB Lebrecht.

[57] Tauchnitz to Smith, Leipzig, 4.9.1839, ESP, ABC 60 (80).

[58] Fleischer letters to Smith, Leipzig, 24.4, 14.10.1847, ESP, ABC 60 (24).

[59] Robinson to Smith, Berlin, 1.12.1838, 10.1.1839, ESP, ABC 60 (63).

[60] In other places he uses 'our Goettinger Oriental periodical,' referring to the *ZKM*.

[61] Rödiger to Smith, Halle, 11.11.1838, ESP, ABC 60 (65).

should wait until success was ensured. The majority of the letter deals with Mesopotamian issues, the region that interested Rödiger the most. Smith offers to introduce him to philologist Justin Perkins (1805–69), ABCFM missionary and the first American to settle in Persia. Rödiger promptly did so and, in 1840, began publishing letters he received from Perkins in the *ZKM*.[62] Rödiger protests lightly in the letter to Smith, 'I am taking too great liberty for a stranger. But the interest you have shown in the Modern Syriac, encourages me to proceed, & give you what information I possess respecting the almost expiring remains of that language, which we still have in Syria itself.' 'The question how far a Modern Syriac dialect is used on the Mesopotanian side of the Kurdish Mountains,' Rödiger adds, 'as well as how far it differs from the dialect of (?), requires further investigation.'[63] Rödiger sought information related to the study of Mesopotamia, noting that missionaries living for years in the relevant country and speaking its native language, 'obtain, not only linguistic, but geographical & other information (…). What they need, is to have their information on these subjects inquired for.' Long discussions follow, mainly of Christian sects, and of people, villages, dialects, manuscripts, vegetation, landscape, and water sources, a true and rarely used treasure for any study of that region.[64] The letter also includes a report of a visit to Berlin of a group of Chaldean Christians, in which Julius Heinrich Petermann (1801–76), professor of Oriental literature in Berlin (and Consul in Jerusalem 1868–69),[65] Ritter, Robinson, and Kiepert were involved. This rather long and complicated story, connected with the Nestorians, will not be dealt with here.[66]

After Robinson's initial introductions, it was Tauchnitz who steered Smith's connections with the German scholars. In early January 1839, Tauchnitz took Smith in his carriage to a 'very interesting visit' in Halle, where 'the Orientalists took a great deal more matter of me than I deserved.' Tholuck organized a dinner for Smith, where, Smith writes to Anderson, 'you would have been amused of the Babel of languages. With

[62] Smith to Anderson, Leipzig, 9.1.1839, ESP, ABC 60 (105) 1839; Rödiger (1840). Cf. Kawerau (1958), mainly 276–90; Phillips (1939), 147.

[63] Rödiger to Smith, Halle, 11.11.1838, ESP, ABC 60 (65).

[64] Smith to Rödiger, Leipzig, 31.11.1838, ESP, ABC 60 (105/9) 1838 Jul.-Dec.

[65] Petermann (1865). Cf. Goren (2003), 209–14.

[66] The Anglican mission targeting of the Eastern Christian Churches of the Nestorians and the Assyrians or Chaldeans has been widely studied; see e.g. Kawerau (1958), 218 n. 1 (long citation, Anderson about the Nestorians), 220–54, 310–30. Smith's archive holds some dozens of letters concerning this issue.

professors Gesenius & Roediger, on each side of me, I spoke in English, & they replied in German, with Prof. Tholuck I conversed wholly in Arabic, which he speaks with considerable ease…' The issue of Arabic calligraphy was entirely new to all of them.[67]

In late 1839, Rödiger was the first to publish, in his *ZKM*, a 'short report' of Robinson and Smith's expedition, though only under the name of the former.[68] Following Smith's visit, Rödiger wrote a letter thanking Smith 'for the rich and so friendly instruction,' dedicating most of the communication to the studying and printing of Kurdish and Arabic.[69] The German scholar could not possibly have published such detailed studies of these languages without the vast information that he received from the American missionaries, Smith being one of his central sources.[70]

It is not clear when and how Smith began lecturing in Germany, most likely in private circles in Leipzig. In a letter to Smith, young Franz Delitzsch (1813–90), then a student of Fleischer in Leipzig and later a leading German theologian and Hebraist, calls him 'most learned, beloved friend,' adding his deep sorrow at missing 'two of the precious lectures that I receive from you.' Delitzsch signed his letter 'your grateful pupil.'[71] In March 1839, Smith writes to Robinson, 'In the course of my Arabic reading with Dr Delitzsch (…),' indicating that he and Delitzsch read German and Arabic together, each teaching the other the language of his expertise.[72] This study relationship was short-lived. Tauchnitz wrote Smith in September 1839 that he met Delitzsch, who asked him to send his regards to Smith, or, in the wording of the day, 'to be kindly remembered to you.'[73]

Smith's research was recognized and used by other highly regarded scholars of the Holy Land and neighbouring countries. Of much interest and significance is Carl Ritter's opinion. During the late 1840s, Ritter was busy writing his four volumes of the *Comparative Geography of*

[67] Smith to Anderson, Leipzig, 9.1.1839, ESP, ABC 60 (105) 1839.

[68] Robinson (1839b).

[69] Rödiger to Smith, Leipzig, 18.02.1839, ESP, ABC 60 (65); see also Robinson on the visit, citing Rödiger's letter: Robinson to Smith, Berlin, 10.1.1839, ESP, ABC 60 (63).

[70] Rödiger (1840–2).

[71] Delitzsch to Smith, Leipzig, 29.12.1838, ESP, ABC 60 (20). For Delitzsch as Orientalist cf. Wokoeck (2009), 146–63; ADB Delitzsch.

[72] Smith to Robinson, Leipzig, 18.1.1839, 16.3.1839, ESP, ABC 60 (105/10) 1839 Jan.-Mar.

[73] Tauchnitz to Smith, Leipzig, 4.9.1839, ESP, ABC 60 (80).

Palestine and the Sinaitic Peninsula, basing his research, of course, on everything he could get from Robinson, as well as other relevant contemporary studies.[74] Ritter openly admired Smith's scholarship. In a letter to Robinson, he wrote about his use of Smith's *Ortsverzeichnisse* (gazetteers) for the countries east of the Jordan River, and his regret over the loss of Smith's diary from his excursion in the Hauran.[75] In his *Comparative Geography*, while discussing at length the *Biblical Researches* and their value, Ritter notes that 'the service of Mr Smith (...) was indispensable.'[76]

What about Smith himself? His assessment of his own significance for the expedition and contribution to the *Biblical Researches* is worth examining. As early as 1839, due, in part, to Berghaus' reaction upon seeing the material for the maps, Smith became aware of the value of his compass bearings and data. Berghaus, Smith writes, 'finds in my [emphasis added] bearing by compass, one exactness, which astonishes me; & they are so full, that from them alone, he is able to construct a net of triangles over almost the whole of Judea.'[77] This letter may be the only proof of the fact that Smith was the one to execute the measurements throughout the expedition!

Robinson also admired Smith's scientific talents and expertise in locating 'Arabic treasures,' or rare books and manuscripts. Taking into consideration Smith's planned travel from Leipzig to the US before returning to his mission in Beirut, Robinson suggested that he should look en route for worthy Oriental published material, in Paris, London, and Oxford.[78]

5.a.3. The Buckingham rivalry

While in Leipzig, Smith engaged in intensive and bitter discussions with James Silk Buckingham, the well-known British journalist and traveller who became his and Robinson's 'archrival.' Their dispute

74 Ritter (1842); (1844); (1848–55); (1866, trsl. to English). Cf. Goren (1999); (2003), 68–83.

75 Ritter to Robinson, Berlin, 18.3.1850, ERP, L.4 Ritter. See Smith to Rödiger, Leipzig, 31.11.1838, ESP, ABC 60 (105/9) 1838 Jul.-Dec.: 'I am sorry that I must (of course leave?) my account of this charter of villages. My journal of that whole tour, was unfortunately lost, in a shipwreck I suffered in the year 1836, at the coast of Caramania.'

76 Ritter (1866), II, 74.

77 Smith to Thomson, Leipzig, 20.3.1839, ESP, ABC 60 (105/10) 1839 Jan.-Mar.

78 Robinson to Smith, Berlin, 10.1.1839, ESP, ABC 60 (63).

encompassed places, facts, and theories from Buckingham's book, initially published in London in 1821, following his 1816 Holy Land travels.[79] Smith mentions Buckingham repeatedly in his personal correspondence from early 1839, and a series of Smith's letters critiquing Buckingham was published in *The New York Observer* in 1839–40.[80] The letters were published anonymously under the pseudonym 'Palestinensis' during summer 1839, and Buckingham's responses from October were collected and published in 1840.[81] The identity of the writer was revealed to Buckingham by distinguished Yale Professor of Chemistry and Natural Sciences Benjamin Silliman, related by marriage to Robinson and to his distant cousin, artist Colonel John Trumbull (1756–1843), who is also mentioned in the letter.[82] Smith, who had just arrived in the US from Leipzig, regretted the fact that he had not received Silliman's 'remarks upon my answer' before sending his reply. He notes that 'Mr Buckingham has written me a very gentlem[an]ly letter, & I have endeavored to give him a simi[lar] one in reply, so far as I could, & still decline his request.' However, Silliman's comments, had Smith received them on time, would have greatly improved his response. Smith adds that since Buckingham 'did not give me his address, you, Silliman, are my only way to communicate with him.'[83] Robinson was unable to avoid the conflict. Writing to Smith from Berlin in spring 1840, using Solomon's Proverbs he calls 'Buckinghams strife' truly 'the lottery of vote,' and asks Smith to send him 'the whole controversy.' Robinson reassures Smith that he 'took just the right ground' and that he could use his name if necessary, though 'I felt too a great repugnance at having my name mixed up with Buckingham.'[84] In response to Buckingham's New York lecture series there, Smith continues his critique in a long letter to Bird, this time somewhat cynically:

79 Buckingham (1821); cf. Tobler (1867), 143; Shepherd (1987), 46–75; Rée (2001).

80 Smith to Robinson, Leipzig, 18.1.1839, 19.1.1839 (the longest and most detailed letter, discussing all points in dispute); Smith to Messers. S. E. Morse & Co. Eds. N. Y. Observer in Rochester, N. Y., Leipzig, 28.1.1839, ESP, ABC 60, (105/10) 1839 Jan.-Mar.; Robinson to Smith, Berlin, 11.2.1839, ESP, ABC 60 (63).

81 Buckingham (1840).

82 Smith to Silliman, Boston, 9.11.1839, ESP, ABC 60 (105/11) 1839 Apr.-Dec. Cf. Williams (1999), 23.

83 Smith to Silliman, Boston, 9.11.1839, ESP, ABC 60 (105/11) 1839 Apr.-Dec.

84 Robinson to Smith, Berlin, 15.4.1840, ESP, ABC 60 (63): 'The beginning of strife is as when one letteth out water,' Proverbs 17:14.

> I received yours on Saturday requesting my opinion of the character of Mr Buckingham's lecture (...). You are aware that it is only in regard to a portion of them, viz. to such as relate to Palestine and particularly Mt Lebanon, that I am at all qualified to pass an opinion, and to these my remarks will be chiefly confined. Mr B seems to think that because no man in America undertook to rectify his mistakes, therefore they could not be of any great importance. But he might be expected to know that public controversy is a work in not which every man, however conscious he may be that he is defending the truth, is arbitrary to engage.[85]

In order to explain their positions concerning Jerusalem, Robinson and Smith even published a map of the city with explanations.[86] The discussion tapered off in the spring of 1840, and consequently Buckingham disappears from Smith's correspondence.

5.a.4. Conclusion

In 1842, only a short time after publication of the *Biblical Researches*, Robinson received one of the highest geographical honors of that time, the Gold Medal of the Royal Geographical Society (RGS), one of two medals awarded annually by the Queen. Robinson was awarded the medal for his 'very valuable, important and interesting work.'[87] The address of RGS President William Richard Hamilton and the congratulations of Robinson's acquaintances, such as Lord Prudhoe (Algernon Percy, 4th Duke of Northumberland, 1792–1865), to whom Robinson dedicated his London edition of the *Biblical Researches*,[88] express the high honour given Robinson:

> (...) by unanimous vote, for your valuable work on the Holy Land. Their vote will prove to you how justly that work is appreciated by them & the English public. It is a well merited compliment to your talents & research. Your volumes now take their place in many libraries among our standard works, by the side of Milton & Shakespeare.[89]

85 Smith to Bird, Gilmanton, 16.3.1840, ESP, ABC 60 (9).

86 Robinson (1840a).

87 Formal announcement, J. R. Jackson [secretary of the RGS] to Robinson, RGS 29.5.1842, ERP, I, 1. See chap. 4.b.1.

88 See Ruffle (1998). The American edition had been dedicated to Moses Stuart, the German to Carl Ritter.

89 Prudhoe to Robinson, Whitehall Gardens, 23.4.1842 (Citation), W. R. Hamilton to Robinson, London, 2.5.1842, ERP, I, 1, Hamilton (1842); Everett to Robinson, London, 28.5.1842, ERP, I, 2; Announcing the decision: J. R. Jackson to Robinson, RGS

Edward Everett, Robinson's longtime acquaintance who served from 1841–45 as American ambassador to Great Britain, accepted the medal for Robinson on May 23, 1842.[90]

As noted throughout, the expedition's superior scientific results were dependent upon Robinson having Smith at his side. It is difficult to understand how Robinson could have adhered to the 'traveling policy' that was so critical to the expedition's success without Smith. They strictly avoided the 'classical' lodging places, preferring the open air or houses of locals. They also knew exactly the right questions to ask and how to ask them: 'Nor did they ask direct questions, which usually get the answer which the Arab thinks the questioner wants; but the most indirect interrogations and cross questions, and by comparing the answers gained from different persons, they at last felt, in most cases at least, that they had in some measure attained the actual facts.'[91]

Naturally, not only to understand but to uncover the real intention of their sources, they needed Smith's fluency in Arabic and its dialects, to the deepest nuances. Moreover, the impressive list of their Arab travel companions and guides, indicating that they knew not only how to avoid the mistakes of their predecessors but to find the best available people for each different region, bears testimony to Smith's contribution. This, and more, was beyond Robinson's expertise and was clearly the work of Eli Smith. In his speech, Hamilton recognizes Smith's integral contribution: 'Dr Eli Smith, his former pupil, long a resident in the East, not only in Syria and Palestine, but also in Armenia, Persia and Egypt, and already familiar with the inhabitants and their language, to the advantage of whose society and assistance Dr Robinson is on all occasions happy to bear the fullest testimony.'[92]

A detailed search of Robinson's entire correspondence connected with the presentation failed to uncover all but brief mention of Smith

29.5.1842, ERP, I, 3; RGS-IBG, to Robinson, Letter Book 1841–44, 27.5.1842; Robinson thank you letter to the RGS, RGS-IBG, R.G.S. Corr. Block 1841–50 RGS/CB3/655.

[90] Hamilton (1842), 6. Cf. Williams (1999), 298–99. As early as 1826, Robinson received from Everett a letter of introduction to His Excellency, the historian Baron Bathold Georg Niebuhr, Minister of his Prussian Majesty in Rome, describing him as 'a gentleman, distinguished among the scholars of America for his philological attainments;' Everett to Niebuhr, n.p., 13.5.1826, ERP, 4 k 3 (in Weigel Williams (1994)). The connection is even more interesting, as Niebuhr was the son of Carsten N., the pioneer Arabia-Felix traveller who visited the Holy Land years earlier in August 1766; see Hansen (1964); Goren (2003), 30–36. Robinson used his *Travels* intensively (Niebuhr, C. (1969); cf. Robinson (1841), III, 1st Appendix, 21).

[91] Ritter (1866), II, 74.

[92] Hamilton (1842), 3. Hamilton also noted the contributions of Kiepert and Ritter.

subsequent to that time. Surely the medal should have been awarded to both Robinson and Smith! Smith's public peer recognition was not long in coming. In recognition of his achievements, the scientific community elected him member of the American Oriental Society (1843) and the German Oriental Society (DMG, 1847).[93] During the latter half of the 1840s, Smith published in the periodicals of both societies and received letters of appreciation from their leading scholars.[94] It is also not surprising that Smith's lists of Arabic names merited separate publication in Halle, in addition to the German version of the *Biblical Researches*.[95]

The news of Smith's illness caused great concern in the missionary community. 'Truly God's ways are not as our ways. It seems as though the life and health and continued labors of Dr Smith are absolutely indispensable to the completion of the Arabic version of the Scriptures. No one else is qualified now to take his place...'[96] Eli Smith died on January 11, 1857. Two days prior, he delivered several messages to friends and colleagues. His letter to fellow missionary Henry Alfred de Forest (1814–58, missionary physician in Syria from 1842–54) and his wife reveals how he saw himself: 'Indeed "I have been a man of few words," and have never said much of my own feelings toward my friends.' Smith begs forgiveness from them, for every misunderstanding during their journey together as missionaries and, as his last will, he demonstrates his loyalty to the ABCFM by asking his colleagues not to 'throw their great influence at home against the Board,' and declaring that he 'would give anything could Dr De F. feel more friendly toward Dr Anderson.'[97] For this chapter, no ending is more fitting than Robinson's words to Smith's widow: 'Perhaps with no friend on earth have I stood in nearer & more intimate relations. Why he should thus be taken at the very moment of highest usefulness, - and I, comparatively a broken reed, be left, - is one of the inscrutable mysteries of Divine Providence.'[98]

93 Greenough (corresponding secretary) to Smith, Boston, 6.10.1843, ESP, ABC 60 (31); Fleischer to Smith, ESP, ABC 60 (24).

94 Rödiger to Smith, Halle, 12.10.1848, ESP, ABC 60 (65); Edward Elbridge Salisbury (corresponding secretary) to Smith, New Haven, 8.5.1847, ESP, ABC 60 (67).

95 Smith (1842).

96 Henry Harris Jessup (1832–1910, ABCFM missionary to Syria; DAB Jessup) to Rev. Dr Stearns, Duma, Mt. Lebanon, 16.8.1856, ESFP-YD, Box 2, Folder 4. See also Mehitable (Hetty) Butler Smith, 'In Memoriam: My Husband,' [Beirut], n.d., ESFP-YD, Box 2, Folder 15a.

97 Dying words of Dr Eli Smith to Dr De Forest, 9.1.1857, ESFP-YD, Box 1, Folder 6. De Forest died in 1858.

98 Robinson to Hetty Smith, New York, 29.7.1857, ESFP-YD, Box 2, Folder 4.

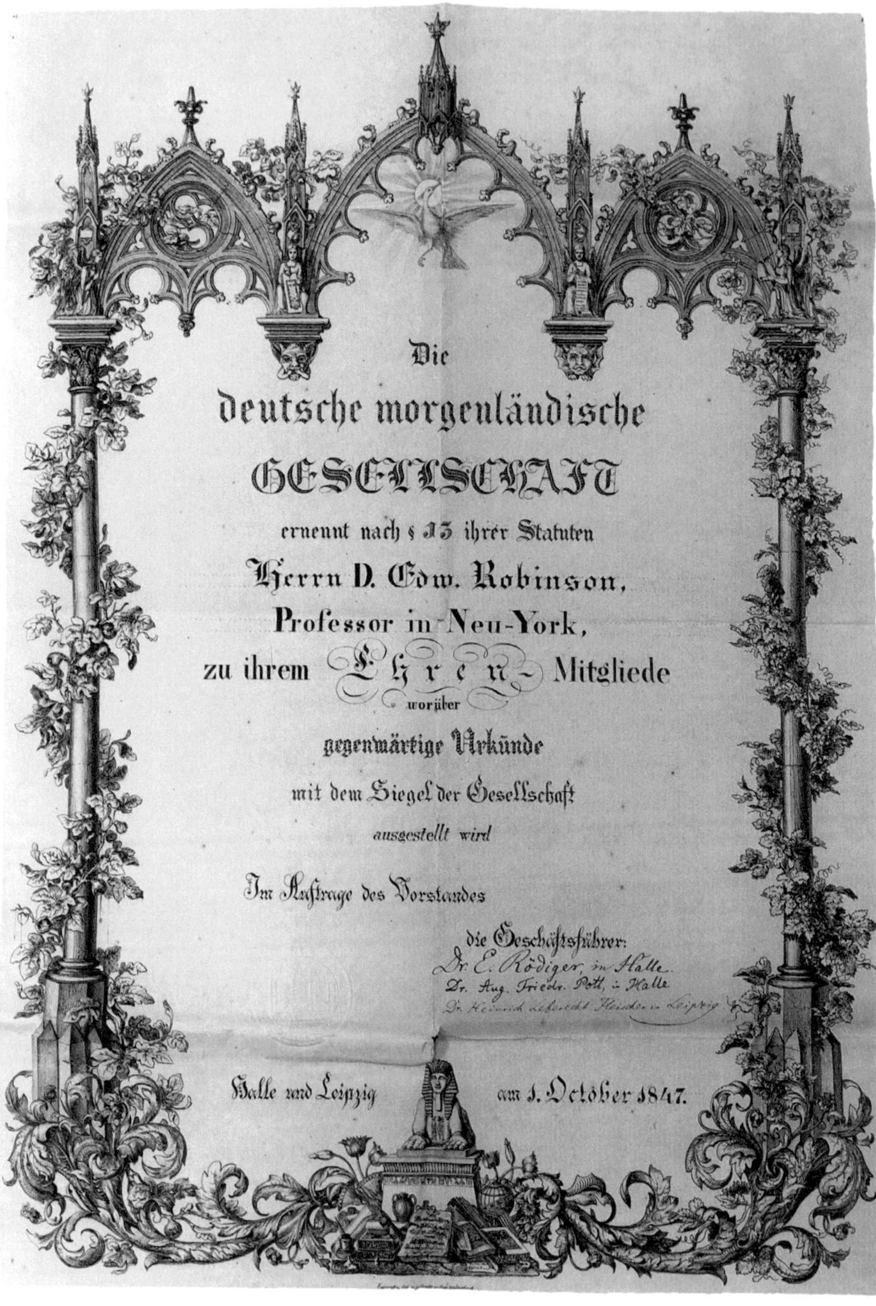
Die
deutsche morgenländische
GESELLSCHAFT
ernennt nach § 13 ihrer Statuten
Herrn D. Edw. Robinson,
Professor in Neu-York,
zu ihrem Ehren-Mitgliede
worüber
gegenwärtige Urkunde
mit dem Siegel der Gesellschaft
ausgestellt wird
Im Auftrage des Vorstandes
die Geschäftsführer:
Dr. E. Rödiger, in Halle
Dr. Aug. Friedr. Pott, in Halle
Dr. Heinrich Lebrecht Fleischer in Leipzig
Halle und Leipzig am 1. October 1847.

Ill. 26. German Oriental Society membership, October 1, 1847, signed E. Rödiger in Halle; Aug. Friedr. Pott, in Halle; Heinrich Lebrecht Fleischer in Leipzig.

5.b. Therese Albertine Luise von Jakob-Robinson, 'Mediator of the Balkan Slavs,' and Holy Land Studies

A scientific conference, dealing with the comparison of Slavian, mainly east-Slavian literature, should not forget, to remember this pioneer of these efforts [that] today is largely forgotten.

Eine wissenschaftliche Konferenz, die sich mit dem Vergleich der slawischen, vor allem der Ostslawischen Literatur beschäftigt, sollte nicht vergessen, sich einer heute weitgehend vergessenen Bahnbrecherin dieser Bestrebungen zu errinern.[99]

Eduard Winter, 1962.

While studying in Halle and Berlin from 1826 to 1830, Robinson, 'in steady, plodding diligence (…) became a German among Germans (…), the most German between the Scholars of English tongue.'[100] In 1828, he married Therese Albertine Luise von Jakob, youngest daughter of the political economist Ludwig Heinrich von Jakob (1759–1827) of Halle, who was deeply involved in early nineteenth-century German academic, social, and cultural circles.[101] Talvj (von Jakob-Robinson's pen name) honed her literary abilities in Halle's cultural '*salons*' and other *litera-*

[99] Winter (1962), 313.

[100] His daughter cited in her father's eulogy of his friend and colleague Hitchcock: Robinson, M. A. (1862–64), 150; Licht (2009), 21. Cf. also for his 'Germanisation': Goren (1999), 30–31; (2003), 83–91.

[101] Much had been written about von Jakob, his career, works and importance. See e.g. ADB Jakob; Krause (1982); Licht (1999); Abašnik (2008); Dehrmann (2010), 352–53; Stiehler (2013). See also, in connection with the current study: Weigel Williams (1997), 1–3; (2007), 7–53. Already in 1880, Talvj had been praised in an early study of the German element in the USA (Körner (1880), 133–36). Several biographies of Talvj have been published, the earliest in 1874 by her daughter, Mary Robinson (Robinson, Talvj) and the next in 1897 by Ludwig Wagner (Wagner (1897)), also ADB Robinson, T.; DAB Robinson, T. The best and most complete to date is that by Hermine Weigel Williams (Weigel Williams (2007)), which includes a detailed, probably full bibliography of Talvj's publications (247–53) and all published studies on Talvj, with the leading works of Friedhilde Krause (Kruse (1982); (2001); (2001a)); Gisela Licht (Licht (1999), some mistakes in her description of Robinson's voyage, 27; (2009)); Franz von Löher (von Löher (1886)); Nikola R. Pribić ((1973); (1985); (1999a)) and Martha Kaarsberg Wallach (2001). Every biographer emphasized the problem of lack of relevant material (e.g. Licht (2009), 10–11). A fire in the New York home of Edward W. Robinson, grandson of Edward and Therese, destroyed most of Therese's papers, articles, and books held by the family (Weigel Williams (1994), II).

Ill. 27. Dr L. H. von Jakob, Russisch Kaiserl. Staatsrath und Ritter des St Annen Ordens. Perofessor der Staatswissenschaften in Halle [Russian imperial state-advisor and Knight of the Order of St. Anne. Professor of Political Sciences in Halle].

turzirkel.[102] Robinson entered, it appears, happily and successfully into the social position of Talvj and her family, of which her father was the dominating force until his death in June 1827.

Talvj facilitated, to a significant degree, Robinson's integration into German academic circles. Leading contemporary German scholars found in Robinson a competent and most interesting and innovative scientist, who contributed much to their knowledge and research. Career backing was not one-directional; researchers hinted quite early at Talvj and Robinson's mutual support, which, upon reading their early biographies, appears inevitable, almost natural. In 1897, Ludwig Wagner writes in his biographical sketch that Talvj assisted her husband

[102] Licht (1999), 18–30; (2009), 13–14.

with his literary works until his death.[103] Gisela Licht writes that the 'Robinsons cooperated as a team for the mutual uses.'[104] Talvj provided constant support, translating German papers into English for Robinson's periodical and translating his papers for publication in Germany. Irma Voigt also notes that Talvj interpreted and translated many of the German contributions to Robinson's periodicals.[105] Williams describes Talvj's and Robinson's mutual influence and support in two sections of his book. Recounting their early period from letters and other documents in the Robinson Archive,[106] Williams writes that 'it is clear that though he learned intellectually from Stuart and Gesenius and Neander, it was Therese who may have been the greatest intellectual influence upon his life.' In a later section describing Talvj's publications, he states that 'one can hardly understand Robinson without realizing that his wife, while caring for their children and their home, was producing intellectually, in her own way, at least as much as he.'[107]

These studies are general and do not provide the reader with primary sources that establish Talvj's yet uncredited role in her husband's expeditions and publications. The highly interesting connection between Talvj's German-Balkan network and German study of the Holy Land, accomplished by the marriage of a New England minister-scholar to a German scholar of Slavic literature, deserves elucidation. Talvj had a significant influence on her husband's travels and research, in which her social and academic circles, including the strong Balkan connection, played a leading role.

Indeed, in her day, Talvj was well known, published quite profusely, and later became the subject of numerous biographies and studies.[108] Today, however, Talvj is all but forgotten, maintains Licht in 1999, which is not altogether true. In 1997, a scientific conference dedicated to Talvj was held in Jena. The conference produced an impressive and rich volume published in 2001; one of its papers is a literary report about Talvj's studies and others were consulted for this study.[109] Licht is probably

[103] Wagner (1897), 12.

[104] Licht (2009), 25.

[105] Voigt (1914), 35, cited by Pribić (1973), 496.

[106] Today housed in the Special Collections, Burke Library, Hamilton College.

[107] Williams (1999), 165 (citation), 198, 304.

[108] Krause (2001b): 79 of Talvj' publications (including after her death), 90 publications on Talvj.

[109] Schubert, G. (2001), Vorwort, (2001a); Krause (2001). Cf. n. 855.

right in concluding that, similar to other women writers of the nineteenth century, it was destined that Talvj's works, at least those in the German language, would exist today solely 'in scarce stocks of few German university libraries.'[110] Well, they also exist in Robinson's archive...

5.b.1. The beginning of a literary career

Talvj wrote poetry even as a young child; yet the most intriguing question concerning her literary interests is how she began translating Serbian poetry, which Hermine Weigel Williams notes is 'a rather complex but interesting saga in and of itself.'[111] Jacob Grimm was probably responsible for Talvj's shift from Russian to the totally unknown and unused Serbian language.[112] The result was, as demonstrated by various scholars, her role in the 'conceptualization of the Serbian image during the nineteenth century.' This conceptualization began with Johann Gottfried Herder (1744–1803), who promoted the idea of *Volkslied* in the German-speaking world, and was continued by Jacob Grimm, who recognized the rising contemporary Slavian national awareness based on 'textualization of the foreigners' through translation and discourse.[113] Talvj was positioned as the 'successor of Herder, Grimm and Goethe;' yet 'her performance had been concentrated in one point, and this one point is: Goethe (...) *dem König in dem Reich des Schönen* [the king of the empire of beauties].'[114]

At the age of ten, Talvj and her family left Halle after the French conquest and spent nine years in Russia's 'exile' (Kharkov, today Kharkiv, Ukraine, and Sankt Petersburg, 1807–16). There, Talvj's interest in the Serbian language was triggered through reading the leading philologist and mythologist Jacob Grimm who, as early as 1815, praised German interest for *Volkskunst* and *Volksdichtung*. She soon mastered the language and began translating Serbian folk songs.[115] Talvj transformed into a

[110] Licht (1999), Vorwort.

[111] Weigel Williams (2007), 26.

[112] Pribić (1999a), 12–13.

[113] Ellermeyer-Životić (2001), 95–97; Schubert, G. (2001a), Einleitung, 15–16; Merchiers (2007), 51; Dehrmann (2010), 354–56. For 'Herder and the Serbian folk song': Ćurčin (1905), 20–37.

[114] Mojašević (1976), 166–68, cited also by Ellermeyer-Životić (2001), 97; Grasshoff (1966), 482. See also Talvj's dedication to Goethe: Wagner (1897), 10.

[115] Winter (1962), 313–14; Grasshoff (1966), 477; Pribić (1979), 140. Cf. for Grimm's connections with the Serbs Vuk and Milutinović and the Russian von Köppen: Vasmer (1939), XII–VII, XX, and the many letters in this collection; Krause's detailed study of the Russian experience in Talvj's deeds, Krause (1982); Licht (1999),

leading linguist and translator of Slavic languages into German and later into English. She was encouraged by Johann Wolfgang von Goethe (1749–1832), a close, personal friend who showed much interest in her work and from whom she received a letter at least every four weeks.[116]

Talvj became a key figure in what one might call 'the Slavic circle,' centred around the linguist Vuk Stefanović Karadžić, an 1823 graduate of Jena University [today Friedrich-Schiller University]. With Karadžić were Grimm and the leading Slovene Slavicist Jernej Barthol (Bartholomäus) Kopitar, who from 1808 had been living in Vienna and working in the Austrian Imperial Court Library. From her letters, it is clear that Talvj relied, more than anything else, on the judgment and suggestions of these two 'masters of Slavic culture, in general, and Slavic folklore, in particular.'[117] When Grimm and Kopitar sought a translator of Slavonic poems into German, it was Karadžić who suggested Talvj, whom he first met in 1824.[118] Other members of this Slavic circle were the Bosnian Serb poet, rebel, and adventurer Simeon ('Sima') Milutinović 'Sarajlija' (of Sarajevo, 1791–1847), who from 1815 was an associate of Vuk's and who was a frequent guest at von Jakob's home in the 1820s,[119] and the Russian writer and publisher Petr Ivanovič Köppen (1793–1864), one of her closest acquaintances.[120]

10–16; Abašnik (2008), 901–05. Wagner (1897), 6–7, claimed that her interest in folk songs began as a child, when she used to hear them in the markets of Kharkov.

116 Ćurčin (1905), 121–86 ('Geothe and the Serbian folk song,' includes subchapters concerning Goethe and Grimm, Talvj, and Milutinović). Cf. Krause (1982); Weigel Williams (1997), 2–3; Licht (1999), 21–23; (2007), 11–17. For their correspondence: Steig (1891), and the discussion 62–65.

117 Pribić (1973), 503.

118 Dobrašinović (2001), 69. Gabriella Schubert and Friedhilde Krause edited in 2001 an outstanding collection of papers concerning Talvj's *Balkan-Slaven* connections and works (Schubert, G. (2001)), including a biography titled 'Talvj, Deutsche und Südslaven' (14–18), an extensive publication of Talvj's letters (Krause (2001a)), and a bibliography to Talvj's publications as well as her studies (Krause (2001b)). Most of the papers include long discussions of the three, their relations, and collaboration (e.g. Dobrašinović (2001); Delić (2001); Ellenmeyer-Životić (2001)). Another important study is Merchiers (2007), where she discusses 'Kopitar's network building or extensive circle of acquaintances' (48–50), and on 222–87 a detailed study of the Kopitar – Karadžić network. Weigel Williams added to her 2007 publication a detailed bibliography of Talvj's works (Weigel Williams (2007), 247–53).

119 Ćurčin (1905), 163–86; Pribić (1979); Weigel Williams (2007), 35. He probably fell in love with Talvj, who rejected him, exactly at the time when her relationship with Robinson was developing: Pribić (1979), 148–50.

120 Grasshoff (1966). The detailed development of Talvj's Slavic connections and her journey from young amateur to central figure in the Slavonic literature of her time is discussed in numerous publications dedicated to these narratives: Steig (1891); (1894); Milović (1941); Coleman (1944); Winter (1962); Grasshoff (1966); Pribić (1979); (1985).

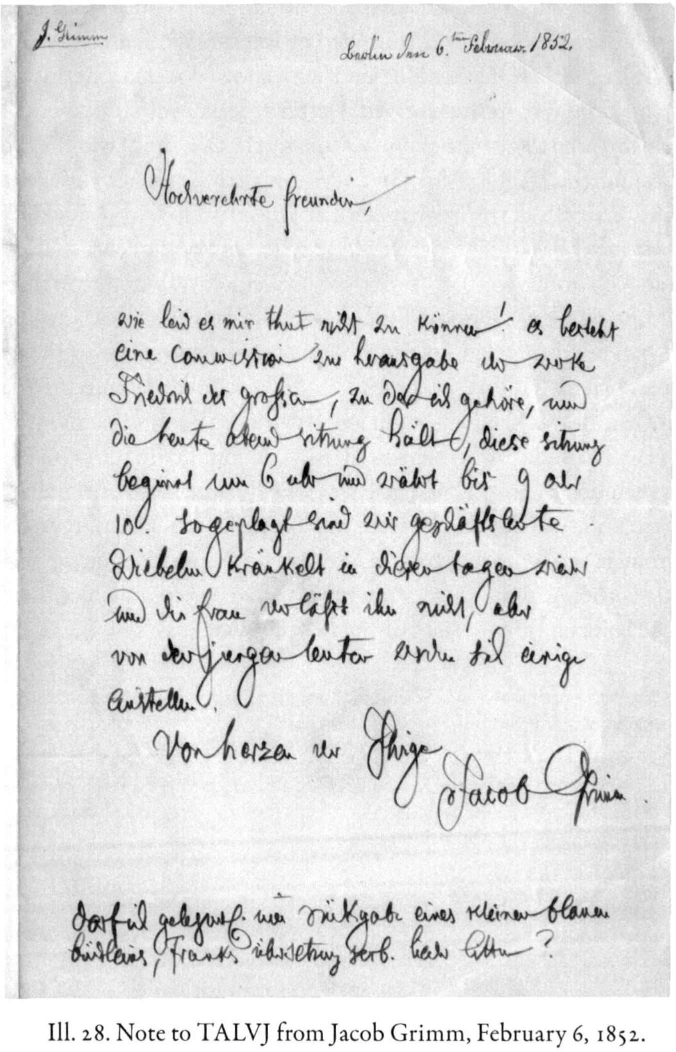

J. Grimm

Berlin den 6. Februar 1852.

Hochverehrte freundin,

wie leid es mir thut nicht zu können! es besteht eine commission zur herausgabe der werke Friedrich des grossen, zu der ich gehöre, und die heute abend sitzung hält, diese sitzung beginnt um 6 uhr und währt bis 9 oder 10. so geplagt sind wir geschäftsleute. Wilhelm kränkelt in diesen tagen wieder und die frau verläßt ihn nicht, aber von den jungen leuten werden sich einige einstellen.

Von herzen der Ihrige
Jacob Grimm.

darf ich gelegentlich um rückgabe eines kleinen blauen büchleins, Franks übersetzung serb. lieder bitten?

Ill. 28. Note to TALVJ from Jacob Grimm, February 6, 1852.

Grimm studied German folklore and oral traditions together with his younger brother Wilhelm (1786–1859), a prolific author. They worked in the Kassel library from 1816 to 1828, moved to Göttingen where Jacob served as professor and librarian until 1840, and then continued to the 'real centre,' to Berlin.[121] The newly married Robinsons vis-

[121] Naturally, there are numerous studies on 'the Brothers Grimm,' many dedicated to Jacob Grimm's *Volkslied* and Slavonic connections. With regard to this research, see discussions by Weigel Williams (2007), passim; Williams (1999), 153, 165. For the following discussion cf. Ćurčin (1905), 65–121; Kropej (2013).

ited Jacob Grimm in Kassel on their honeymoon tour. Robinson writes, 'I had expected to see an old man, covered with the dust and lumber of learning,' yet they met 'a man in the prime of life, some 40 years old, with the air & warmth of a gentleman.'[122]

Karadžić is known as the leading reformer of the Serbian language and author of the first Serbian dictionary. His collections of Serbian national songs stimulated 'the great vogue in Germany for Serbian popular poetry' and his studies of Serbian folklore led him to collect songs, fairy tales, and riddles.[123] As for Kopitar, 'the building of foreign and domestic networks was very important to him,' writes Ingrid Merchiers, 'since he wished to discuss academic matters with as many experts as possible, in order to enable interaction in the form of mutual communication, feedback, criticism, approval, introduction of new ideas etc.'[124] Clearly, Talvj adopted this social-academic approach, and happily entered and integrated into Kopitar's circles. For Talvj, crossing accepted social borders was natural. From childhood, she exploited every opportunity to acquire knowledge and pave her way to becoming a writer.[125]

5.b.2. The beginning of a partnership

Talvj and Robinson's successful partnership began in 1826 when Robinson, who had just reached Halle, was introduced to the father and daughter. 'Her name is Therese, the youngest daughter of the late Prof. von Jakob,' writes the 34-year-old Robinson in spring 1828 to his uncle:

> between 4 & 5 years younger than myself (...) could not but think at [text missing] with her superiority to those around her. Her eager desire to improve herself in the knowledge of English [led] me to be frequently in the family & with her; & the more intimately I became acquainted with her, the more was I unstrained to [admire] her talents & acquirements, the fruits & elevations of her soul, the kindness, gentleness, & excellence of her whole character.'[126]

Robinson elaborates on Talvj's linguistic abilities, noting that she knew 'some other Slavonic languages, particularly that of Servia. [text miss-

[122] Robinson to Stuart, Halle, 5.8.1828, Weigel Williams (2003), 102.

[123] Coleman (1944), 85 (citation); Schubert, G. (2001a), 16–17.

[124] Merchiers (2007), 48.

[125] Licht (2009), 10. Cf. Steig (1891), 66–67.

[126] Robinson to Norton, Halle, 4.4.1828, ERP, ER 1.17. For additional citations see Weigel Williams (1997), 5.

ing] national poetry of this nation she has translated & published under the patronage of Goethe, who has shown her much kindness. She speaks also French & English, & reads Spanish & Italian.'[127]

With regard to her character and social behaviour, Robinson writes, 'Her character is serious & earnest; – she has all the strength of mind & curiosity of Eliza,' comparing her to his first wife, Eliza Kirkland (1779–1819).[128] Robinson then adds that 'her heart & home are in the private social circle, where she is the life & soul of all.' He was absolutely right; Therese and Talvj were one and the same. Throughout her life, she never expended effort on differentiating between the private (housewife and mother) and public person.[129] Therese had also been a woman, writes their daughter, 'her true womanliness afforded him a devoted helpmeet.'[130]

In the letter to his uncle, Robinson cites Charles Hodge's favourable opinion of his potential match with Talvj. Like Robinson, Hodge was Moses Stuart's student at Andover, sharing Robinson's theological tradition attributing strong authority to the Bible. He later served as long-term president of the Princeton Theological Seminary. Stuart was influential in their development, sending both to Germany for advanced studies. Hodge followed Robinson to Halle and became his roommate for an extended period, and shared these thoughts with Robinson:[131]

> I always thought you could not be so much in the society of the Fräulein von Jakob without being more deeply interested in her than you yourself were aware of. – I sincerely congratulate you: she is the most interesting female I have met within this side of the Atlantic, and her superior attainments render her a prize of which you may will be proud. May God's blessing rest upon your connexion & all your plans.[132]

[127] Robinson to Norton, Halle, 4.4.1828, ERP, ER 1.17.

[128] Eliza Kirkland was the daughter of Samuel Kirkland (1741–1808, DAB Kirkland, S.), missionary to the Oneida Indians and founder of the Hamilton Oneida Academy, forerunner of Hamilton College, and younger sister of John Thornton Kirkland, president of Harvard College (1810–28). Robinson married her in 1818 but she died in childbirth within less than a year. Williams (1999), 59–77; Oliver (2014), 148–54.

[129] Licht (1999), 6, 18; (2009), 18–19.

[130] Robinson, M. A. (1862–64), 150.

[131] Dearman (1991). Williams (1999) points out the interesting fact that Robinson and Hodge became good friends despite the fact that Hodge, in contrast to Robinson, 'was to become famous as the Princetonian defender of Old School Presbyterianism and the perpetuation of slavery in the South' (149).

[132] Robinson to Norton, Halle, 4.4.1828, ERP, ER 1.17.

Robinson married 'a German woman who had travelled the length and breadth of Europe, knew the most modern poets and philosophers of Germany, and corresponded with Johann Goethe and Jacob Grimm,'[133] 'young as she was beautiful, and beautiful as she was talented.'[134] 'No man was ever more fond of his wife,' writes Hitchcock in his biographical sketch of Robinson, 'or more considerate of the happiness of his children.'[135]

Robinson found a partner who took charge of her own development and acquired knowledge in fields of interest particular to her, paving her own career path. 'No less prominent than the spiritual gifts of this rare woman, of which she in such a many-sided way made use for the world,' writes her daughter, 'were the qualities of her heart and character. With a lively imagination, clear judgment and sharp practical understanding, she united a deep sincerity of feeling and a pure, chaste nature (...). For her, it was definitely unpleasant to be considered only as an "educated woman;" she wanted to raise love and confidence...'[136] And Goethe? Following their farewell visit to him during their long honeymoon, he wrote to a friend that he recently was visited by Robinson and his 'most favorite (*allerliebsten*) Wife.'[137]

The Robinsons' relationship was characterized by mutual understanding and respect, and Talvj shared her thoughts and feelings generously. 'Fortunately,' she intimates to Grimm in 1828, 'Robinson belongs to those few men who know how to respect a lively interest in art and knowledge even among women, and he would rather encourage me than hold me back in my literary work.'[138] Six years later Talvj again wrote how fortunate she was to have married Robinson.[139] Robinson could not have chosen better!

133 Williams (1999), 181.

134 Steig (1891), 75, cited the Austrian playwright and poet, Franz Grillparzer (1791–1872); copied by Mojašević (1976), 164–65; Weigel Williams (1997), 2.

135 Cited by Robinson, M. A. (1862–64), 160.

136 Robinson, M. A. (1874), xxvi–vii.

137 Licht (2009), 20.

138 Talvj to Grimm, Töplitz, 9.7.1828, Staig (1894), 357, cited by Krause (1999), 51, and Weigel Williams (1997), 7.

139 Talvj to Grimm, Pforta, 10.4.1834, Staig (1894), 362.

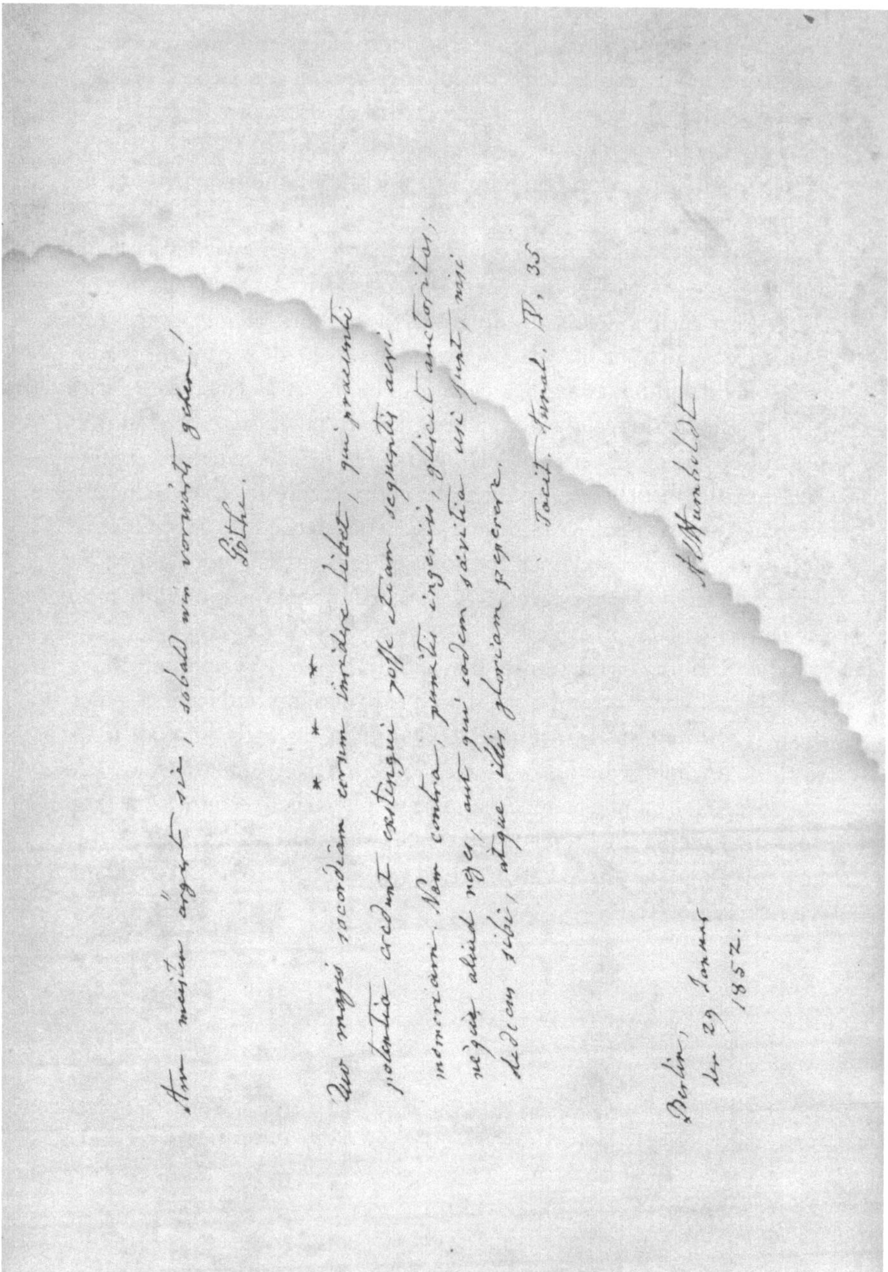
Am meisten ärgert sie, sobald wir vorwärts gehen!

Göthe.

* * *

Quo magis socordiam eorum inridere libet, qui praesenti potentia credunt extingui posse etiam sequentis aevi memoriam. Nam contra, punitis ingeniis, gliscit auctoritas; neque aliud reges, aut qui eadem saevitia usi sunt, nisi dedecus sibi, atque illis gloriam peperere.

Tacit. Annal. IV, 35

AlHumboldt

Berlin, den 29 Januar 1852.

29. Göthe and Humboldt, dedications in Talvj's autograph book, 1852.

5.b.3. Becoming a recognized linguist and Slavist

Talvj's extensive correspondence is testimony to her status in social and scientific circles ('*Therese hat viel und gern geschrieben*'[140]) and, fortunately for contemporary scholars, has been catalogued and published nearly in its entirety. The literary historian from Berlin, Reinhold Albert Steig (1857–1918), published her *Briefwechsel* with Göthe as early as 1891 and with Jacob Grimm in 1894.[141] In 1941, Jevto M. Milović (1908–91), a Montenegro-born lecturer in Serbokroatian languages in Berlin, published in Leipzig excerpts from Talvj's letters to Kopitar.[142] Three years later, Arthur Prudden Coleman (1897–1974), an American expert on Slavic languages, published three letters of hers addressed to the 'talented, but temperamental' New Haven poet (and physician and geologist) James Gates Percival (1795–1856). Coleman claims that 'no good biography of this woman' had been written to date, and that the letters 'constitute probably the first serious correspondence on Slavic matters to be carried on in the United States.'[143] In 1966, the German Slavist Helmut Grasshoff (1925–83) published letters from among Talvj's correspondence with Köppen.[144] The East Berlin Slavicist and librarian Friedhilde Krause (1928–2014) devotes more than 50 pages to her correspondence in a 2001 volume of papers on Talvj (*Aus Liebe zu Goethe: Mittlerin der Balkanslawen*) co-edited with Gabriella Schubert (retired Slavicist from Jena university).[145] Many others, usually single letters, were published over the years.[146] The vast literature attributed to Talvj, her history, correspondence, publications, and connections are certainly the best proof of her scientific importance and position!

Talvj's acquaintance with Köppen began when Köppen's father, a German physician, was called to Russia by Catherine II, and then in 1808, began studying with Professor von Jakob at the University of Kharkov. Through his mother's friendship with Mrs von Jakob, the two youngsters became closely acquainted, and their friendship deepened

[140] Krause (1999), 50: 'Therese wrote much and gladly.'

[141] Steig (1891), 12.4.1824 to 13.12.1826, always Therese's letter and Goethe's answer; (1894).

[142] Milović (1941). Cf. Müller (2011), 485.

[143] Coleman (1944), 83–84 and n. 2. The three letters, dated 1835 and 1836, deal solely with translations of Slavic poetry.

[144] Grasshoff (1966).

[145] Krause (2001a); see also her discussion of 'Talvj's studies': Krause (2001).

[146] Krause (2001b).

when Köppen moved to Sankt Peterburg in 1814.[147] Köppen's trip to the Crimea and the Caucasus Mountains, encounters with Russian personalities, and exposure to the Russian nationalistic idea left deep impressions, generating an interest in Russian culture and history that led him to embark on a search for the earliest Slavonic traces and origins. During a second trip through Europe between 1821 and 1824, he met with many leading Slavists, including Kopitar and Vuk Karadžić. In Halle, where he spent ten days, Köppen met Vuk, who was working with Jacob Grimm on the latter's translations of Serbian poems.[148] Vuk met often with Goethe and was a frequent visitor of von Jakob. Talvj's letter exchange with Köppen, from 1824–52, holds much material concerning her 'Slavic circle' as well as her private life.[149] Talvj's letters attest not only to the scientific but to the deep, personal nature of the friendships in the Slavic circle, which included a long friendship between Minna, Vuk's daughter (Wilhelmine Karadžić, later Vucomanović, 1828–94, 'a gipsy girl') and her 'bosom friend' Mary Augusta Robinson (1829–1905), with whom she continued exchanging letters for many years.[150]

Talvj became interested in Serbian poetry when she read 'Jacob Grimm's favourable review of a recently [1823–24] published collection of them,' a German translation of Vuk Karadžić's volume, published at the same time. Köppen gave Talvj a copy of the original, enabling her to compare the translation.[151] 'I still did not think about printing,' writes Talvj, who adopted this pen name for the first time in her review of Vuk's work, 'I wanted only to occupy myself and to please Goethe, from him I experienced all encouragement.' Supported by her father, Talvj had published poems and short stories; however, her work in Serbian truly expanded as she developed her unique method of 'metric translation,' feeling for the rhythm of the text. She sent Kopitar some of the text for correction and, 'cleaned and improved,' the two volumes of *Volkslied-*

[147] Grasshoff (1966), 478.

[148] Grasshoff (1966), 478–79. Weigel Williams (2007), 26–31, discusses in detail Talvj's entrance into this group, until her publication of the *Volkslieder der Serben* in 1825.

[149] Grasshoff (1966); see also Coleman (1944), 84–86; Licht (2009), 16–17.

[150] Milović (1941), 85 n. 1; Mary Robinson to Minna, New York, 5.11.1853, Krause (2001a), 272–74.

[151] Robinson, M. A. (1874), xiii–iv; Dehrmann (2010), 353, and 370 n. 10; Grasshoff (1966), 479–81; Weigel Williams (1997, 12 n. 7. The opening pages, up to xviii, in the 'Einleitung' of M. R.'s published novels by her mother (Robinson, M. A. (1874)), are given in Talvj's words.

er der Serben came out in 1825 and 1826.[152] Talvj broke conventional gender boundaries when, as such a young woman, she began writing and publishing reviews (in the beginning, still anonymous: *Briefe eines Frauenzimmers über mehrere Erscheinungen der Litertur*: 'Letters of a women's room about many literary appearances ').[153] Her newly adopted pen name, which sounded more masculine, attributed to her uniqueness and most likely made it easier for the literary community to accept her.

5.b.4. Adapting to the New World

Talvj continued her literary career when she moved with her husband and family to live in 'the new world,' a difficult move, which for her meant a 'painful separation from all, who had until now been dear to my heart, and the loss of my mother language.'[154] On the other hand, according to Martha Kaarsberg Wallach, Talvj might have experienced a kind of 'freedom' from Grimm's rather constant criticism.[155] She did have mastery of the English language (as well as Latin, Greek, French, Spanish, and various Scandinavian languages[156]) and had, in 1823, translated two novels of Walter Scott.[157] Rather, she was concerned that the uprooting from her motherland to a strange world, and recurrent illness and personal loss (only two of four children born between 1829 and 1836 survived) would be the factors that 'prevented also further literary employments.' She was worried that in the United States she might not have the time to resume her literary work.

It seems that Talvj does not completely fit Stuecher's model of German-American women writers in the nineteenth century. Certainly, Talvj did share 'in the disorientation, economic insecurity, and privations of immigrant life from the very outset,' and it is also correct that the difficulties during her transitional period 'militated against the leisure time necessary for pursuing literary interests.' However, she came to the United States with an American, not German husband, to his

[152] Talvj (1825–26): Robinson, M. A. (1874), xiv; Coleman (1944), 85; Grasshoff (1966), 478–82; Mojašević (1976), 168–70. For her way into metrical translation see Weigel Williams (2007), 28–31.

[153] Wagner (1897), 8; Licht (2009), 15–16.

[154] Talvj to Kopitar, Töplitz, 9.7.1828, Milović (1941), 81; Grasshoff (1966), 479–80; Licht (1999), 28.

[155] Kaarsberg Wallach (2001), 248.

[156] Licht (2009), 13.

[157] Wagner (1897), 8; Krause (1999), 52; Weigel Williams (2007), 24–25.

country, culture, and way of life.[158] Her difficulties derived mostly from the unique and hard reality she had been 'thrown' into in Andover, which, in fact, became her 'second exile.'

'Since the beginning of November [1829] we live in Andover, a city (you know it the word cannot be translated; village, small town, whatsoever) with a theological seminary,' Talvj writes to Kopitar in June 1831. This land of Congregationalists, successors to the old Puritans, was governed, Talvj writes, by 'much diligence and true effort for the Science; but also an Orthodoxy, which I could not believe as possible in todays Christian world.' The Unitarians, with their independent thinking, faced endless difficulties with which Talvj identified, 'I live here, in Andover, **totally apart** from the society.'[159]

Talvj describes living in 'spiritual loneliness,' an inner feeling so pervasive that she writes, 'therefore is nothing left for me, but to feel that I do not in any way fit here.' 'Indeed, marriage and motherhood combined with immigration' led to a break from her impressive and intensive literary career.[160] After having been married for some three years and living over a year and a half in the States, in the same letter, she makes a rare, direct reference to her role in Edward Robinson's work, 'Perhaps, dear friend, will I always have to restrict myself to writing **my husband's** literary productions, instead of my own.'[161] Did Talvj repeat this sentiment? With all the difficulties she faced (Licht claims that she also had to work on Robinson's farm[162]), no similar reference has been found in all her letters. Time does not appear to have improved her situation or her feelings. In her next letter to Kopitar, dated February 1832, she writes, 'I could not have probably found a place in the whole world' less suited to me. Consequently, she preferred to set aside her own needs and concentrate on those of her children. She went to lectures and observed the country and its society critically, 'it is the country of freedom, but surely not of freedom of thoughts.'[163]

[158] Stuecher (1990), 24–25, 28–29.

[159] Talvj to Kopitar, Andover in Maßachusetts, 17.6.1831, Milović (1941), 84–85, emphasis in original; Kaarsberg Wallach (2001), 249; Licht (1999), 25.

[160] Stuecher (1990), 29–31.

[161] Talvj to Kopitar, Andover in Maßachusetts, 17.6.1831, Milović (1941), 84–85, emphasis in original.

[162] Licht (2009), 24.

[163] Talvj to Kopitar, Andover in Maßachusetts, 21.2.1832, Milović (1941), 87–89; see also Krause (1999), 53–55, where she cites extensively from these letters concerning Talvj's early impression of American society.

It seems that the situation did not improve significantly after leaving Andover and moving to Boston in 1833, at least in the beginning. In July 1834, Talvj wrote to Kopitar, explaining the long break of over one and half years from her last letter, 'I faced many, many heart-breaking misfortunes.' She lost two of three children and nursed Edward, who suffered repeatedly from severe illnesses. Talvj wrote that she found it difficult to muster the strength to continue. Yet having always found literary activity to be the best remedy, she was determined to re-immerse in the Slavic languages.[164] Talvj requested that Kopitar forward information about any new, relevant publications for translation in the area of Slavic people and culture. She sent Kopitar examples of her work in English published in *The Biblical Repository* and suggested exploiting existing connections between Robinson's publishing houses, the Orphanage (*Weisenhaus*) in Halle, Barth and Tauchnitz in Leipzig, Gould in Andover, and Crocker-Brewster in Boston.[165]

Though Talvj disliked Bostonian social life, the fact is that the 'Andover situation' did slowly change as she resumed writing.[166] As early as 1834, her first 'American' manuscript was published, a German translation of an English language publication on American Indian languages, with her original comments added.[167] With the aim of improving her English, Talvj submitted a rather long paper in 1834 to the new periodical *The Biblical Repository*, founded and edited by her husband, titled 'the first general view of Slavic languages and literatures available to the English reader.'[168] In 1836, she submitted two anonymous articles to 'the leading literary journal of the day,' *The North American Review*, also a forum for her husband's publications. The first article was a review of four books dealing with popular poetry of the Germans, Dutch, Danes, and Swedes and, for the following issue, an article concerning popular Slavic poetry. Talvj even pointed out, that she ordered 50 copies of her paper 'Historical View of Slavic Languages,' as it did not reach the book shops, probably for distribution to her colleagues and interested acquaintances.

[164] Talvj to Kopitar, Boston, 17.7.1834, Milović (1941), 90–91.

[165] Milović (1941), 92–94; Pribić (1999a), 13.

[166] Krause (1999), 55–58, based on Talvj's letters to the singer Auguste Loewe, second wife of the composer Johann Carl Gottfried Loewe (1796–1869), who had been married to Talvj's sister Julie who died in 1823, cf. Licht (1999), 20; Kaarsberg Wallach (2001), 249–50; Weigel Williams (2007), 20–22, 26.

[167] Talvj (1834a). Cf. Robinson, M. A. (1874), xv–vi; Wagner (1897), 12; Pribić (1973), 496.

[168] Talvj, (1834); cf. Wagner (1897), 13; Pribić (1973), 496–97 (citation).

Coleman contends that Talvj undertook a mission to supply the American people with 'intellectual nourishment and spiritual stimulation.'[169]

5.b.5. *The return to Germany*

Sailing from Boston, the Robinson family arrived in Germany in September 1837. Talvj, who had travelled quite a bit with her husband in Europe and North America, wrote that she would have happily accompanied him to the East, but there were family and children who 'needed me more than him.' Moreover, as the primary destinations were the Sinai Peninsula and the 'Land Eden [*sic*], the poor children will find themselves rocking on Camel's backs.'[170] Talvj and the children spent the winter with her sister, brother, and brother-in-law in Stettin. There she wrote a letter to Kopitar dated November 17, 1837, that Robinson was to hand deliver upon continuing his travels in Europe alone before sailing to the East. Talvj and the children then moved to Berlin from where she wrote, in a letter dated March 23, 1838, that they planned to continue to Dresden the following week and wait there for Robinson's return.[171] Shortly thereafter, in Dresden, Talvj received a visit from her long-time correspondent and close acquaintance, Jacob Grimm.[172] From Talvj's letters we also learn that she followed Robinson's expedition through his letters to her and shared details with her correspondents. Almost needless to say, Talvj intensified her work now that she was able to easily locate everything that had been published since 1834 and to expand her circle of acquaintances.[173] 'Nobody is more than her responsible for the fact, that now the whole world, including Kings, is going to Montenegro (& Serbia),' writes Kopitar to Vuk, upon learning that Talvj had a long translating 'romance' with Friedrich August II (1797–1854), King of Saxony (1836–54).[174]

169 Talvj (1836); (1836a); cf. Coleman (1944), 86; Krause (1982), 518–22; Pribić (1999).

170 Licht (2009), 26; Talvj to Köppen, Berlin, 23.3.1838, Krause (2001a), 293.

171 Talvj to Köppen, Berlin, 23.3.1838, Krause (2001a), 294–95; Weigel Williams (2007), 105–08.

172 Weigel Williams (2007), 108. Cf. Steig (1894); Talvj to Kopitar, Dresden, 28.7.1838, Milović (1941), 95.

173 Talvj to Köppen, Berlin, 23.3.1838, Krause (2001a), 294. See, for example, the detailed discussions in her letters to Kopitar: Dresden, 28.7 and 16.8.1838, Milović (1941), 94–97.

174 Milović (1941), 96 n. 1; Talvj to Kopitar, 16.8.1838, Milović (1941), 95–96.

Kopitar, who lived in Vienna, was naturally one of the first persons from whom Smith sought help for his gravely ill companion upon their return from the Holy Land.[175] Smith had the names of two of Robinson's acquaintances, Dr Julius and Mr Kopitar, though 'the residence of the latter I was ignorant of.' Kopitar, who, as mentioned, served as Kustos of the Imperial Library in the Austrian capital, was already aware of Robinson's expected arrival through a letter sent by Talvj in July 1838.[176] Smith could only have acquired Talvj's address in Dresden from Kopitar, which he did in order to send her the alarming letters concerning Robinson's critical situation, causing her to leave everything and hurry to Vienna. The planned short stop in Vienna turned into a long and rather difficult saga. Talvj writes to Grimm, that *Schmerz und Angst gaben mir die Kräfte, das fast Unmögliche möglich zu machen* ('pain and fear supplied me with the forces, to make possible [what seemed to be] the almost impossible).' As noted in chapter 3.c., where the saga is presented in detail, the whereabouts of the various parties were later reported by Smith in a nineteen-page letter to Robinson written before they left Vienna. The details of Talvj's involvement in her husband's recovery derive from a reply she sent from Dresden on September 21, 1838, to Smith in Vienna, describing her plan to leave for Vienna immediately, and from a much later, short account of the experience she wrote to Grimm in March 1840, that opens, 'some months after I had the pleasure to welcome you, a storm-wind cast me away to Vienna.'[177]

Robinson recovered, but they were forced to stay in Vienna until he was able to travel to Germany.[178] Smith left Robinson in Vienna with Talvj, arriving in Leipzig in late October 1838, and, of course, immediately started writing letters. On October 25, 1838, Robinson and his family left Vienna travelling through Dresden, Leipzig, and Halle before reaching Berlin on November 15, 1838.[179] On December 1, 1838, Robinson, now in Berlin, sent Smith a letter with a package containing 'one of

[175] Only half a year earlier, on February 8, 1838, the US appointed its first diplomatic envoy to Vienna. See chap. 3.c.

[176] Milović (1941), Einleitung; Krause (2001a), 265–97 (letters). On Kopitar: BLKO Kopitar; NDB Kopitar; Merchiers (2007); Weigel Williams (2007), 106. I assume that Julius was the physician's first name.

[177] Talvj to Grimm, Berlin, 16.3.1840, Schubert, G. (2001), 269–71 (original letter in Grimm's legacy, National Library, Berlin); Therese Robinson to Smith, Dresden, 21.9.1838, ESP, ABC 60 (63); Weigel Williams (2007), 106–07. For detail see chap. 3.c.

[178] Weigel Williams (2007), 106–07; Homes to Smith, Constantinople, 12.11.1838, ESP, ABC 60 (37).

[179] Robinson to Smith, Berlin, 23.11.1838, ESP, ABC 60 (105/9); Talvj to Kopitar, Berlin, 14.12.1838, Milović (1941), 97.

Kopitar's pamphlet[s]' that he had brought from Vienna, intended for the German philologist Moritz Haupt (1808–74).[180]

The Robinson family stayed in the Prussian capital for the next 22 months, during which Robinson would be extremely busy writing the *Biblical Researches*. As mentioned above, this stay was unplanned. Talvj even complained lightly that they could have prepared themselves better had they known in advance they would be in Berlin for so long. However, for Talvj, it turned out to be a heavenly gift, as there was no place like Berlin for meeting fellow scholars and finding material, an ideal centre for productive and enjoyable scientific work.[181] Even with a household to run and a very busy husband whom she was reluctant to disturb, Talvj displayed unprecedented productivity. Working with superior materials, her first publication appeared in 1840. Hinting at her work in the US, she wrote, 'no translation, but a totally new work.'[182] Linked to this first publication was a second manuscript dealing with the unauthenticity of Ossian's cycle of epic poems, as translated from Gaelic by the Scottish writer Jack Macpherson (1736–96).[183]

It is no surprise that Talvj published her first Berlin manuscript with Brockhaus in Leipzig. Renewing early social connections with the Brockhaus Leipzig publishing house that originated between her father and Brockhaus founder Friedrich Arnold Brockhaus (1772–1823), she developed a close relationship with his son Heinrich B. who ran the firm. Though they initially experienced some difficulties in communication and Smith had been forced to intervene,[184] the connection quickly turned into a professional association as well as personal friendship, lasting for many years and producing numerous publications.[185]

As described in Chapter 4, Robinson, who received permission from his superiors at UTS to delay his return to summer 1840, started work-

[180] Robinson to Smith, Berlin, 1.12.1838, ESP, ABC 60 (63). Cf. ADB Haupt.

[181] Talvj to Kopitar, Berlin, 14.12.1838, 12.8.1839, Milović (1941), 97–101.

[182] Talvj (1840a). Cf. Robinson, M. A. (1874), xvi–vii; Pribić (1973), 500.

[183] Talvj (1840); cf. Pribić (1973), 500–01; Weigel Williams (2007), 113–15. In 1760, Macpherson published the English-language text of ancient poetry, claiming he collected them in Scotland and translated from the Gaelic. Later he claimed to have found an epic on the subject of the hero Fingal, written by Ossian.

[184] Smith to Robinson, Leipzig, 11.3.1839, ESP, ABC 60 (105/10). Cf. NDB Brockhaus, F. A.; NDB Brockhaus, H. For H. Brockhaus and his connection to German Orientalism see Wokoeck (2009), mainly 134–36.

[185] Talvj (1840); (1840a) (see Vernahgen von Ense (1840)). Cf. Hexelschneider (2001); Krause (2001a), 308–09; (2001b), 5; Licht (2009), 14, 26; Weigel Williams (2007), 110–11.

ing intensively on various shorter publications in addition to the book,[186] diving into the treasures of Berlin's libraries and many scientific meetings and talks. Talvj was able to renew, maintain, and even improve her social connections, establishing a 'Salon,' at first against her husband's wishes (Wagner: *welcher Abneigung vor Geselligkeit zeigte*, 'which showed aversion to sociability').[187] Talvj's assessment shared with Kopitar was certainly correct, 'that overall there was no finer place where a scholarly life could be had than in Berlin.'[188]

5.b.6. Translation of the Biblical Researches

The question of who translated the three volume (over 600 pages each) *Biblical Researches* into German is somewhat of a mystery. The *Weisenhaus* publishing house in Halle accepted it under the strong recommendation of the Orientalist Emil Rödiger, who actively followed its translation and production. But who was the actual translator? The German translation of this pioneering work, wrote Ludwig Wagner in 1897, had been supervised by Talvj.[189] 'There has been much speculation as to who the translator of the German edition might have been,' writes Hermine Weigel Williams, adding that 'a number of writers have assumed Therese was the translator.'[190] Jay Williams writes that 'the German edition had been corrected if not wholly translated by Therese.'[191] Mary Robinson, Talvj and Edward's daughter, states that Robinson was the actual translator, with Therese assisting in revising and language editing.[192] She did recall that Therese translated Robinson's volume after his second expedition simultaneously 'as he wrote it.'[193]

Robinson's letter exchange with Smith illuminates this process. In spring 1839, Robinson initially considered not hiring a translator due to financial considerations.[194] In May, it appears that Robinson had

[186] For example: Robinson (1839); (1839a); (1839b).

[187] Wagner (1897), 15.

[188] Talvj to Kopitar, Berlin, 14.12.1838, Milović (1941), 97; Weigel Williams (2007), 108–10.

[189] Wagner (1897), 14.

[190] Weigel Williams (2007), 118.

[191] Williams (1999), 261.

[192] Robinson, M. A. (1862–64), 154.

[193] Robinson, M. A. (1862–64), 155; (1874), xxi; von Löher (1886), 465. The book: Robinson (1856); (1857).

[194] Robinson to Smith, Berlin, 17.4.1839, ESP, ABC 60 (63).

hired a translator, yet had been tempted to give up the whole idea as the translation was becoming 'exceedingly volatile and unscrupulous.'[195] However, a December 1839 letter makes it clear that he hired a translator, a young theologian recommended by the theologian August Neander, one of Robinson's closest supporters.[196] A careful examination of letters to Smith and Rödiger, among others, did not reveal the translator's identity. It is only natural to assume that Talvj followed the translation closely. On this matter and others relating to Edward and Talvj Robinson's scholarly collaboration, their correspondence is silent.

5.b.7. New York

The Robinsons returned to the US only in November 1840, clearly intending to settle in New York; however, they had difficulties finding a proper house.[197] The *Biblical Researches* was still in progress, Robinson being extremely busy correcting the sheets. His inauguration into the UTS faculty took place on January 20, 1841.[198] As one of the three full-time faculty members of the UTS as well as its librarian, Robinson was extremely busy (as Talvj writes, with '*die fatallen Geld- und Konstitutionsgeschäfte des Seminars* (fatal money- and constitution-business of the seminary).'[199] Moreover, since Robinson was an ordained Congregational minister, on November 15th of that year, he met UTS requirements and was ordained as a Presbyterian minister. With little spare time, the household and two young children became the almost exclusive responsibility of Talvj and the maid. Nonetheless, Talvj's literary career continued to flourish. New York offered her much better social and academic conditions (including many German-speaking *Einwanderer*) than either Andover or Boston. During the winter months, her house became 'the assembly place for New York's aristocracy, Ameri-

[195] Robinson to Smith, Berlin, 21.5.1839, ESP, ABC 60 (63).

[196] Robinson to Smith, Berlin, 24.12.1839, ESP, ABC 60 (63).

[197] Robinson to Ritter, 21.12.1840, SBB-PK, Nachl. 133 (Carl Ritter), K 5, Q – SCH, Bl. 32: 'But owing to a very (?) & absurd custom in this city, which has come down from the Dutch, all houses are rented only from the 1st of May in each year, – so that at any other season, it is next to impossible to find a house;' Weigel Williams (2007), 116–17, see the letter exchange between Robinson and his sister Elisabeth.

[198] ERP, AL COLL R65 B43 1841: Edward Robinson, *The Bible and Its Literature; an Inaugural Address*, New York: Office of the American Biblical Repository and The American Eclectic, 1841. The lecture was given on January 20, 1841.

[199] Talvj to Kopitar, New York, 5.7.1842, Krause (2001a), 267.

cans as well as Germans,' and the summer was used for travelling. The result was an impressive list of new publications, books, and articles.[200]

Throughout Talvj's impressive career, she remained opposed to any publicity of her personal matters.[201] Stuecher reflects on Talvj's reputation as an 'uneducated woman,' an image she cultivated as a wife dedicated to elevating the wisdom of her husband, who 'avoided politics and chided women's participation in the public eye, kept a low professional profile, and insisted on her own basic domesticity.'[202] This rather long citation written by her daughter is demonstrative:

> The world knows Talvj from her creations within the literary fields; everybody who was in touch with her remembers her wise and warm-hearted politeness in social intercourse, and her untiring activity as helper to the poor and the distressed placed her a monument in many hearts. But her full value derived first from her rule and work in the narrow circle. Here she appeared as sacrificing, always participating friend, as the truest, loving wife and mother, and in all intellectual efforts as caring, prudent house-woman, whose pride was that she had never started her written works before preparing her house for the day.[203]

In contrast with Europe, in the US, Talvj was challenged by a lack of relevant research material, being limited to the resources in her own library.[204] This fact may have affected her choice of genre, and she began to write novels. Her novels, which were influenced by her experience and reflected her world and personal narrative, 'reveal an underlying preoccupation with the vagaries of cultural displacement and isolation.' Certainly, the autobiographical element was central.[205]

5.b.8. Conclusion

Much has been written about the Robinsons, and though it is clear that they lived and worked together for 35 years, many of the specifics of their scholarly collaboration are unknown. Robinson wrote a long introduction to his wife's pioneering article on Serbian Folk Songs published in

200 Robinson, M. A. (1874), xviii–xxi; Wagner (1897), 15–18; Weigel Williams (2007), 119–20.

201 e.g. von Löher (1886), 451–52, 464–65.

202 Stuecher (1990), 31–32.

203 Robinson, M. A. (1874), xxviii. See also Wagner (1897), 22–23, who uses many of Mary Robinson's words for describing Talvj.

204 Talvj to Köppen, Berlin, 16.4.1852, Grasshoff (1966), 487–88.

205 Stuecher (1990), 53–58; Kaarsberg Wallach (2001), 251–53.

The Biblical Repository as early as 1834.[206] It is not clear whether Talvj already had command of the English language to such an extent that she could write a long and detailed study. It seems logical to assume that Robinson read and edited, as needed, this article and her later English publications. In addition, it was Robinson, with his network of connections, who opened doors to relevant publication forums. It is no coincidence that some of Talvj's early articles written in the US were published in *The Biblical Repository* and the *North American Review*.[207] Some years later, Robinson added a long preface to Talvj's two volumes summarizing her *Historical View of the Languages and Literature of the Slavic Nations*. The book, highly praised by Ritter in a letter to Robinson,[208] 'represents a continuation of Therese's ideas expressed in her lengthy essay published in *The Biblical Repository*.' She wrote it in English, and a German edition came out two years later, and there the title included the translator's name.[209]

What was the extent of Talvj's assistance to Robinson? In addition to the *Biblical Researches* discussed above, Robinson wrote many articles for which Talvj's collaboration must have been crucial. Notable are the scientific works that Robinson wrote and published in German. Following their marriage, the Robinsons embarked on a rather long honeymoon, travelling for 41 weeks throughout Europe, ending on May 24, 1829. On June 25th, their first daughter was born and, one year later, on July 2, 1830, they arrived in New York. Before leaving Europe, Robinson had already been appointed editor of a new journal for Biblical studies[210] and may have already started preparing material in Germany.

Robinson published many of these articles in his periodicals with the addition 'translated by the Editor.' This attribution also appears on articles by an impressive list of German scholars, theologians, Orientalists, and historians, published by Robinson in his *Biblical Repository*. The volume of 1831 includes no fewer than nine [!] translations of essays written by German scholars, 'only' three in 1832, seven in 1833, and two in 1834, the last year that Robinson edited this periodical.[211] A good example is the translations of four papers of the Leipzig theologian Johann August

206 Robinson (1859); Pribić (1973), 496; Mojašević (1976), 178–79.

207 e.g. Talvj (1825–26); (1836); cf. Weigel Williams (2007), 248–49.

208 Ritter to Robinson, Berlin, 27.1.1851, ERP, L. 4 Ritter.

209 Talvj (1850). See Weigel Williams (2007), 138–41, 250.

210 Weigel Williams (2007), 82–83; Williams (1999), 171–74.

211 Williams (1999), 185, 351–53 (Robinson's list of publications).

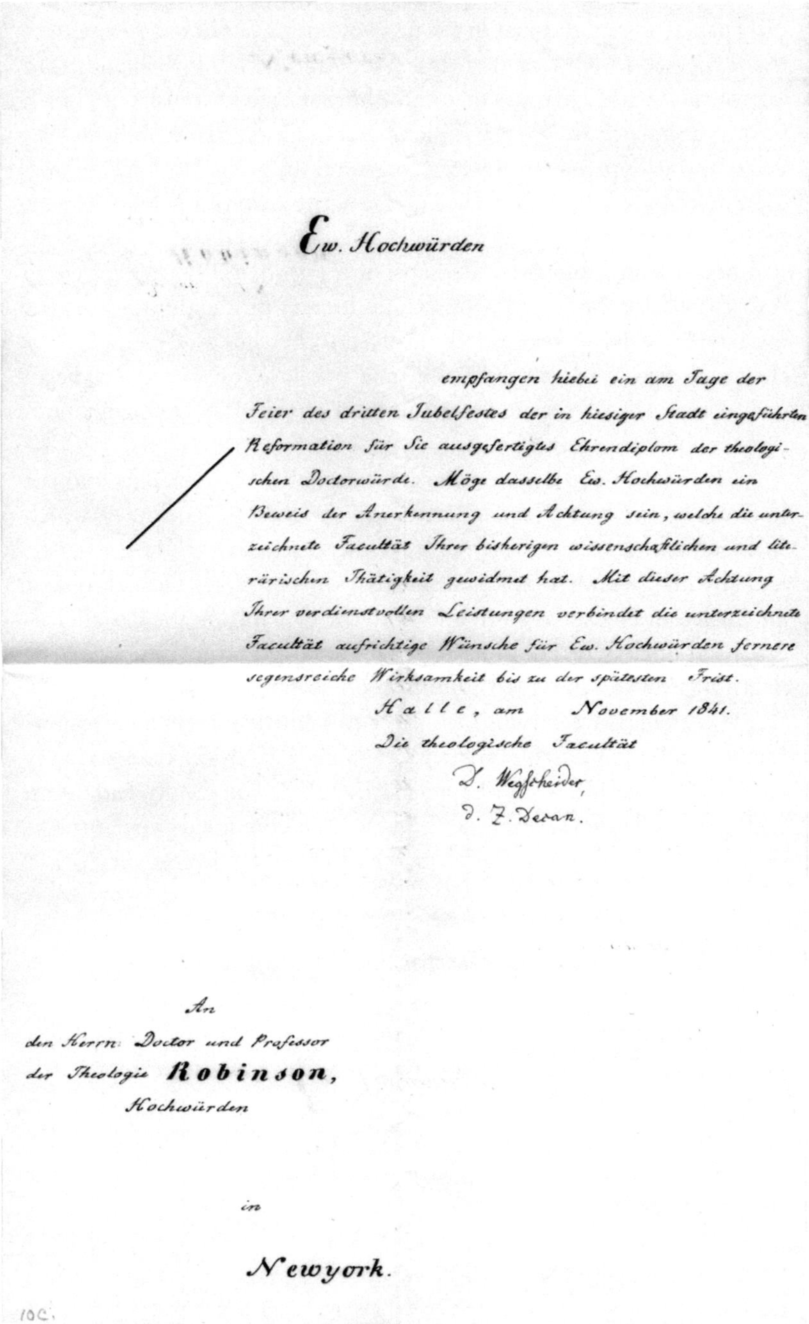

Ew. Hochwürden

empfangen hiebei ein am Tage der Feier des dritten Jubelfestes der in hiesiger Stadt eingeführten Reformation für Sie ausgefertigtes Ehrendiplom der theologischen Doctorwürde. Möge dasselbe Ew. Hochwürden ein Beweis der Anerkennung und Achtung sein, welche die unterzeichnete Facultät Ihrer bisherigen wissenschaftlichen und literärischen Thätigkeit gewidmet hat. Mit dieser Achtung Ihrer verdienstvollen Leistungen verbindet die unterzeichnete Facultät aufrichtige Wünsche für Ew. Hochwürden fernere segensreiche Wirksamkeit bis zu der spätesten Frist.

Halle, am November 1841.

Die theologische Facultät

D. Wegscheider,
d. Z. Decan.

An
den Herrn Doctor und Professor
der Theologie **Robinson**,
Hochwürden

in

Newyork.

Ill. 30. Robinson's Honorary Doctor of Theology, University of Halle, November 1841.

Heinrich Tittmann (1773–1831), who died on the last day of 1831. The early papers were published in the first volume of *The Biblical Repository* that appeared in January 1831. It is not clear whether Robinson had to ask for the author's permission or how he managed to translate so much in such a short time. Weigel Williams's contention that Robinson 'was fully capable of preparing his own English translation'[212] is logical; however, it seems less probable without the active help of his wife. This is, perhaps, the correct way to interpret Talvj's rare sentence in her letter to Kopitar from June 1831, 'Perhaps, dear friend, will I always have to restrict myself to writing **my husband's** literary productions, instead of my own.'[213] Talvj was so deeply involved in her husband's scholarship that, as Weigel Williams states, 'she may not have considered it to be an overwhelming task to put the manuscript of the "Biblical Geography" into a publishable format.'[214]

Talvj was, of course, proud of her husband's achievements and the honours he received after publication of the *Biblical Researches*, which included an honorary doctorate from the University of Halle ('but he had already been a Doctor; it seems that they probably did not look at an American Doctor as a full one') and the medal from the Royal Geographical Society of London.[215] There is no one better to describe the depth of Edward and Talvj Robinson's personal and professional relationship than Talvj herself. In a letter to Heinrich Brockhaus following her husband's death, Talvj writes, 'more than thirty years of living together in a definitely fully happy marriage based on love and attention and against all differences in views and morale, combined two natures so deep that by tearing of the bond, only half a life is left for the survivor.'[216]

5.c. James Adger, the anonymous fellow traveller

'Throughout all the journey from Cairo to Beirût, the Rev. James Adger of Charleston, S.C. was our companion and fellow traveler, except on the

[212] ADB Tittmann; Weigel Williams (2007), 82.

[213] Talvj to Kopitar, Andower in Maßachusetts, 17.6.1831, Milović (1941), 84–85, emphasis in original.

[214] Weigel Williams (2007), 214.

[215] Talvj to Kopitar, New York, 15.7.1842, Krause (2001a), 267–68; Weigel Williams (2007), 119. Robinson's archive holds the honorary doctorate certificate, the medal, and, of course, all the relevant correspondence, as described above.

[216] Cited by Licht (2009), 32.

excursion from Jerusalem to Gaza and Wady Mûsa.'[217] Though he travelled with Robinson and Smith for nearly the entire expedition, this 'secret companion' received only a brief mention in the *Biblical Researches*. Evidently, he made no contribution to the publication and it seems that he was not consulted. Williams did not mention Adger even once in his book. However, the availability of new primary sources provides a fresh opportunity to delve deeper into the expedition narrative and to clarify the connection between the Adger family and Robinson and Smith.

James Adger II (1777–1858), head of a well-to-do merchant family of Charleston, South Carolina, was a banker and owner of a shipping company. John Bailey Adger (1810–99) was the eldest of his eleven children. James Adger III (1812–82), who accompanied Robinson and Smith on their expedition, was the fourth son. His older brother John was a long time ABCFM missionary to the Armenians and later professor at Columbia Theological Seminary (1857–74). John Adger's autobiography was published in the year of his death; accordingly, much more is known about this elder Adger brother.[218] He studied at the Princeton Theological Seminary under Charles Hodge and also spent some years in Germany.[219] The young John Adger was introduced to the world of missionary activity by his classmate, the northwestern Presbyterian William McLure Thomson who, Adger wrote, was the person who most 'affected my future life.'[220] On August 2, 1834, Adger sailed with Elizabeth Keith, his bride, from Boston to Smyrna. Smyrna, with its relatively large Armenian community, would become his center for the next twelve years. Here, he served as missionary with all attendant duties, including managing the printing house, directing translation of the Bible into Armenian, printing the Bible in Armenian letters, and more.[221]

As active and central among ABCFM missionaries in the East, John Adger naturally was strongly connected to Eli Smith. In addition to their regular missionary duties, both were managing a printing house

[217] Robinson (1841), I, Preface, xiv.

[218] Adger (1899).

[219] Adger (1899), 76–80; Dearman (1991), 164–65; Williams (1999), 149–50 (citation), 158–79.

[220] Adger (1899), 76–82; Kawerau (1958), passim; Finnie (1967), 171–72, 187–88; Oren (2007), 122–24. Cf. Also DAB Thomson; the Adgers did not receive references in this biographical dictionary.

[221] Adger (1899), 90–103; Kawerau (1958), 268–72, 369–74, 392–97. Smith to Thomson and Hobbard, Leipzig, 10.12.1838, ESP, ABC 60, (105/9) 1838 Jul.-Dec.: 'I found a large printing establishment in the Armenian convent, (?) types are now making for our Smyrna press.'

and overseeing the translation of the Holy Scriptures into an Oriental language for publication.[222] But beyond their professional ties, fate led them to share a very personal and tragic experience that Adger recounted in his autobiography. Adger's first son was born on June 1, 1835, some eight months after the couple's arrival in Smyrna, and died on April 15, 1837. His second son was born in the interim on June 2, 1836, and died at the age of one year and two days, shortly after the death of his older brother. Both were buried 'in one grave alongside of the one where we had shortly before assisted in depositing the remains of the wife of the Re. Eli Smith, missionary to Beirut.'[223] The story of Mrs Sarah Smith, who 'had spent the last days of her suffering life with us in our little Turkish cottage' and her untimely death on September 30, 1836, is told elsewhere in this book.[224] The narrative includes the heartrending episode of Adger and Smith buying a house and secretly turning it into a chapel and reburying the bodies of their loved ones there one night. Smith, who lived with Adger in Smyrna, baptized Adger's third child, a girl, born on September 4, 1837.[225]

Smith's archive holds numerous letters exchanged with John Adger or referring to him. In late 1837, expecting Robinson to come to Smyrna on his way to Egypt, Smith wrote that he should come directly to Adger's house where Smith was now living.[226] Subsequent letters, from early 1838 and before Smith left for Egypt, deal mainly with professional issues such as the printing house and with Adger's wife's deteriorating health.[227] A worried Adger writes to Smith that if the plague were to reach Smyrna forcing them into quarantine, 'I shall feel it my duty to leave Smyrna for the summer and may go to Paris!' Later, when it became clear that the threat was averted, he writes, 'Mrs Adger however will probably try the effects of a change of air & going up to Constantinople.'[228]

[222] Adger (1899), 103: 'My chief business, as missionary to the Armenians, being the management of the press in modern Armenian (...), I was consequently very much confined to my desk...' This, in contrary with the always travelling Smith.

[223] Adger (1899), 104–07.

[224] Adger to Smith, Smyrna, 19.1.1839, ESP, ABC 60 (1); 'Death of mrs. Smith,' Isaac Bird's diary, IBP-YSM, Box 2, Folder 25: Journal 1827–60; Hooker (1845), 320–60; see chap. 2.d.

[225] Adger (1899), 107.

[226] Smith to Robinson, Smyrna, 8.12.1837, ESP, ABC 60 (105/7 1837 Aug.-Dec.).

[227] Hallock to Smith, Smyrna, 10.2.1838, ESP, ABC 60 (32); Adger to Smith, Smyrna, 18.2 and 6.3.1838, ESP, ABC 60 (1).

[228] Adger to Smith, Smyrna, 6.3.1838, ESP, ABC 60 (1).

Adger's plans changed, however, after being informed in early summer 1838 that his parents, two sisters, and his brother William had arrived in England. Adger tells of the trip to meet his family, 'in July we started with our babe' and 'her Greek nurse.' Both his wife and the baby were very sick throughout the trip. They were quarantined in Malta and had a long journey through France by carriage, reaching England only in early September. The baby died on September 9 shortly after their arrival and Adger's wife returned with his family to the US, 'there again, (...) we were left childless.'[229] Continuing the narrative, John Adger recounts that 'My brother James, having just arrived from his travels in Egypt, my father took him and me to his native country Antrim' (Ireland). James then accompanied John to Paris[230] from where John returned to his mission in Smyrna.[231]

In contrast to his brother John, James Adger is barely mentioned in the letters of Robinson and Smith or in their correspondence with his brother. John Adger's letter to Smith of February 18, 1838, hints at the fact that James was in Smyrna. In that letter, he mentions posting a letter to his brother in Smyrna that he invites Smith to read.[232] In a letter from Smith to his parents, written exactly a month later 'in the Desert between Suez & Mount Sinai,' Smith describes the sleeping arrangement in their tent, 'at one side is Dr Robinson & on the other Mr Adger similarly situated.'[233] A few days later, Smith writes that after 'two days wandering on foot over the deserts of Sinai (...) my companions, exhausted, have both fallen asleep.'[234]

Another letter of John Adger to Smith, dated early 1839, hints at a detail barely touched upon elsewhere regarding the fact that the travellers collected specimens *en route*. As described above, missionary Adger had been away meeting his family when Smith visited Smyrna after the expedition. Adger writes, 'I am a little curious also to know whether you brought our copper plates for me from my Brother & whether you know anything of a box of my seals collected in Egypt of which he spoke to me & which I thought you were to bring.' Adger did find a box addressed

[229] Adger (1899), 111.

[230] Adger (1899), 111–13.

[231] Adger (1899), 107–13; Homes to Smith, Constantinople, 12.11.1838, ESP, ABC 60 (32): 'Mr Adger to arrive the last of this month – childless!'

[232] Adger to Smith, Smyrna, 18.2.1838, ESP, ABC 60 (1).

[233] Smith to parents, Sudr, 18.3.1838, ESP, ABC 60 (105/8) 1838 Jan.-June. See ill. 9.

[234] Smith to Hoadly, Mount Sinai, 27.3.1838, ABC 60 (105/8) 1838 Jan.-June.

to Smith and opened it, discovering that its contents were 'all Beyroot specimens' and not the copper plates belonging to his brother, so he replaced the specimens carefully and closed the box.[235] In July 1839, Smith, in London on his way to the US, was surprised to discover that their fellow traveller, James Adger, was a shipmate. Smith's expression of his discontent is notable: 'I found him a burden in this house, the same man unchanged.'[236]

In contrast to his older brother, it appears that James Adger III did not write an autobiography and, as noted, Robinson and Smith left few clues about the man. But fortunately for contemporary scholars, a number of American researchers have published studies of their countrymen's Middle East travels. An outstanding book in this collection is Andrew Oliver's *American Travellers on the Nile*, published in 2014. The volume covers 'Early U.S. Visitors to Egypt, 1774–1839,' and, in a way, is a continuation of David Finnie's 1967 *Pioneers East*.[237] Oliver read every American travelogue, checking carefully 'who mentioned whom' and 'who met with whom' and when and where, enabling a thorough cataloguing of the whereabouts even of 'secondary' persons. This list includes James Adger, who is mentioned in the chapter dealing with 'Edward Robinson, Biblical Archeologist.'[238] 'On December 1 [1837], in Trieste, Robinson met, by chance, two younger clergymen, Rev. James Adger of Charleston and Rev. George B. Cheever [1807–90], a controversial pastor of Salem, Massachusetts.'[239] Adger, then 24 years-old, sailed from New York in late July 'with his mother and two of his sisters,' probably only shortly after Robinson's family set sail and, as Oliver supposes, 'perhaps intending to reach Smyrna where one of his brothers, John, was stationed as a missionary.'[240]

Without further evidence, it is impossible to establish James Adger's original objectives for his journey to the East or why or how he joined Robinson and Smith on their expedition. From their meeting in Trieste and until their arrival in Beirut in July 1838, James Adger accompanied Robinson on his brief trip to Egypt (together with Cheever) and then

[235] Adger to Smith, Smyrna, 19.1.1839, ESP, ABC 60 (1).

[236] Smith to Robinson, London, 8.7.1839, ESP, ABC 60 (105/11) 1839 Apr.-Dec.

[237] Oliver (2014); Finnie (1967).

[238] Oliver (2014), 235–41.

[239] Oliver (2014), 236. On Cheever: DAB Cheever; Mackey (1973).

[240] Oliver (2014), 236. Their father, who was with them (Adger (1899), 107), is not mentioned. Their brother William was also present but it is not clear when he arrived (Adger (1899), 111).

Robinson and Smith from Cairo through the Sinai to Jerusalem and throughout the Holy Land. For some unknown reason, Adger did not join the excursions to Gaza and Wadi Musa, foregoing the opportunity to visit Petra. On July 8, 1838, they sailed from Beirut to Alexandria (where they were quarantined on board) and set sail for Syra and Smyrna a week later on July 17. As noted above, James Adger subsequently set out for England, arriving there in September 1838. After travelling with his brother John to Ireland (together with their father), back to London (to escort their brother William who sailed home), and on to Paris (from where John returned to Smyrna), James Adger returned to the US in late July 1839.[241]

[241] Adger (1899), 111–12, 236–41, including additional details about their Egyptian travel.

CHAPTER 6
REVISITING THE DISCOVERY OF ROBINSON'S ARCH

Robinson and Smith's epoch-making publication, the *Biblical Researches*, generated numerous scientific discussions over Robinson's approach and his interpretation of findings. Many of the discussions were contentious, and the list of contemporary opponents counted by Ritter is quite impressive. It includes George Williams and Ernst Gustav Schultz, the Bonner theologian Wilhelm Ludwig Krafft (1821–97), and even the leading German Egyptologist Karl Richard Lepsius (1810–84).[1] Williams (1814–78), Church of England clergyman, served for more than two and a half years (1841–43) as chaplain of the church of the first Protestant Bishopric in Jerusalem. He conducted most of his studies in the company of the young Orientalist Ernst Gustav Schultz, first Prussian Consul to the city.[2] Late eighteenth- and early nineteenth-century Holy Land scholars also critiqued Robinson's studies.[3] Scholars debated most vehemently, however, over Stephen Olin's (1797–1851) challenge of the attribution to Robinson of the discovery of an initially unimposing group of stones jutting from the western wall of the Temple Mount in Jerusalem. 'Robinson's Arch' is the reason that the name Edward Robinson is well known to scholars and the non-academic public alike, discussed in nearly every guidebook, tour, and research publication concerning the connection between the Temple and other parts of the an-

1 Ritter's letter to Robinson, Scientific Proceedings (1847). See Krafft (1846); Williams, G. (1849). For Krafft's involvement in Holy Land studies: Krafft to Ritter, Bonn, 26.7.1848, SBB-PK, Nachl. 133 (Carl Ritter), K 3, J – L, Bl. 96–97.

2 Williams, G. (1849); cf. ODNB Williams. For Schultz: Goren (2003), 194–201.

3 Stinespring (1939).

cient city.[4] However, Olin claimed that others discovered the arch and even connected the arch to the bridge described by Flavius before Robinson.[5]

John Price Durbin (1800–76), a Methodist clergyman, taught at Wesleyan University and served as Chaplain of the United States Senate and president of Dickinson College. He visited the Holy Land in 1843. The forward to his *Observations in the East* contains this neutral passage:

> My principle guidebook in the Holy Land, besides the Bible, was the *Biblical Researches* of Messrs. Robinson and Smith. Their exceeding accuracy was a matter of daily surprise to me in my travels through Palestine; and I must express a deep sense of obligation to such indefatigable and successful observers. I had not then seen Dr Olin's excellent volumes, which convey a great amount of information, also remarkably accurate.[6]

Using notes on Catherwood's 1835 map of the Temple Mount, Durbin describes the arch in detail:

> To the south of the place of the wailing is one of the most interesting relics of antiquity to be found in Jerusalem – the remains of the ancient bridge which was thrown over the Tyropœn, to connect mount Zion with the Temple buildings. About fifty feet from the southwest angle of the temple area, in the western wall, three courses of immense stones are springing from the wall, obviously forming part of a large arch.[7]

In the first volume of their *Biblical Researches*, Robinson and Smith write that upon initial observation, they and their companions thought that a strong natural force such as an earthquake caused the stones to jut out from the wall. Describing his tour on Wednesday, April 18, 1838, only four days after reaching Jerusalem, Robinson notes that 'we paid little attention to this appearance at the time; but subsequent examina-

[4] For a relatively recent summary of excavation and views with regard to the role of the arch see Reich (2015); Gibson (2020), 6–8 [temporary pages, in press]. Onn (2010) and Bdolach claimed in 2010 that the real bridge connection was Wilson's Arch, which had actually been first discovered by Tobler (Gibson (2020), 7 n. 5).

[5] Flavius (1895), Book 15, [410]–[411]; cf. Ben-Dov (1982), 115, 121–23.

[6] Durbin (1845), iv.

[7] Durbin (after 1838), 26.

tion led to one of our most interesting discoveries.'[8] In the next section, 'Jerusalem: Topography and Antiquities,'[9] Robinson describes returning to the arch to study it methodically. After discussing it 'in a circle of our friends, we found that they had also noticed it; and the remark was incidentally dropped [Robinson did not mention by whom!] that the stones had the appearance of having once belonged to a large arch.' Taking his time and checking the stones carefully, Robinson lists the reasons for his decision that 'they form the commencement or foot of an immense arch, which once sprung out from this western wall in a direction towards Mount Zion...'[10] He notes that he measured the stones, calculating 'the proximate length of the ancient bridge,' and concluded decisively that 'the existence of these remains of the ancient bridge, seems to remove all doubt as to the identity of this part of the enclosure of the mosque with that of the ancient temple.' He concludes with an interesting thought, 'how they can have remained for so many ages unseen or unnoticed by any writer or traveler, is a problem.'[11]

Robinson's words were prophetic. Quite soon thereafter, Robinson was confronted with Olin's claim that he was not the first to discover the arch. Discussion of this dispute, its background, stages, and participants, will illuminate whether this special monument should be attributed to Robinson or to someone else. Olin, a Methodist Episcopal Church minister from South Carolina, became a theologian, educator, and professor at the University of Georgia. He was the first president of Randolph Macon College (1834–37), founded in Boydton, Virginia (which later relocated north to avoid Civil War fighting), and then president of Wesleyan University from 1839 until his death in 1851. In spring 1837, following his term at Randolph Macon College, Olin left for Europe 'broken and exhausted by protracted sickness.' He remained 'more than a year' in Paris 'for medical advice,' reaching Rome in the winter of 1838–39 where he regained his health. In May 1839, he lost his wife who had been accompanying him, and only in 1840 travelled to the East. Olin believed strongly that going on pilgrimage was akin to having been born in the Holy Land; only there does the Bible become real.[12] He first published his book *Travels in Egypt* in 1843 using Catherwood's

8 Robinson (1841), I, 351.

9 Robinson (1841), I, 424–25.

10 Robinson (1841), I, 424–25.

11 Robinson (1841), I, 426.

12 Olin (1860), Preface, vii–xiv.

map of Jerusalem.[13] By 1860, several editions had been printed, and the book received almost unprecedented praise from the very critical expert Titus Tobler. Tobler describes the book as 'detailed, rich in contents, noteworthy. After Robinson and Smith, I think that no American had better written.'[14]

In his forward, Olin notes two prominent contemporary American travellers who raised new issues and contributed fresh perspectives to the well-worn yet, in America, still exceedingly popular Palestine pilgrimage literature: 'After enjoying favourable opportunities for forming an opinion upon the subject, I bear the willing testimony to the high and peculiar merits of both these authors; of Mr Stephens (...); and of Dr Robinson, who has contributed so largely to its gratification.'[15] Why Robinson? Because 'he has entitled himself to the gratitude of biblical scholars and of the religious public,' as his work is 'rich in the fruits of extensive erudition, and in the reports of faithful and shrewd observation.'[16] The fact that Olin mentions Robinson in various sections of his book indicates that he did read the *Biblical Researches*. Moreover, he opines, 'I have since examined the opposite hypothesis, and have bestowed especial attention upon the very able and learned argument of Dr Robinson, without, however, being able to adopt his conclusions.'[17] It is not surprising that the Methodist southerner disputed Robinson, his theories, and conclusions; his objections might have also stemmed from their opposing attitude towards slavery and abolition.

After publication of Olin's book, the issue of 'who discovered the arch' was raised in various publications: a review of the book, Olin's letter to the editors of the review, two rebuttals by Robinson, and more. The discussion took place in 1843–44, primarily in the *North American Review*, the first American literary magazine. The magazine was found-

[13] Olin (1860), Preface, xi–xii for his explanation of the delay in publication. Map: II, facing title. For Olin's biography: DAB Olin. Wesleyan University in Middletown, Connecticut, was founded in 1831 as an all-male Methodist college.

[14] Tobler (1867), 164.

[15] Olin (1860), Preface, viii. For detail on the lawyer John Lloyd Stephens, later also famous as 'the Maya Explorer': von Hagen (1948); for his part in reconstructing Costigan's 1835 Dead Sea exploration: Goren (1997a), 67–70.

[16] Olin (1860), Preface, viii. Robinson met Stephens in New York, the latter also promised him to send Ritter a copy of his 'Travels in Central America' (Robinson to Ritter, 1.10.1841, SBB-PK, Slg. Darm. Asien 1841), which he actually did (Robinson to Ritter, New York, 31.12.1841, SBB-PK, Nachl. 133 (Carl Ritter), K 5, Q – SCH, Bl. 34.

[17] Olin (1860), II, 277.

ed in 1815 in Boston, and from 1820 became a quarterly focusing on social and cultural issues.[18]

The reviewers of Olin's book, who were most likely the editors of the New England periodical, did not veil their strong criticism of his *Travels in Egypt*. Their critique does not stop at his failure to consult Robinson's book; they clearly accuse Olin of plagiarism. 'That Dr Olin should have availed himself of its statements is matter neither of complaint nor surprise,' they write somewhat cynically, 'but we regret to find observations, for which he was manifestly indebted to it, appearing in his work as original.'[19] Their examples include 'the pointing out and determination of the true character of the arch (...),' the 'most important and striking of all of Dr Robinson's discoveries.'[20] They note that Robinson made his discovery in 1838 and published it immediately when, in fact, the book was published only three years later. Additionally, they note that Olin visited the city in the spring of 1840, and it took three more years until his book was published:

> He describes the same remains, and quotes the same authorities for their character. And not only does he make no acknowledgement of Dr Robinson's discovery, but he adds the following statement. 'I could not learn, that the most interesting and unquestionable of these remains – the massive arch of the ancient bridge – had been so much as mentioned by any modern traveler, though its existence had long been well known to European and other residents, as well as visiters [*sic*].' The first part of this sentence is to us wholly inexplicable; the reminder we know to be a mistake.[21]

Olin could not remain silent in the face of such an extreme accusation. On November 30, 1843, he sent a letter to the editors that was published in January 1844. His aim was 'to offer a full and unqualified denial of the charge of plagiarism,' which he did by a detailed chronology of the events and participants. First, Olin notes that his visit took place between Robinson's visit and the publication of the *Biblical Researches*. Second, Olin claims that a 'large portion' of his own book was already in press when he saw Robinson's work. It is true, Olin admits, that on

[18] Eds of the *NAR* (1843); Olin (1844); Robinson (1844). Robinson (1844a) was published in the *BS*.

[19] Eds of the *NAR* (1843), 492.

[20] Eds of the *NAR* (1843), 492–93.

[21] Eds of the *NAR* (1843), 492–93.

his way to the East in 1839, he learned from 'Mr Wheaton, in Berlin' about a synopsis published by Robinson but was unable to obtain it. Olin notes additional reasons for the fact that he was aware of Robinson's writing only in the beginning of 1842 – too late, he claims, to make changes in his own book.[22]

Henry Wheaton, scholar of international law, served from 1835 to 1846 as US envoy, then called 'Minister,' to the Prussian capital.[23] Robinson mentions Wheaton in his January 22, 1839 letter to Smith: 'I put the mss.[24] as far as Jerusalem into Mr Wheaton hands, who has read it & expresses a much more favorable opinion than I expected. He thinks the topographical notices cannot well be omitted, & that there is enough of personal narrative & interest attached to the subject, to overbalance any objections.'[25] The diplomat did help Robinson and Smith deliver their letters (Robinson: 'Whenever you are writing to the United States, your best way perhaps would be to send them to me, & I will forward them through Mr Wheaton').[26] Robinson's and Wheaton's paths were due to cross again, almost ten years later. 'Not a year ago,' writes Robinson in another letter, dated September 27, 1848, 'Lynch & Dale & Mr Wheaton just returned from Berlin, sat around my table, all in high health & spirit & full of hope & expectation; now two of them are in their graves...'[27]

It is only natural, Olin continues, that there will be 'numerous coincidences' between the two travelogues, both written 'amid the scenes which they describe' or composed from notes written on the spot, relying upon the same guidebooks and information 'from natives, or

22 Olin (1844), 253.

23 DAB Wheaton; Wheaton (1889); Kellen (1902); Findling (1989), 543. In 1827, President Adams sent him as Chargé d'Affaires to Denmark, in 1835, he was transferred to Berlin, and in 1837, promoted there to 'Envoy Extraordinary and Minister Plenipotentiary' by Van Buren. He returned home in 1847. See also chap. 4.b. 4.

24 ERP-UTS: Edward Robinson Papers, 1836–38. Series 1: Writings. Box 1. Robinson dated his manuscript of the *Biblical Researches*. On 14 December 1838 he started writing the second part of section X, 'Excursion to 'Ain Jidy.' 'As far as Jerusalem' are sections I–V.

25 Robinson to Smith, Berlin, 22.1.1839, ESP, ABC 60: Robinson, M. A. (1862–64).

26 Robinson to Smith, Berlin, 13.12.1838, ESP, ABC 60 (63).

27 Robinson to Smith, New York, 27.9.1848, ESP, ABC 60 (63). Robinson's involvement in William Francis Lynch's (1801–65) Jordan and Dead Sea Exploration (1848) should be further studied. John B. Dale (1814–38), naval officer, painter, and Lynch's second in command, died in Beirut before returning home. Cf. Lynch (1849); Jampoler (2005); Bain (2011); Goren (2011), 149–61.

resident foreigners [i.e. Smith], or prevalent traditions.' Following this explanation, Olin answers each accusation in detail. The first accusation involves identifying a larger mountain in place of 'the monkish Sinai' as the site for the giving of the Ten Commandments, big enough to contain 'the situation of the encampment.' This issue, he writes, was discussed previously by Carne, Laborde, and Lord Lindsay, all of whom preceded both him and Robinson and had varying influence. In this, 'my account of Mt. Sinai is no more indebted to Dr Robinson's, than his to mine.'[28] The English traveller John Carne (1789–1844) visited the East in 1821 and published the first edition of his *Letters from the East* in 1826. Tobler was unimpressed by his book, writing that it demonstrated 'no thorough research.'[29] Carne's countryman Lord Lindsay, 'nobleman, traveller, and writer on art' and former student at Trinity College, Dublin who held two earl titles, spent his life in studious pursuits, in the collection of a magnificent library, and in travel. In 1836–7, he travelled in Egypt and Palestine accompanied by friends. Tobler describes his letters, published in 1838, as 'für Jerusalem werthlos, [for Jerusalem worthless]' but they are of much use for this study.[30] Laborde's major role in Robinson's routes and publications was discussed above.[31]

In the next paragraph of his response, Olin confronts the charge that he unjustly disputed Robinson's claim of discovery of 'the true character of the arch.' Olin refers to missionary John Nicolayson, his guide 'to the monument,' who claimed that the existence of the protruding stones (not of the arch itself) was a well-known fact, identified by Catherwood seven years prior to Robinson.[32] Olin adds that Catherwood, not Robinson, was the first to use the term 'monument' for the arch and, importantly, Catherwood measured the Temple Mount in 1833, five years prior to Robinson's visit.[33]

Frederick Catherwood became famous as companion to Maya explorer John Lloyd Stephens on his travels in Central America between 1839 and 1841; both also travelled in the East.[34] 'A man of many lives,'

[28] Olin (1844), 253–54.

[29] Carne (1826); Tobler (1867), 148. For Carne: ODNB Carne; Dawson, W. R. (1995), 213.

[30] Lindsay (1838a); Tobler (1867), 159; cf. ODNB Lindsay; Paxton (1839), 211–12.

[31] Cf. mainly chap. 4.c.

[32] For detail on Nicolayson, see chap. 2.b.

[33] Olin (1844), 254.

[34] von Hagen (1948); An American (1858); Davis, J. (1996), 59–72.

Stephens' biographer describes Catherwood as an 'architect, explorer, archaeologist, draughtsman, artist, daguerreotypist, surveyor, and railroad builder.' Romanticism strongly influenced his architecture, and he sketched archaeological remains whenever possible. Earlier, in 1822, Catherwood left for Greece with a group of English artists, sculptors, and architects, but due to the Greek war (war of independence or civil war, or both) was forced to flee through Syria to Egypt dressed as an Arab. Returning to London, Catherwood met in Malta with the wealthy Scot, Robert Hay (1799–1863), whose dream was to underwrite a systematic study of Egypt, its ruins and culture, in competition with the French project of Napoleon's scientists.[35] In 1824, Hay hired artist Joseph Bonomi, whom he had met in Rome and who was fluent in Arabic, to accompany him to Egypt. They stayed in Egypt from November 1824 until 1828, and again from 1829 to 1834, recording monuments and inscriptions and drafting a large number of architectural plans.[36] Following Hay's invitation, Catherwood joined his team of artists and architects and left for the East in 1829. Dressed as a Turkish officer, he sketched many ancient sites and became architectural adviser to Muhammed 'Ali. In 1833, Catherwood, accompanied by Bonomi, visited Jerusalem in his familiar disguise, claiming he had been sent by the Egyptian ruler. 'From a combination of fortunate circumstances (...) while engineer to the Pasha of Egypt,' Catherwood sketched and measured the Temple Mount for six weeks until he received news that Ibrahim Pasha, 'Ali's son and governor of Syria, was approaching the city. In 1834, Catherwood returned to London.[37]

In a declaration dated April 23, 1840, published only in his 1844 letter to the *NAR* ('my declaration, that I could not learn that this monument had been mentioned by any modern traveller'), Olin asserts that '(...) I never saw or heard the name of Dr Robinson connected with this

[35] The first full biography of Hay, a 'forgotten Egyptologist,' was published only in 1984 (Tillett (1984)). The study includes much information about the artists he recruited. See for Hay, Catherwood and Bonomi, Simpson (1879); von Hagen (1968), 27–32; Searight (1979), 201–03, 238–41; Tillett (1984).

[36] Bartlett, W. H. (1844), 151; Dawson, W. R. (1995), 53–54, 194; Thomson (2010), 112–16. For an exact timetable of the Egypt expeditions: Tillett (1984), preface.

[37] Durbin (1838), 1 (Catherwood's citation); Dawson, W. R. (1995), 53–54; Cooke (1998), 86–93; Thompson (1998), 131–35; von Hagen (1948), 61–62; (1968), 32–40; Bartlett, W. H. (1844), 148–65 (Catherwood's detailed letter concerning his work in Jerusalem); Ben-Arieh (1973), 65–69; (1974); Gibson (1996), passim.; (2020), 6–8. Cf. also Dabach (2016), 28–40. For Catherwood as 'Holy Land painter' see Ben-Arieh (1997), 42–47 (on 47 his 'Interior of the Dome of the Rock' from 1833); for Bonomi: Khatib (2003), 83–84; For Catherwood's Panorama in London: Burford (1835).

subject in Jerusalem or elsewhere.'[38] In the 1844 letter he continues that upon reading the *Biblical Researches* in early 1842, he could find no reason to distrust the information included and accepted that Robinson discovered the arch. Olin then presents his argument against Robinson. He argues that given that architect Catherwood was also 'the author of Dr Robinson's plan of Jerusalem,' and whose own map depicts the valley lying between the Temple and Mount Zion, Catherwood 'could hardly have doubted or been mistaken with regard of its design.' Olin met Catherwood several times and the latter 'regarded and spoke of this monument' as the remains of an ancient bridge.[39] Hence Olin's conclusion that Catherwood, or even someone predating Catherwood, such as Nicolayson, should be accepted as the real discoverer of the arch and not Robinson.

In his rebuttal letter, Olin responds to additional accusations of plagiarism of Robinson, including identification of 'the Tower of David,' which contemporary scholars assign to Josephus Flavius (the Hippicus tower), as well as the pool of Bethesda. He refers continuously to his guides in the city, LJS missionary Nicolayson and ABCFM missionary Lanneau,[40] and the fact that he wrote his book based solely upon his own observations: 'I have not knowingly derived either facts, arguments, or opinions from Dr Robinson.' The editors of the *North American Review* note that they were publishing Olin's letter untouched, and that it was up to the readers 'to judge (…) how far the explanation is a satisfactory one.'[41]

A thorough investigation of Catherwood's earliest issue of the map, signed 'London: Published Aug. 1st 1835, by F. Catherwood, 21, Charles Square, Hoxton,' failed to locate any notation or mention of the arch. However, The Risa and Richard Domb Collection in the Israel Museum houses another version of the map, undated, but clearly originating much later, this time showing an 'ancient bridge' outside the southwestern corner of the Temple Mount. The title of the later map is 'Plan of Jerusalem by F. Catherwood, Architect, with the improved measurements of Professor Robinson.' Clearly, the second map could not have been

38 Olin (1844), 255.

39 Olin (1844), 254–55.

40 For detail on Lanneau, see chap. 3.a.

41 Olin (1844), 255–56. Olin answers the editor's accusations, one to one: the true Mount Sinai (253–54), the arch (254–55), the Dead Sea (255), the Tower of David (255), the pool of Bethesda (255–56), and more.

Ill. 31–32. Catherwood's Map of Jerusalem, 1835 and n.d. (1839-40?).

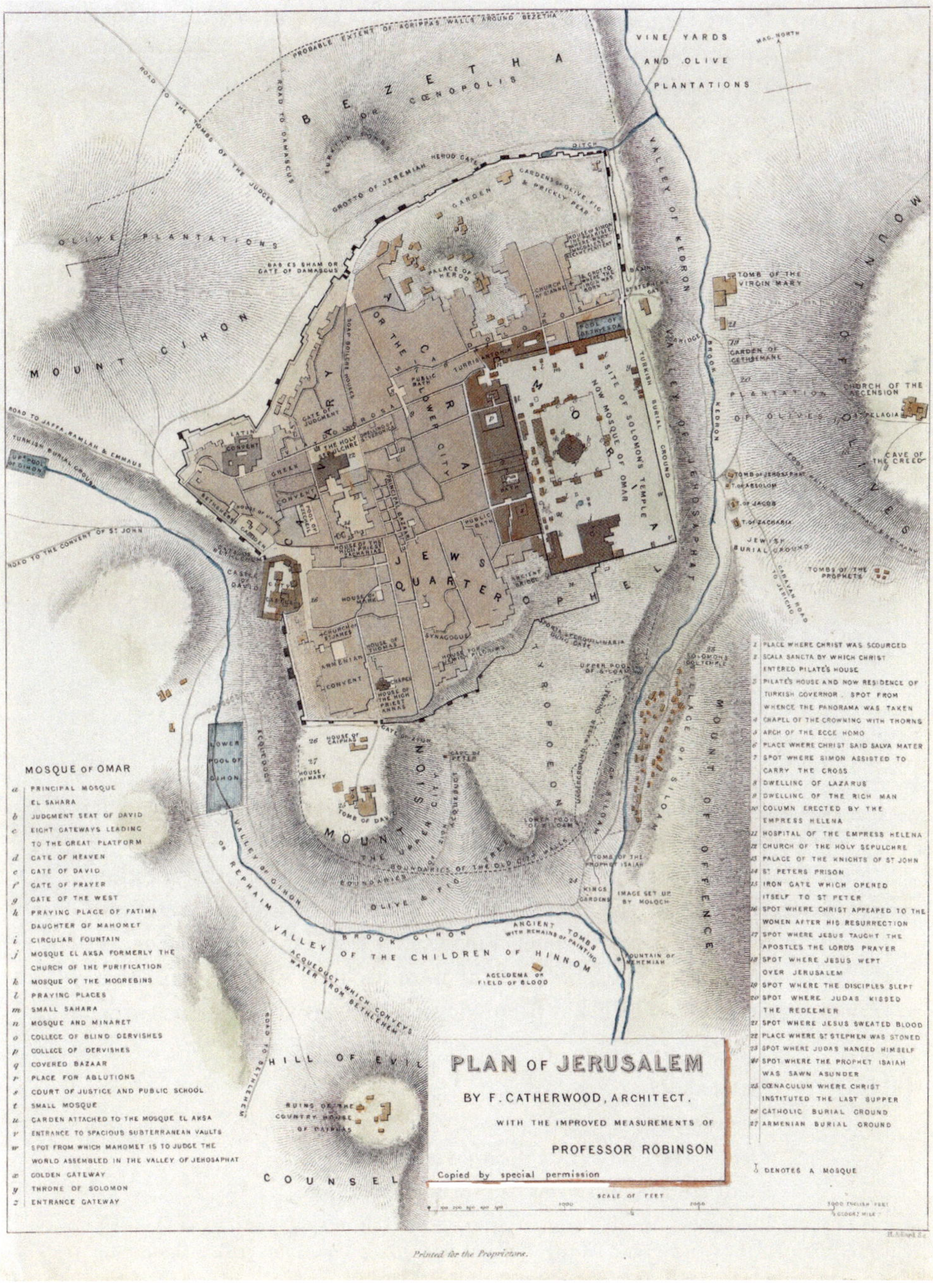

PLAN OF JERUSALEM
BY F. CATHERWOOD, ARCHITECT.
WITH THE IMPROVED MEASUREMENTS OF
PROFESSOR ROBINSON
Copied by special permission
SCALE OF FEET
MOSQUE OF OMAR
a PRINCIPAL MOSQUE
EL SAHARA
b JUDGMENT SEAT OF DAVID
c EIGHT GATEWAYS LEADING
TO THE GREAT PLATFORM
d GATE OF HEAVEN
e GATE OF DAVID
f GATE OF PRAYER
g GATE OF THE WEST
h PRAYING PLACE OF FATIMA
DAUGHTER OF MAHOMET
i CIRCULAR FOUNTAIN
j MOSQUE EL AKSA FORMERLY THE
CHURCH OF THE PURIFICATION
k MOSQUE OF THE MOGREBINS
l PRAYING PLACES
m SMALL SAHARA
n MOSQUE AND MINARET
o COLLEGE OF BLIND DERVISHES
p COLLEGE OF DERVISHES
q COVERED BAZAAR
r PLACE FOR ABLUTIONS
s COURT OF JUSTICE AND PUBLIC SCHOOL
t SMALL MOSQUE
u GARDEN ATTACHED TO THE MOSQUE EL AKSA
v ENTRANCE TO SPACIOUS SUBTERRANEAN VAULTS
w SPOT FROM WHICH MAHOMET IS TO JUDGE THE
WORLD ASSEMBLED IN THE VALLEY OF JEHOSAPHAT
x GOLDEN GATEWAY
y THRONE OF SOLOMON
z ENTRANCE GATEWAY
1 PLACE WHERE CHRIST WAS SCOURGED
2 SCALA SANCTA BY WHICH CHRIST
ENTERED PILATE'S HOUSE
3 PILATE'S HOUSE AND NOW RESIDENCE OF
TURKISH GOVERNOR. SPOT FROM
WHENCE THE PANORAMA WAS TAKEN
4 CHAPEL OF THE CROWNING WITH THORNS
5 ARCH OF THE ECCE HOMO
6 PLACE WHERE CHRIST SAID SALVA MATER
7 SPOT WHERE SIMON ASSISTED TO
CARRY THE CROSS
8 DWELLING OF LAZARUS
9 DWELLING OF THE RICH MAN
10 COLUMN ERECTED BY THE
EMPRESS HELENA
11 HOSPITAL OF THE EMPRESS HELENA
12 CHURCH OF THE HOLY SEPULCHRE
13 PALACE OF THE KNIGHTS OF ST JOHN
14 ST PETERS PRISON
15 IRON GATE WHICH OPENED
ITSELF TO ST PETER
16 SPOT WHERE CHRIST APPEARED TO THE
WOMEN AFTER HIS RESURRECTION
17 SPOT WHERE JESUS TAUGHT THE
APOSTLES THE LORD'S PRAYER
18 SPOT WHERE JESUS WEPT
OVER JERUSALEM
19 SPOT WHERE THE DISCIPLES SLEPT
20 SPOT WHERE JUDAS KISSED
THE REDEEMER
21 SPOT WHERE JESUS SWEATED BLOOD
22 PLACE WHERE ST STEPHEN WAS STONED
23 SPOT WHERE JUDAS HANGED HIMSELF
24 SPOT WHERE THE PROPHET ISAIAH
WAS SAWN ASUNDER
25 CŒNACULUM WHERE CHRIST
INSTITUTED THE LAST SUPPER
26 CATHOLIC BURIAL GROUND
27 ARMENIAN BURIAL GROUND
DENOTES A MOSQUE
BEZETHA
MOUNT GIHON
MOUNT OF OLIVES
MOUNT OF OFFENCE
MOUNT ZION
HILL OF EVIL COUNSEL
VALLEY OF JEHOSAPHAT
VALLEY OF THE CHILDREN OF HINNOM
VINE YARDS AND OLIVE PLANTATIONS
OLIVE PLANTATIONS
JEWS QUARTER
MORIAH
OPHEL
Printed for the Proprietors.

drawn before 1839, and most likely even later, in the early 1840s.[42] When Robinson and Smith published a map of the city in 1840, as part of their rivalry with Buckingham, they did not depict any object or structure in the southwestern corner of the Temple wall.[43]

Verifying whether Kiepert used Catherwood's map and data for the map of Jerusalem he drew for the *Biblical Researches* can shed light on the question of the arch's discoverer. In his *Memoir*, Kiepert describes all his sources in detail.[44] As early as January 1838, Smith promised Robinson to acquire 'Catherwood's plan of Jerusalem' for use on the expedition.[45] However, it is unclear if and when Robinson saw the map. In a February 1839 letter, Robinson notifies Smith:

> In the London Booksellers' circular, I see advertised from America, a new 'Topographical Map of Jerusalem & the Surrounding Country. In the style of Wilkinson's Thebes. By Catherwood. Price 2. 12. 4. warranted.' To judge from the model & price, this must be a very large & splendid map; but what materials can he have to justify such a work? I have put matters in trace here among the literary men, to obtain a copy as soon as possible; & we shall then see. His plan is not accurate in all respects; he makes the wall from the Jaffa gate near W.N.W. whereas it is N. 40 W. Dr Parthei who was there in 1825 has put into my hands, the angles etc. taken by his companion Westphal, with a sextant. Perhaps something will be made of them.[46]

In a letter written in early April, Robinson updates Smith that he sent for Catherwood's large map of Jerusalem, intending to make a present of it to Carl Ritter; however, this letter does not answer the question of whether he saw the map.[47] Kiepert, nonetheless, is very clear: Robinson and Smith's measurements of the city were far from sufficient to draw a new map. Thus, he used the most accurate existing map, 'that of Catherwood, which has been adopted as a basis,' and corrected it.

[42] Catherwood (1835); (n.d.).

[43] Robinson (1840a).

[44] Kiepert (1841); (1841a). Cf. Robinson (1844), 254 note.

[45] Smith to Robinson, Smyrna 6.1.1838, ESP, ABC 60 (105), 1838.

[46] Robinson to Smith, Berlin, 11.2.1839, ESP, ABC 60 (63). For Johann Heinrich Westphal (1794–1831) and Parthey [actual spelling] see Goren (2015a). For the leading and world-famous British Egyptologist, Sir John Gardner Wilkinson (1797–1875), see, e.g., Searight (1979), 200–41; Dawson, W. R. (1995), 443–45. For Catherwood's maps: Ben-Arieh (1973), 66–67; (1974).

[47] Robinson to Smith, Berlin, 3.4.1839, ESP, ABC 60 (63).

Catherwood's plan, he writes, is almost identical to the 1828 German version published in Jakob Berggren's *Reisen*, and both maps 'appear to have been reduced from Sieber's large plan of Jerusalem.' Because it was accepted that Catherwood measured the Temple Mount, Kiepert did not change his plan of that area, and mainly corrected the city walls, their directions, angles, and length.[48]

The first of these earlier maps referred to by Kiepert was that of the Swedish clergyman, traveller, and doctor of theology Jacob Berggren (1790–1868), who was ordained in 1818 and went to the Swedish embassy in Constantinople where he worked until 1822. In August 1820, he travelled to Syria, Egypt, and Palestine and later journeyed into the interior of Russia. In addition to *Reisen*, Berggren wrote about the relationship of works of Josephus and Philo to the Gospels.[49] The second map referred to was that of German physician Franz Wilhelm Sieber (1789–1844) from Prague. Sieber was a frequent traveller whose primary interest was in building natural history collections, mainly herbaria, presenting and selling them to finance further travels. Following his 1818 six-week visit to Jerusalem, Sieber issued the first map of the city and its surroundings based upon at least partial geographical and topographical data from his own observations.[50]

No mention of Olin or discussion of refuting his claims was found in Robinson and Smith's extensive letter exchange; however, Robinson's archive holds a copy of his inevitable response to the editors of the *NAR*. Signed June 10, 1844, and published in the same month, Robinson requested space in the journal to clarify only one point, that of the arch and the bridge.[51] His response to Olin seemed of tremendous importance to Robinson, as he duplicated it, with minor corrections, in the November volume of the *Bibliotheca sacra*. Robinson found it necessary, 'partly for the purpose of introducing that letter; and partly as a matter of literary history relating to an interesting point in Jewish antiquities.'[52] Perhaps Robinson thought the *NAR* was insufficiently popular.

[48] Kiepert (1841), 39 and n. 1. See also Robinson to Smith, Berlin, 15.4.1840, ESP, ABC 60 (63): '(...) Catherwood's small plan of Jerusalem is a mere copy from a German one published before 1830; his large one I have not been able to get sight of; but it cannot have been made from actual survey, as claimed.' See Ben-Arieh (1973), 69.

[49] Berggren (1826); (1832).

[50] Sieber (1818); (1823); Goren (2001a); (2003), 62–63.

[51] Robinson (1844); ERP, AL COLL R65.

[52] Robinson (1844a), 794. He did add here that these remarks appeared in 'the north American for January, 1844,' probably a mistake as it was in the June volume.

E. Robinson
from the author
June 1854

FLAVIUS JOSEPHUS

DER

FÜHRER UND IRREFÜHRER

DER PILGER

IM

ALTEN UND NEUEN JERUSALEM.

MIT EINER BEILAGE

JERUSALEM DES ITINERARIUM BURDIGALENSE ENTHALTEND

HERAUSGEGEBEN

VON

JAKOB BERGGREN,

D. THEOL. DR., PROBST UND PFARRER ZU SKAELLVIK BEI SOENDERKOEPING.

LEIPZIG,

T. O. WEIGEL.

1854.

Ill. 33. Jakob Berggren, *Flavius Josephus der Führer und Irreführer der Pilger im alten und neuen Jerusalem*, Leipzig: T. O. Weigel, 1854. Berggren's dedication to Robinson.

LETTER FROM DR. ROBINSON.

The following letter refers to the article in review of Dr. Olin's "Travels in Palestine" in the North American Review for October, 1843, and to the letter from Dr. Olin in reply to that article, which we published in a note to the number for January, 1844.

To the Editor of the North American Review.

Dear Sir, — I am very reluctantly induced to ask a place in your Journal for the following statement, respecting a single point in the matter recently at issue between the North American Review and the Rev. Dr. Olin. It relates to *the bridge* which anciently connected the Jewish temple with Mount Zion. As to the other matters in question, I suppose the issue to lie, not with me, but between Dr. Olin and the public, whom the reviewer represents.

In my work on Palestine, after recounting the manner in which I was led to recognize the remains of the ancient bridge so often mentioned by Josephus, and after describing those remains, there is subjoined the following note.

"Since the above was written, I have been informed by both Messrs. Bonomi and Catherwood, the well known artists, that they likewise remarked these large stones in 1833, and recognized in them the beginning of an immense arch. They regarded them, too, as probably among the most ancient remains in or around Jerusalem; but had no suspicion of their historical import." *Biblical Researches in Palestine,* Vol. I., p. 427.

This note was first written in London, in October, 1840, after an interview with Mr. Bonomi. He spoke of the remains as being the fragment of an arch; but frankly added, "We could make nothing more of them." The note was afterwards submitted in manuscript to Mr. Catherwood, in New York; who kindly showed me his very beautiful drawing of the remains in question, and corroborated the general statement of Mr. Bonomi. The note was printed with his sanction. My work appeared in July, 1841. The facts respecting the recognition of the bridge had been published in this country in October, 1838; and, before the middle of 1839, they had been further spread before the world as widely as the public presses of England, Germany, and the United States could give them currency.

Dr. Olin was in Jerusalem in April, 1840; and in his Travels, published in April, 1843, after describing the ancient remains around the mosque, he has the following passage.

"I could not learn that the most interesting and unquestionable of these remains — the massive arch of the ancient bridge — had been so much as mentioned by any modern traveller, though its existence has long been well known to European and other residents, as well as visiters. At least, this is the impression which I derived from my conversation with Mr. Nicolayson; who told me that Mr. Catherwood had examined the remains of the bridge seven years before." Vol. II. p. 268.

To this passage the reviewer took exceptions; and Dr. Olin, in his reply, uses the following language.

"Mr. Nicolayson was my guide to this monument; and I recorded his statement and my own measurement at the time. I now declare, that I never saw or heard the name of Dr. Robinson connected with this subject in Jerusalem or elsewhere, until I read the 'Researches' nearly two years after my visit. Having no reason to distrust my own information, I of course presumed Dr. Robinson was in an error, in regarding himself as the original discoverer. Mr. Catherwood, who is a professional architect, and the author of Dr. Robinson's plan of Jerusalem,* as well as the one always in my hand, in which he had laid down the Temple, Mount Zion, and the valley between them, across which the arch looks directly, could hardly have doubted or been mistaken with regard to its design. Mr. Catherwood has often told me since, that my account is strictly true; and that he, as well as several other gentlemen with whom he conversed in Jerusalem, regarded and spoke of this monument as the remains of an ancient bridge, that connected the Jewish temple with Mount Zion."

It is the testimony here ascribed to Mr. Catherwood, to which I would invite the reader's attention. On comparing it with my note in the Biblical Researches quoted above, the discrepancy is seen to be so striking, that one of three things must necessarily follow, namely: either I was wrong in my statement respecting Messrs. Bonomi and Catherwood; or Dr. Olin was here in the wrong; or Mr. Catherwood at different times had made different statements. It seemed due to Mr. C. as well as to myself, to call his attention to the matter. The article appeared on the 1st of January last; and by the packet of January 6th, I wrote to him on the subject. His reply is so full of frank and honorable feeling, and so creditable to his candor, that I should not feel justified in withholding it from the public. I therefore give it here entire, subjoining a few remarks. The date of the postscript shows why it could not appear in your April number.

* Mr. Catherwood was not the author of the plan of Jerusalem in the Biblical Researches; but Mr. H. Kiepert, of Berlin. He made use of the same original as Mr. C. appears to have done, viz., the earlier plan of Sieber, and introduced all Mr. C.'s corrections in and around the haram, which were made from careful measurements. But Kiepert's plan varies very materially from Mr. C.'s, on the south and west, and in the shading of the hills within the city; all these being corrections derived from measurements made with my own hands. — E. R.

"London, 9th February, 1844.

"DEAR SIR, — Your favor of January 6th reached me but a few days before the sailing of the February steamer, and I was too much engaged to answer it at the moment. I had also to make some inquiries, to refresh my memory, which is not very good, in regard to conversations held many years ago. I am sorry that any thing I should have said, or omitted saying, should have produced an apparent discrepancy in my testimony regarding the bridge; but I will endeavour to recall to mind and relate all I know of the matter in question.

"Before going to Jerusalem, I was furnished with a manuscript map of the city by Mr. J. J. Scoles, architect, who made it on the spot, and at that time it was the best extant. I also had conversations with Mr. Barry and Mr. Scoles regarding the most interesting points that still remained for investigation. Among other directions, Mr. Scoles told me to 'look out for the remains of a bridge which joined Mount Moriah to Mount Zion.' He had been unsuccessful in finding it himself; and did not mention to me whence he derived his information respecting it. (I have a note from Mr. Scoles to this effect, dated a few days ago.) This direction was a verbal one, not written down, and which I afterwards forgot altogether; and when I discovered the arch, it was not from purposely looking for it, but casually, in making my survey of the walls surrounding the mosque of Omar.

"I therefore was in error when I stated to Dr. Olin, that others (meaning Messrs. Barry and Scoles) were previously acquainted with the arch in question. Mr. S. has set me right on this point; and I thus rather unexpectedly find myself to have been (so far as I know) the discoverer. I had no doubt, from the moment I saw it, that it had formed part of a viaduct and aqueduct; but I was totally ignorant of its historical importance. I merely looked at it, as an architect, with reference to its position, both to *the water-course from Bethlehem* and the deep ground between it and Mount Zion. I do not recollect whether I spoke on the subject to Mr. Bonomi or Mr. Nicolayson; but when Dr. Olin told me that Mr. N. mentioned my name in connection with it, I concluded I had spoken to him on the point; which is very probable, from my having had almost daily intercourse with Mr. N.

"This, therefore, will explain my not having made any observation, when I read your note. I was in doubt, and therefore said nothing; at least, so far as I can recollect. I was, moreover, desirous, that you, who have labored so diligently and successfully in the field of Jewish antiquities, should have the full merit (as is justly due to you) of being the first to publish and bring to light the historical importance of this monument. I have stated the facts to the best of my recollection and present knowledge; and hope the explanation will prove satisfactory.

"I am, dear Sir, very truly yours,

"F. CATHERWOOD.

"P. S.—March 2d. About a week ago I received a letter from Dr. Olin, asking some explanation on this subject; and I thought the best plan would be, to send him a transcript of my letter to you; which I have accordingly done. F. C."

REMARKS.

I. The question here at issue is not, whether these remains have ever been noticed before; for they must have been seen by thousands upon thousands, in the long course of seventeen centuries, and especially in the age of the Crusades. Nor is it, whether they have before been recognized as an arch; for among the multitudes who have looked upon them, it is scarcely possible to suppose, that some one should not have detected their true character in this respect. Yet there is no known testimony extant earlier than that of Mr. Catherwood; so that, in this sense, he is the discoverer, as I have stated in the Researches. Nor is it here the question, whether any one had, or had not, before speculated upon the purpose of such an arch in this place; for among the multitudes of learned men and artists who have visited the city, as, for instance, during the Crusades, we can hardly suppose, that such speculations would not have arisen in some minds; and then nothing would have been more natural than to refer these remains to a bridge or an aqueduct. Yet here, too, there is no recorded testimony in behalf of any one before Mr. Catherwood. The true question at issue is simply this: Had any person, before my visit to Jerusalem, in April, 1838, in any way brought these remains into connection with the important historical fact, made known to us by the Jewish historian, that a bridge anciently existed over the valley between the Temple and Mount Zion? I know of no such person. Had Mr. Scoles found the spot, he very probably would have brought out the result years ago. Or had Mr. Catherwood published his own observations, it is hardly to be supposed, that scientific inquirers would not have quickly perceived their identity with the bridge of Josephus. But he did not do this; and he frankly says of himself, "I was totally ignorant of its historical importance." There is no other person, so far as I know, who can in any way be brought forward in derogation of my right to this very casual honor; and this is all that I have ever claimed for myself in the Biblical Researches or elsewhere. I went to Jerusalem knowing nothing of the existence of any such remains; my attention was called to them there; and their identity with the ancient bridge instantly suggested itself to my mind.

II. I sincerely regret, that Mr. Catherwood, in his conversations with me, in January, 1841, did not mention that he had, at the time, regarded the arch as having "formed part of a viaduct and aqueduct." Had he done so, I certainly should have stated the fact in connection with my note; both because my only object was, and is, the truth, the whole truth, and nothing but the truth; and also because, in a matter of so much archæological importance, it is interesting to scholars to be able to trace the progress of discovery. The idea of an aqueduct and viaduct would very naturally present itself to Mr. C.'s mind, not only as an architect, but also from the analogous fact, that the water-course from Bethlehem is actually carried over the Valley of Hinnom, on the west of the city, in this very manner. But how little stress Mr. Catherwood himself laid upon this hypothesis, as also upon the whole matter, is obvious, not only from his having thus left it to sleep for so many years; but also from the fact, that, on his plan of the city, instead of bringing the said water-course into the haram in any possible connection with the arch, he actually brings it in from the south, at a point where the ground is nearly a hundred feet below the level of the mosque and of the ancient bridge. The real place of its entrance is along the eastern precipice of Zion, and across the Tyropœon, at an elevation considerably less than that of the bridge.

III. It may be asked, What is here the difference between *a* bridge and *the* bridge? Why is not the inference of a scientific architect just as conclusive and important as the testimony of a historian? The reply is, that, while such an inference brings out no result beyond or more important than itself, the identification of the arch in question with a known ancient bridge is at once an immense step gained in the archæology and topography of the Temple and city. For example: travellers have for ages gazed upon many courses of huge stones in the external substructions of the haram; and perhaps all have conjectured, and many have believed, that these had been in some way connected with the Jewish Temple. This, however, was merely matter of credence, and not of demonstration; and it is not too much to say, that not a single point in the topography of the city had ever been certainly and indubitably settled. But the moment we identify this arch with the bridge described by Josephus, the conclusion follows irresistibly, that these courses of stones, with which it is thus connected, are parts of the identical wall existing in the time of Josephus. Similar reasoning applies to the southern and western walls; and hence is demonstrated beyond

258 *Letter from Dr. Robinson.* [July,

cavil the general identity of the present area of the mosque with that of the ancient Temple. Further; the same bridge connected the Temple with Zion; and the hill now opposite on the west is thus proved to be Zion. In this way is shown the falsity of Dr. Clarke's theory, who held Zion to be the hill south of the Valley of Hinnom; — a theory which, heretofore, scholars might disbelieve, but could not disprove. Thus we may go on through the whole city; and, as one point after another is gained, each may be referred back, for all the certainty it can claim, to the identification of the arch in question with the ancient bridge.

IV. The sum of the whole matter may, therefore, be stated as follows: Mr. Scoles was aware, doubtless, from Josephus, that an ancient bridge had existed; but was unable to find any remains. He suggested to Mr. Catherwood to look for such remains, but the suggestion was forgotten; and when Mr. C. found the arch in question, he was not aware that there had been an ancient bridge. That is to say; the former knew there had been a bridge, but found no remains; the latter did not know that a bridge had existed, but found the fragment of an arch, which he referred to an aqueduct and viaduct. Thus far there was no gain to history or topography. It was left to a third person, five years afterwards, knowing nothing of what the former two had done, to connect and identify on the spot the said remains with the ancient bridge; and thus to fix a definite and imperishable landmark, from which to trace out and settle beyond controversy many most important points in the archæology and topography of the Holy City.

V. In accordance with these remarks, the statement of Mr. Bartlett is to be modified, in his recent work, entitled, "Walks about the City and Environs of Jerusalem," London, 1844, p. 149. — I may here add, that the engraved view of the arch given in that work (p. 150) does not strike me as very correct. The wall in that part does by no means present an appearance so regular and well defined. If my memory is right, the drawing made by Mr. Catherwood is far more truthful.

In conclusion, I would repeat, that it is with very great reluctance that I thus come before the public. It has, however, seemed to me, that justice to Mr. Catherwood, as well as to myself, demands it.

Yours, with great respect,

EDWARD ROBINSON.

New York, June 10th, 1844.

Ill. 34. 'Letter from Dr Robinson,' *North American Review*, 59 (1844), 253–58.

Both texts are composed of two parts: the facts according to Robinson, followed by his remarks. Robinson refers first to the note in his book where he explicitly mentions that 'Messrs Bonomi and Catherwood' told him that they also 'remarked these large stones' during their visit in 1833. They regarded them 'as probably among the most ancient remains in or around Jerusalem,' but 'had no suspicion of their historical import.'[53] In the original note in his book, Robinson also mentions Henry Maundrell, chaplain of the British Levant Company factory in Aleppo who visited Jerusalem in Easter 1697, and the British prelate Richard Pococke who visited the city in 1737, as examples of prominent and trustworthy travellers who mentioned observing the stones while visiting the site.[54]

In fact, Robinson, on his way to New York, met Joseph Bonomi in London in October 1840. Bonomi did tell Robinson that he and Catherwood thought the stones were part of an arch, 'but they could make nothing more of them.' As noted, Robinson wrote to Smith regarding Catherwood's maps in April 1840, that 'his large one I have not been able to get sight of.' Catherwood was based in New York and Robinson met him there in December 1840, where he was able to view 'the very beautiful' painting of the stones and 'corroborated' Bonomi's statement.[55]

Robinson continues his response to Olin by noting that he presented the facts 'respecting the recognition of the bridge' prior to the July 1841 publication of his book, both to the American readership as early as October 1838, and to 'public presses of England, Germany and the United States in mid-1839.'[56] A thorough check of Robinson's list of publications, as presented by Williams, casts doubt on this assertion. The only publication in 1838 was in German in the second volume of the *ZKM*, founded the year prior. It is a rather long paper titled 'short report,' a text identical to that of his book.[57] This publication raises a question of timing, as Robinson arrived in Berlin on November 15, 1838 and wrote, eight days later, that although most of his time was spent looking for

[53] Robinson (1841), I, 427 n. 1; (1844), 253.

[54] Maundrell (1703), 99: 'Here we were shewn [*sic*] several large Vaults, (...). This might possibly be some under ground work made to enlarge the Area of the Temple. For *Josephus* seems to describe some such work (...).' Pococke (1745), 14: 'The whole [Temple] was supported by walls and buttresses towards the vallies...'

[55] Robinson to Smith, Berlin, 15.4.1840, New York, 28.12.1840, ESP, ABC 60 (63); Robinson (1844), 253; (1844a), 794.

[56] Robinson (1844), 253; (1844a), 794–95.

[57] Robinson (1839b), 347.

lodging for his family for the next six months, he had already visited Ritter and Neander who both offered 'the use of their libraries in the kindest manner.'[58] Perhaps Robinson began writing already on his way to Germany or while sick in Vienna. This question is addressed in chap. 4.b.3. and 4.

Robinson's next reports, both much shorter than the German one, were published in English the following year and are the two mentioned by Robinson. In April 1839, his 'Brief Report' was published in the *American Biblical Repository* (*ABR*, Robinson founded it in 1831 in Andover and was its editor until 1835; in 1839 it was published simultaneously in New York, Boston and London), and his journal 'Extracts' was published in the *JRGS*, which certainly came out later than June 1839.[59] In all three reports, Robinson wrote about the arch and its apparent function in the beginning of his discussion of Jerusalem, demonstrating his recognition of the importance and significance of this discovery.

These reports were published prior to Olin's visit to Jerusalem, and certainly prior to his book, published only in 1843. Robinson cites Olin's book as well as his response to the review in the *NAR* where Olin lists Nicolayson and Catherwood as his main corroborators. Robinson, fighting for his integrity and recognition, wrote to both. In early January 1844, he writes to Catherwood, 'either I was wrong in my statement respecting Messrs. Bonomi and Catherwood; or Dr Olin was here in the wrong; or Mr Catherwood at different times made different statements.'[60] Subsequently, Robinson decided to expose his entire letter to the public, a body to which he refers as the judge in the dispute.

In his highly sympathetic answer, Catherwood, who made some inquiries 'to refresh my memory,' admits that there was an 'apparent discrepancy in my testimony.' The first discrepancy involved the map, and Catherwood's correction is rather surprising. To date, historians have accepted, as noted above, that Catherwood must have used Sieber's 1818 map and based his detailed description of the area of the Temple Mount upon it. Sieber is not mentioned in the letter. Instead, Catherwood writes, 'I was furnished with a manuscript map of the city by M. J. J. Scoles, architect, who made it on the spot, and at the time it was the best

[58] Robinson to Smith, Berlin, 22 January 1839, ESP, ABC 60, Robinson, M. A. (1862–64).

[59] Robinson (1839), 411; (1839a), 299. For the latter see Goren (2015).

[60] Robinson (1844), 254–56; (1844a), 795–97.

extant. I had also conversations with Mr Barry and Mr Scoles regarding the most interesting points that still remained for investigation.'[61]

A 2016 M.A. Thesis in the Department of Geography, The Hebrew University of Jerusalem, checked the accuracy of nineteenth-century measuring maps of Jerusalem using computerized tools. One of the prominent conclusions of the thesis involves Catherwood's map. The accuracy of the area of the Temple Mount is relatively high, while that of the other parts of the city is much lower. Comparing it to Sieber's map, 'it is easy to establish without much doubt that Catherwood copied Sieber's city map (...) except, of course, the Temple Mount.'[62]

Joseph John Scoles (1798–1863), a Gothic Revival Catholic architect, designed many Roman Catholic churches, studied with Bonomi, and in 1822 they set out to further their architectural studies in Italy, Greece, Egypt, and Syria. Scoles also travelled with architects Catherwood and Henry Parke (1792–1835). He drew a map of Jerusalem and of the Holy Sepulchre, published only by Robert Willis in 1849.[63] Ben-Arieh cites Scoles' map as a cartographical work that was never published but that was used by other scholars. He derived this information from a single source, George Williams' book *The Holy City*, published in 1845.[64] It seems that Williams' main purpose in publishing his book was to contest Robinson's discrediting of local Palestinian traditions, waving the flag of conservative Anglican reactions to the so called 'innovations' of the American. Williams' references to the *Biblical Researches* can only be described as wild and furious attacks.[65] Catherwood's second conversant, architect and later Sir Charles Barry (1795–1860), who was most well known for his role in the rebuilding of the Palace of Westminster, went on an extensive grand tour of the Mediterranean and Middle East from 1817 to 1820.[66] A study of the connection between Sieber's and Scoles' maps still awaits.

[61] Robinson (1844), 255.

[62] Dabach (2916), 36–39.

[63] Dawson, W. R. (1995), 319, 382; Willis (1849). See Arundale (1839), 54; Ben Arieh (1973), 67; Khatib (2003), 168–69.

[64] Ben Arieh (1973), 64; (1974), 154.

[65] Williams, G. (1849), I, vi–ix, and as examples: II, 345–57 (Outer Court of the Haram), 387–403 ('The Ruined Arch did not belong to the Bridge'); Ben-Arieh (1979), 133, 137; Lipman (1989), 35–42. Cf. Robinson letters to Smith, New York 22.12.1845 and 12.1.1846, ESP, ABC 60 (63).

[66] For detail on Barry as 'painter of the Holy Land' see Ben-Arieh (1997), 40–43, 70–71.

Among others, Williams mentions Scoles and Catherwood as rendering him 'particular services.' Scoles helped him with many drawings, plans, and suggestions, including of the Holy Sepulchre and Absalom's Tomb.[67] The important issue here, however, is his reference to Catherwood's map. The latter had insufficient time to complete his survey of the city, so 'his plan (...), with the exception of the Haram, is based on an unpublished survey of Mr J. J. Scoles in 1825, executed under highly difficult and dangerous circumstances, and though infinitely superior to any that existed prior, could not be safely considered authoritative (...).' In contradiction to Robinson's statement in the *Biblical Researches*, Williams opines that Catherwood's map 'has (...) been the standard Plan from that time to the present,' and that it 'had introduced (...) some additions and corrections (...) yet are the plans essentially the same (...).'[68]

On his map of Jerusalem, Williams marked the remains of the arch in brackets as 'Dr Robinson's Bridge.'[69] Moreover, he noted that this was the arch 'which Dr Robinson imagines' could have been a part of the bridge mentioned by Josephus.[70] Given the prominence of Josephus, Williams could not possibly 'pass over, without a word of description or admiration, such an astonishing performance.' But the greater difficulty, according to Robinson, amounting in his mind to an absolute impossibility, derives from the different ancient street levels, which he discusses at length.[71]

Returning to Catherwood's revised recollection of events in his response to Robinson, he writes that although Scoles told him 'to look out for the remains of a bridge which joined Mount Moriah to Mount Zion,' he forgot about this recommendation and discovered the remains of the arch casually, while surveying the walls of the Haram.[72] The letter continues with a rather strange assertion, 'I therefore was in error when I stated to Dr Olin, that others [meaning Messrs Barry and Scoles] were previously acquainted with the arch...' Catherwood thought the arch belonged to 'a viaduct or aqueduct' that was part of the city's water supply system and examined it as an architect, 'totally ignorant of its his-

[67] Williams, G. (1849), I, xv, II, 138, 152, 158–60, 293–94, plans for which he is indebted to Scoles; Willis (1849), 157.

[68] Williams, G. (1849), 7.

[69] Williams, G. (1849), I, xxx, map.

[70] Flavius (1895), Book 15, [410]–[411].

[71] Robinson (1841), I, 388–91.

[72] For this version see also Smith, H. B. (1863), 83 note *.

torical importance.' Catherwood did not recollect speaking about the arch with either Bonomi or Nicolayson, though this is quite probable. Catherwood was quite positive, that of all the scholars, travellers, map makers, and guides mentioned here, Robinson, who has 'labored so diligently and successfully in the field of Jewish antiquities, should have the full merit (as is justly due to you) of being the first to publish and bring to light the historical importance of this monument.'[73]

There is one important difference between Robinson's two published letters of response to Olin. The letter published in the *Bibliotheca Sacra* includes Nicolayson's reply to Robinson written 'On the board of the Austrian Steamer, Off Cyprus, May 31 1844,' so it could not have been included in the earlier *NAR* publication. After spending six months in Constantinople, Nicolayson's steamer was expected to stop in Beirut, where he planned to give the letter to Smith for delivery to Robinson. Nicolayson's answer supports Robinson's claim, as he recollected clearly that he never paid attention to the stones until Robinson pointed them out to him. Robinson was also the first person who told him about the connection of the stones to the bridge mentioned by Josephus.[74] Nicolayson explains Olin's error in this way, as 'for the slight mistake into which Dr Olin seems to have fallen (...), that I had referred to Mr Catherwood on this subject,' derived from the fact that Nicolayson had extensive talks with Olin concerning Catherwood's measurements on the Temple Mount. Nicoylayson also mentions the 'measurement of the supposed spans of the arch, conducted by a certain Brettell, an English engineer, to whom I showed it after Robinson's visit.' Joseph C. Brettell, an English mining engineer in the service of Muhammed 'Ali, contributed articles on Acre in 1841 to *Nautical Magazine*, which was issued in London. Brettell's sketch was used for a map titled 'Bombardment and Capture of Acre.'[75] The paper was published anonymously, but it was a letter sent to Palmerston by Major General Charles Felix Smith (1786–1858), commander of the allied forces of Syria, dated 5.11.1840 following the bombardment and capture of Acre.[76]

To his recitation of the facts as he viewed them, Robinson adds, in his letters to both the *NAR* and the *Bibliotheca Sacra*, pages of 'remarks' reinforcing his point that '...the true question at issue is simply this: Had any person, before my visit to Jerusalem, in April, 1838, in any way

[73] Robinson (1844), 255–56; (1844a), 797–98.

[74] Robinson (1844a), 796.

[75] Brettell (1840); Röhricht (1890), 641.

[76] Smith, H. B. (1841).

brought these remains into connection with the important historical fact, made known to us by the Jewish historian, that a bridge anciently existed over the valley between the Temple and Mount Zion? I know of no such person.'[77] Robinson states that Catherwood should be credited with identifying the stones as being part of an arch, with being the first to speculate about the arch's function, and with drawing a truthful rendition of the ruins.[78] Robinson summarizes 'the whole matter' in this way: Scoles knew of the bridge from reading Josephus but was unable to locate it and suggested to Catherwood that he look for it. The latter forgot; hence, Robinson surmises that when Catherwood saw the stones, he was unaware that there had been a bridge in that location, and concluded that the stones were part of a viaduct or aqueduct. Resolving the puzzle had to wait for Robinson's arrival five years after Catherwood's 1833 visit, at which time Robinson connected the remains with the bridge recorded by Josephus. 'I had to bring the whole issue before the public,' Robinson ends his letter, as 'justice to Mr Catherwood, as well as to myself, demands it.'[79]

Robinson's letter only appeared to end the discussion. Olin was busy with his position as president of Wesleyan University, a new marriage, and endless abolition and church discussions, and Robinson remained the accepted 'founder' of the arch. But Olin did not give up. In January 1845, under the title 'the whole matter settled,' he published letters he received from two ABCFM missionaries, Homes and Hamlin, that he claimed changed the entire picture. Disregarding Robinson's apparent reputation as an unrelenting opponent, in the words of Robinson's colleague, friend and biographer Hitchcock,[80] 'no man who provoked him to controversy cared to repeat the experiment,' Olin used the letters of these ABCFM missionaries he never met and with whom he had no previous connection in order to have the last word. This time, he chose a New York newspaper, *The Commercial Advertiser* of May 21, 1845, and, twenty days later, the periodical *Christian Advocate and Journal*, published in New York for the Methodist Episcopal Church between 1833 and 1865.[81]

[77] Robinson (1844), 256; (1844a), 798.

[78] Robinson (1844), 256, 258. Robinson compared it to painter William Henry Bartlett's (1809–54) engraving (Bartlett, W. H. (1844), facing 136 (engraving), 148–65 (Catherwood's cited letter)).

[79] Robinson (1844), 258.

[80] Cited by Robinson's daughter: Robinson, M. A. (1862–64), 159.

[81] Olin (1845); (1845a).

Cyrus Hamlin (1811–1900) was appointed in 1837 by the ABCFM to serve in Constantinople. Between 1840 and 1860, he directed the mission school 'Bebek Seminary.' Hamlin's name does not appear to be mentioned in connection with Robinson.[82] Henry Augustus Homes (1812–87), Amherst graduate who also studied at Andover and Yale, was fluent in Arabic and Turkish. Homes 'came in 26. December 1835 to help the station in Constantinople' and was responsible for distribution of printed books.[83]

Hamlin became aware of the dispute between Olin and Robinson when he received the *North American Review* while in Constantinople. He sent an undated note to Homes, asking Homes if he recollected their conversation on the issue, 'did you **previous to meeting Dr R.** in Jerusalem regard the arch in question, not as the remnant of a bridge, but of **the** bridge spoken of by Josephus? And **as such** did you mention it to Dr R.?'[84] The reference is to their meeting with Robinson when he joined the missionary assembly in Jerusalem in April 1838. Hamlin asks Homes, who had also returned from Jerusalem to his position in the Ottoman capital, to answer by a note 'by the bearer of this.' Home's answer, dated March 13, 1845, is clear and unequivocal: 'The discovery of the bridge is a just claim of Dr Robinson's (...) on the great principle (...) "he only discovers who proves".' Homes merely suggested the idea of the arch to Robinson. Homes continues, 'But I would simply tell my story, and you will therein find a sufficient answer to your questions.' The first time that Homes saw the remains was in 1837, 'while residing several months at Jerusalem.' Whiting did call Homes attention to the fact that there was a bridge, probably after the latter returned to his house and mentioned it.[85] 'He said that he and some English travelers, one of whom was Mr Moore, regarded this as the bridge mentioned by Josephus, or perhaps we all three aided each other in the *suggestion*.' Homes continued 'dwelling much of this discovery,' and returned 'to re-examine the localities.' Later, he met Robinson in the spring of 1838, at the missionary council in Jerusalem. As 'we were all anxious to show Dr Robinson all the noticeable places in Jerusalem which might possibly

[82] Hamlin (1893); DAB Hamlin; Kawerau (1958), 269, 310–22, 358–62.

[83] DAB Homes; Kawerau (1958), 236–37, 268 (citation), 272. For Smith's letter to him: Smyrna, 4.8.1838, ESP, ABC 60 (105/5) 1838; his letters to Smith: Constantinople, 13.11.1838, 22.1.1839, 25.5.1839, ESP, ABC 60 (37).

[84] Olin (1845); emphasis in original. This letter is missing from the version in the *Christian Advocate.*

[85] For detail on ABCFM missionary Whiting, see chap. 2.b.

suggest to him facts regarding its ancient topography,' Hamlin told him of the existence of the arch. To his surprise, so he adds, it was totally new to Robinson, and Hamlin even remarked that it might be remnants of the bridge.[86]

Homes introduces a new name into the discussion, the Irish nobleman, adventurer, and Dead Sea Explorer George Henry Moore. Moore and his British companion William George Beek (1804-73) were the first to establish and publish the fact that the level of the Dead Sea 'is considerably lower' than the Mediterranean.[87] In early October 1836, Moore was in Jerusalem and his 'Dead Sea expedition' took place between March 15 and April 28 of the following year, after which he remained in the country, including Jerusalem, until mid-July 1837.[88] He left Jerusalem for the last time about ten months before Robinson and Smith arrived in the city on April 14, 1838. Robinson and Smith stayed in Jerusalem until May 5, and then began excursions using the city as a base. Due to plague and quarantine, these three weeks were the only time that Robinson and Smith spent in Jerusalem before leaving north on June 13; hence, this must be the time when Homes met Robinson.

What about Homes? In his diary, Moore mentions missionaries Whiting (and his wife to whom Moore liked to 'pay a visit,' many times), Nicolayson, and Lanneau, but there is no mention of Homes. He also does not mention any stones, nor does he mention the Temple Mount walls.[89] With regard to Robinson's presence in Jerusalem, Moore writes, 'we found collected all the members of the Syrian Mission (...) from the stations at Beirût and in Cyprus; and one also from the Mission at Constantinople.' The latter must have been Homes. Among the eight missionaries (who came with their wives, 'to consult together on the best measures for promoting the great work in which they were engaged'), Robinson found five 'as former friends and pupils.'[90] Eli Smith's letter from Smyrna to Homes, dated August 4, 1838, after the expedition and before leaving for Germany, deals with mission issues. There are also two

[86] Olin (1845).

[87] Moore (1837). For detail on Moore, see Goren (2011), 158–206, and sources there.

[88] Goren (2011), 168; primarily citing his 'Dead Sea Diary,' housed in the National Library of Ireland: Moore DSD.

[89] Whiting is first mentioned on 22.3.1837, and again during visits on 7.4, 9.4, 10.4, 17.4, etc.: Moore DSD.

[90] Robinson (1841), I, 327–28 (citations), 334–35 (about the missionaries' meetings).

letters from Homes to Smith written from Constantinople. The first is dated November 12, 1838 (don't forget to forward my regards to the 'beloved Dr Robinson') and the second January 22, 1839.[91] Smith mentions Homes again in letters from February to March 1839, mainly in regard to his help showing Muslim leaders Smith's Leipzig types and receiving and forwarding their comments.[92]

On March 14, the day after receiving Homes' reply, Hamlin wrote to Olin from his school in Constantinople, apologizing for being unaware of the dispute earlier. He notes that he remembers 'with perfect distinctness,' hence is 'confident that I had heard Mr Homes affirmed that he informed Dr R. of the existence of that arch as a remnant of the bridge spoken of by Josephus.'[93] As Homes' answer appears to agree with Hamlin's version, and while regretting its delayed arrival, 'it is never too late to establish the claims of truth and justice.' Homes, complains Hamlin, who is mentioned in the *Biblical Researches* with regard to minor issues, never received thanks or an acknowledgement from Robinson. 'The "arch",' he stresses, 'was known previous to Dr R.'s visit to Jerusalem,' as was its relation to Josephus' bridge.

An examination of the sources did not uncover a response from Robinson. One of his obituaries, published a day after his funeral, was dedicated to the arch: 'he is doubtless entitled to the honor of directing the eyes of the world to one of the most interesting relics of the past which can anywhere be found above ground.'[94] Perhaps somewhat exaggerated, Robinson's Arch and its original function continues to be an issue of debate between archaeologists to this day, and its location, within the Western Wall makes it a focus of political discussions as well.

'I would like to correct an assumption frequently made in publications regarding monumental features at the Temple Mount, that they were always named after the explorers who first discovered them,' writes Shimon Gibson in a recent paper. He gives further examples, including Wilson's Arch first noticed by Tobler, 'but was given to Wilson, in lieu of his achievements in Jerusalem, by Sir Henry James, Director-General of the English Ordnance Survey,' and Warren's Gate, whose discoverer, Wilson, 'decided to name the gate after Warren, though Warren said he

[91] Smith to Homes, Smyrna, 4.8.1838, ESP, ABC 60 (105/5) 1838; Homes to Smith, Constantinople, 12.11.1838 and 22.1.1839, ESP, ABC 60 (37).

[92] Smith to Robinson, Leipzig, 16.3.1939, ESP, ABC 60 (105/10); Smith to Whiting, ESP, ABC 60 (105/15) 1839.

[93] Olin (1845a).

[94] Anon (1863a); in ERP, Box 2, MM.

was uncomfortable about this.'[95] I tend to agree with Gibson. Robinson might not have been the first to recognize the arch, and was most likely also not the first to connect it with Josephus Flavius' narrative. But it is due to him and his publications that these facts entered, as accepted narrative, into the scientific and lay discourse of old Jerusalem and its remains. So the stones continue to be named after Robinson, and will probably carry his name forever.

[95] Gibson (2020), 7, n. 5. For the mentioned persons see all PEF historiographies, as Watson (1915); Moscrop (2000).

CHAPTER 7
CONCLUDING REMARKS

Ill. 35. Robinson in the later years, 64 years old.

My husband recommends you in the best way, and indeed actually from Jerusalem, as he lives there more than here [Berlin].

Mein Mann empfiehlt sich Ihnen bestens, und zwar eigentlich aus Jerusalem, denn er lebt mehr dort als hier]

Talvj to Kopitar, Berlin, August 12, 1839, Milović (1941), *101.*

Had I been looking for Robinson in the nineteenth century, declares Jay Williams in the *Hamilton Alumni Review* of 1996, 'no research would have been necessary, for everyone in Hamilton knew of Edward Robinson as one of the College's most illustrious alumni. He was not only the first American to achieve an international reputation in biblical studies but also perhaps the most important scholar to graduate from Hamilton during that century.'[1]

Williams' declaration was well-founded. Immediately upon his death, Robinson was eulogized as a scholar of great calibre:

> now we are called to mourn the loss of one of the most eminent of the Christian scholars of our country. Rev. Edward Robinson, D.D., LL.D., who for nearly half a century had devoted himself to the cause of Biblical learning; who was one of the very few men this side the Atlantic who commanded the respect of European scholars; and indeed, who, in the special department of Biblical Geography was the first living authority – at length is gone.[2]

Robinson's scholarship and his reputation are undisputed. I strove to delve beyond the facts and ask 'who were you, Edward Robinson'? Surely, Robinson was an oft praised luminary, a leading name in biblical scholarship and study of the Holy Land, one of the most influential and cited authorities on the country, its history, and geographical-history, and in the fields of scriptural- and theological-history and geography. What motivated you, Robinson, to launch such an exceptional research expedition and create an even more exceptional *magnum opus*?

Considering that 'there is no retention of memories but in writing,'[3] I set out, not without deep concerns, on this exhausting journey in the footsteps of this unique person and prominent scholar through his archival material and that of his companion Eli Smith. This book, resulting from this intensive archival work, differs significantly, so it is hoped, from most if not all other publications dedicated to Edward Robinson. It examines the background to and progress of his and Smith's spring 1838 expedition and the genesis of the idea and writing process of the three volumes of *Biblical Researches in Palestine, Mount Sinai and Arabia Petraea: A Journal of travels in the year 1838*, published almost simultaneously in London, Boston, and Halle (in German). Over one thousand documents, mainly letters, from Robinson's and Smith's archives, as well as other archives located primarily in Germany and New England, were meticulously examined. The central thesis of this study is that the letters reveal actual memories in real time, the issues not covered by

1 Williams (1996), 19.
2 Robinson (1863).
3 Be'er (2019), 171.

prior research, narratives, and publications, in the past and present and, in many cases, differing from what has been accepted to date in the extensive research dedicated to Robinson.

If a scholar, or any other author, trying to penetrate a particular character wishes to increase significantly the pool of narratives and events from which he describes life in written words, he cannot content himself only with materials that have been published in print, as in the case of Robinson until the publication of Williams' and Weigel Williams' books in the late twentieth and early twenty-first centuries. The writer should also enter the hidden world of the characters, the 'heroes,' as expressed in correspondence, diaries, and any other document he can 'lay his hands on.' Meticulously storing all sections of the documents (cover, back, etc.), deciphering and methodically organizing them, enables the researcher to assemble the relevant pieces, creating a whole narrative. In such cases, the documents accompany the researcher hand-in-hand, as a hidden author.

For this study, the handwritten documents turned out to be a rather large treasure, in many cases holding significant, new information, exceeding initial expectations at quite an early stage. The long process of locating the documents, recording and photographing them, was followed by the even longer, more difficult and in many cases frustrating effort to decipher illegible handwriting. Only then began the stage of close reading and choosing of relevant letter selections according to the planned chapters. The reading had to be conducted very carefully as, in the words of author Haim Be'er, 'you might never know where in the letter lies the hidden gold.' Hopefully nothing was left to chance; accompanying envelopes, addresses, dates, side comments, and more were also carefully examined. In an attempt to establish the writer's hidden motives, careful attention was given to what was not written, to who and what had been omitted. The reading was 'sober,' listening closely, asking questions, putting the letters chronologically into the relevant context, checking the different leads originating in the text. Questions emerging from outer circumstances were asked, such as reasons for long breaks in correspondence, as well as those raised by the text itself.

Yes, 'God resides in the details,' and a letter, unlike a book, presents real time. When intending to sketch the portrait of a 'celebrity,' as in the case of Edward Robinson and accompanying figures, letters are the most fruitful source for accurate and personal details. One might view the reliance upon letters as old-fashioned positivism. I firmly believe that there is no real and reliable narrative absent a positivistic approach revealing the details and putting them logically into context. Accordingly, this book has been created.

The first concrete event in Robinson's decision to set out on an expedition to the Holy Land was his pivotal 1832 meeting with Smith at the Theological Seminary in Andover, MA. In the years that followed, the idea evolved as Robinson developed research goals and awaited the right opportunity to fulfil his dream, realized when he embarked on the expedition in spring 1838. Robinson's and Smith's letters of the turbulent months leading up to their meeting in Cairo are evidence of the challenges the pair faced in planning such an undertaking: communication between Robinson and Smith across continents, with one living in the US and the other moving constantly between Smyrna and Beirut; what equipment they carried with them; how and when they executed the complicated preparations so crucial to the success of the expedition outcomes; financing the expedition (from Robinson's personal resources); and the logistics of securing cash in foreign lands for expenses en route. Importantly, the letters also demonstrate how close Robinson came to setting out on the expedition without Smith, a fact that would have changed dramatically the extremely successful outcome of the scientific endeavour.

Robinson's desire to embark on a scholarly expedition to the Holy Land was deeply influenced by the example of leading German scholars he met while studying in Halle and Berlin from 1826 to 1830. His marriage to Therese von Jacob (Talvj) was also tremendously influential, as she introduced him to the highest German learned and social circles. There he also developed and adopted his unique historic-geographical approach to the Scriptures, embedded in his Congregationalist upraising and education. Together, these factors led him to develop his unique and innovative method for scholarly investigation of the land of the Scriptures. Until Robinson, Holy Land study was characterized by different approaches, for example, accepting Franciscan traditions or arguing with them. Among its central aspects were the geographical 'rediscovery' of the country and its central cities and the first, highly inaccurate cartography. Robinson's approach was based on viewing the Bible as a historic-geographical source, walking in the footsteps of biblical figures to map scriptural geography, and re-identifying biblical place names (toponymy).

The expedition highlights have been recounted elsewhere. Hence, based upon archival material, 'Impressions en route' tells the story of

Robinson's and Smith's life on the road, contact with locals and dignitaries, and deportment at various points. It describes, in Robinson's and Smith's words, previously unknown natural phenomena, local and regional events that influenced their itinerary, and their connections with local Anglican, American, British, and German missionaries who are mentioned extensively in the letters and who contributed significantly to the expedition and its success. When the expedition ended, Robinson (travelling to Berlin) and Smith (travelling to Leipzig) sailed up the Danube to Vienna. During the voyage, Robinson became critically ill and barely survived. Smith's letter tells the quite dramatic tale of Robinson's recovery, aided by Smith's faithful treatment and Talvj's care when she reached Vienna. Robinson's and Smith's choice of travel destinations positively impacted their subsequent research, placing them in centres of lively intellectual activity and leading biblical, Near Eastern language, and cartographical scholarship.

The writing of the *Biblical Researches* was not without its own complications. Robinson initially planned to return to New York and begin his UTS teaching in Autumn 1838, but he actually returned only two years later. His decision to remain in Berlin and delay his return necessitated delicate negotiations with his superiors at UTS, negotiations that he entrusted to Smith. Robinson originally wrote the preface to the *Biblical Researches* as the last chapter of the book. In the preface, he discusses his and Smith's investigatory method as well as his approach to writing, illuminating his initial intentions and the changes in his scholarly approach through the writing process. In their correspondence, Robinson and Smith discuss in detail questions such as whether to write a book, of what nature, what to include and in which order, and what to omit. The letters reveal a decision-making process that developed during the expedition and deepened after their arrival in Berlin. After receiving encouragement from German scholars, Robinson decided to go forward, delving into an intensive, extraordinary writing schedule clearly designed to produce a long and detailed research volume and not a typical travelogue.

In Berlin, using its central library as well as Ritter's and others, Robinson quickly realized that the project would require significantly longer than the planned one year. One of the main innovations of the book is the unprecedented attention given to Arabic place names and orthography, and the combination of toponymy and field survey to rediscover biblical sites. Smith's Arabic expertise was invaluable. He developed a method of writing Arabic names in Roman letters and an entire field of Arabic orthography. Publication and distribution of the *Biblical Researches*, three volumes comprising more than 600 pages each, was also a significant challenge. The archival letters reveal deliberations regarding choice of publishers, difficulties with printing of the maps, and more.

Robinson relied on an outstanding number and diversity of sources to construct his arguments and the historical-geographic picture of the land and its sites. Beginning with Greek, Roman, and Church historians, he continued with medieval pilgrims' reports, and then travellers and scholars of his time, some of whose sources were published while he wrote his book. One reason for his thorough research of prior sources was the expedition's lack of accurate measuring instruments, acknowledged by Robinson as one of their greatest obstacles to executing a scientific study. Accordingly, he planned to equip Smith with necessary instruments to be used when the latter returned to his missionary stations in Beirut and other centers of the Near East. During his visits to Europe, Robinson also served as library emissary, seizing every opportunity to purchase books and library collections for his American institutes, first Andover Seminary, then the UTS, and later also for his own collection, which, after his death, was housed in Hamilton's Burke Library. This activity turned out to be one of his important contributions to the advancement of American science and learning institutions.

Two personalities, in particular, had important roles in the expedition and publication. ABCFM missionary Eli Smith was the expedition team Arabic expert and authority on local customs, people, and dignitaries in the East; he was the contact person with the locals, officials, guides, muleteers, and servants. He was also a close partner in the writing process. Yet, contemporary as well as modern research arguably has not sufficiently credited Smith for his crucial role in the expedition and publication as revealed in the primary sources, leading me to bestow the title 'the underacknowledged partner.' Smith's contacts with leading German Orientalists developed and strengthened during his nine months in Leipzig following the expedition. He initially went to Leipzig in order to produce Arabic types for his printing house in Beirut, and his stay in Germany led, fortuitously, to an even greater contribution to the *Biblical Researches*. Smith's significance should be understood as approaching that of Robinson!

Talvj, Robinson's second wife whom he married in Halle in 1828, played, until his death, a crucial role not only in his personal and social, but in his academic life as well. Her primary academic work was in translating, mainly poetry, from South Balkan languages to German. Her important status in German social and scientific society opened doors for Robinson that were significant for his academic research. Clearly, Talvj and Robinson were a team. They worked very closely together and supported each other's careers, and their scholarly achievements are testament to strong mutual influence and contribution.

Whosoever mentions Robinson today, generally does so in reference to the group of large stones protruding from the Temple Mount wall called 'Robinson's Arch.' While Robinson and Smith's epoch-making publication has been the source of many scientific debates, perhaps the most vigorous one concerns attributing the discovery of 'Robinson's Arch' to its namesake. Reverend Stephan Olin contended that the architect Catherwood and others discovered the arch prior to Robinson. In response, Robinson claimed that he was the first to connect the arch with the bridge reported by Flavius. The dispute remains unresolved, but the stones – not an arch, as accepted by scholars today – still bear Robinson's name.

Once again, who were you, Edward Robinson? In his 1991 article, J. Andrew Dearman, then professor of Old Testament at Austin Presbyterian Theological Seminary, introduces Robinson as a scholar and Presbyterian educator:[4] 'Although described by Philip Scaff as "the most distinguished biblical theologian whom America has produced," Edward Robinson (...) has received less attention from North American historians than other contemporaries (e.g. Moses Stuart, Charles Hodge), perhaps due to his reserve in public and his consistency in staying aloof from partisan squabbles.'

Here one touches upon another issue, Robinson's personality and character. What did he look like physically? What was the impression he left on people whom he met? Bliss, describing Robinson over forty years after his death, cites his first biographers, Smith and Hitchcock, who knew Robinson personally, and probably quite well:

> In person, he was built upon a large and even massive scale; with broad shoulders and muscular limbs, that denoted capacity for great endurance and toil; crowned with a head of unusual volume, a broad and open forehead, with perspective powers predominant; a shaggy eyebrow, a full, bright, piercing eye, though usually shaded through infirmity; a firm, yet pliant mouth; and, altogether, giving the impression, even to casual observer, of a man of weight and mark.[5]

4 Dearman (1991), 163.

5 Smith, H. B. (1863), 13; Bliss (1907), 193.

The physical description is impressive and also bears on Robinson's character. Using biblical terms, his biographers write further with regard to Robinson's appearance, 'although some of the features of it were as bold and rugged as the outline of Lebanon itself.' Robinson's contemporaries learned that it was not always pleasant to become his academic rival, 'no man who provoked him to controversy, cared ever to repeat the experiment.'[6]

In contrast, 'other features of it, little known to the world, were as delicate and charming as the rose of Sharon.'[7] This description was of Robinson the family man and of Robinson the social personality, the latter an obligation that he adopted under Talvj's tutelage. But the most important, impressive, and relevant trait, described by his biographers and emphasized here throughout, was his intellect, 'one of great native solidity and vigor.' He sought the truth, his truth, 'the truth of God, as revealed for human guidance in duty;' this was his guiding light. Robinson's correspondence reveals his impressive ability to hone in on an issue, his dedication to find answers, his fortitude, and his deep scholarship. 'We cannot leave the journeying of our Union professor without instancing his powers of endurance, his admirable pluck, and his just as admirable prudence,' writes Bliss.[8] His biographers complete the characterization, 'A man of positive convictions,' he did not hesitate to speak his mind clearly, 'his thoughts came feebly to their birth upon his lips.'[9]

In many aspects, this book continues my 'travels to look for the unknown Edward Robinson,' which began about ten years ago with a study on cartography conducted with Bruno Schelhaas of the IfL in Leipzig.[10] This is the search for 'the unknown Robinson' in all other aspects of his life and Holy Land studies, his background, research travels, and publications. In contrast with most studies, for example, Bliss's magnificent review,[11] and after much consideration, I decided to limit myself to the first expedition and its fruits, stopping at the first half of the fifth decennia of the nineteenth century, and not to include the second expedition and its resulting publications. Surely, some readers will see in this decision a deficiency, making this book a partial, uncomplet-

[6] Smith, H. B. (1863), 86–87.

[7] Smith, H. B. (1863), 84.

[8] Bliss (1907), 220.

[9] Smith, H. B. (1863), 85–86.

[10] Goren (2017); and Goren (2015); (2015a); (2015b); (2018).

[11] Bliss (1907).

ed work. My justification is threefold: 1) the first voyage, as well as its resulting publications, were, I believe, the more important; 2) the enormous quantity of letters, documents, and other archival material is more than sufficient for a deep and comprehensive study; and 3) including the second expedition might have resulted in a superfluous book. I do believe that the reader, scholar or amateur, will, after reading the book, accept this argument.

For me, the anecdotes I uncovered, the gems I unearthed and learned, and the experiences I amassed during this unexpected and impossible to foresee journey into Robinson's world, has been a once in a lifetime experience. 'These miracles, the unexpected ones, that the incident sometimes creates, are probably the most wonderful writers' wages awaiting the author.'[12]

Ill. 36. Robinson Family Graves, Green Wood Cemetery, Brooklyn, New York.

[12] Be'er (2019), 241.

BIBLIOGRAPHY

A. Archives

BP-UTS UTS Board Protocols, Union Theological Seminary, New York City, The Burke Library Archives

ERP Edward Robinson Papers, Hamilton College, Burke Library, Special Collections, Clinton, NY

ESP ABC Eli Smith Papers, 1819–69 (ABC 60), Harvard University, Houghton Library, Cambridge, MA

ABCFMA ABCFM Archives (ABC 6–16.5), Harvard University, Houghton Library, Cambridge, MA

AFSt/N, THOLUCK Friedrich August Gottreu Tholuck Papers, Studienzentrum August Hermann Francke, Archiv (Francke Foundations, Archives)

ERP-UTS Edward Robinson Papers, 1836–38, Union Theological Seminary, New York City, The Burke Library Archives

ESFP-YD RG 124: Eli Smith Family Papers, Yale University Library, Yale Divinity Library Special Collections

FBG SPA ARCH MFV Sammlung Perthes Archiv, Mitarbeiter und Freunde des Verlages Universität Erfurt, Forschungsbibliothek Gotha, Sammlung Perthes Gotha

IBP-YSM MS 82: Isaac Bird Papers, Yale University, Sterling Memorial Library, Manuscripts and Archives

Moore DSD Moore 1837, MS 3509: Dead Sea Diary, National Library of Ireland, Manuscripts reading room

RGS-IBG RGS-IBG Collections, Royal Geographical Society, Collections, London

SBB-PK, Nachl. 133 (Carl Ritter) Carl Ritter Papers, Staatsbibliothek zu Berlin – Preußischer Kulturbesitz, Handschriftenabteilung

SBB-PK, NH Nachlass Hengstenberg, Staatsbibliothek zu Berlin – Preußischer Kulturbesitz, Handschriftenabteilung

SBB-PK, Slg. Darm. Asien Sammlung Darmstaedter, Staatsbibliothek zu Berlin – Preußischer Kulturbesitz, Handschriftenabteilung

B. Abbreviations of periodicals and newspapers (mentioned at least twice)

ABR	*The (American) Biblical Repository*
BS	*Bibliotheca Sacra; or Tracts and Essays on topics connected with Biblical Literature and Theology*
BSTR	*Bibliotheca Sacra and Theological Review*
DUM	*Dublin University Magazine*
GABA	*Gelehrte Anzeigen herausgegeben von Mitgliedern der k. bayerischen Akademie der Wissenschaften*
IM	*Imago mundi*
JBL	*Journal of Biblical Literature*
JHG	*Journal of Historical Geography*
JI	*Jewish Intelligence and Monthly Account of the Proceedings of the London Society for Promoting Christianity Amongst the Jews*
JRGS	*Journal of the Royal Geographical Society*
MVGEB	*Mittheilungen über die Verhandlungen der Gesellschaft für Erdkunde zu Berlin*
MCEHK	*Monatlische Correspondenz zur Beförderung der Erd- und Himmels-Kunde*
MH	*The Missionary Herald*
NAR	*North American Review*
PN	*Pietismus und Neuzeit*
ZKM	*Zeitschrift für die Kunde des Morgenlandes*

C. Published material

Abašnik, V. O. (2008) 'Ludwig Heinrich von Jakob (1759–1827), ein hallescher Professor in Chaŕkov und Sankt Petersburg,' in E. Donnert (ed.), *Europa in der Frühen Neuziet. Festschrift für Günter Mühlpfordt*, 7, Köln, 895–928.

Abel, F.-M. (1939) 'Edward Robinson and the Identification of Biblical Sites,' *JBL* 58/4, 365–72.

Abu-Gazaleh, A. (1990) *American Missions in Syria: A Study of American Missionary Contribution to Arab Nationalism in 19th Century Syria*, Brattleboro, VT.

Abulfeda (1766) *Tabula Syriae: cum excerpto geographico ex Ibn ol Wardii Geographia et historia naturali* [ed. J. B. Köhler], Leipzig.

Adger, J. B. (1899) *My Life and Times, 1810–1899*, Richmond, VA.

Aiken, E. J. (2010) *Scriptural Geography: Portraying the Holy Land* [Tauris Historical Geography Series], London / New York.

Akel, S. (1978) *Der Pädagoge Missionar Johann Ludwig Schneller und seine Erziehungsanstalten*, Tübingen.

Alderson, R.C. (1843) *Notes on Acre and some of the Coast Defences in Syria* (Papers on Subjects Connected with the Duties of the Corps of Royal Engineers, VI), London.

An American (1858) [J. L. Stephens,] *Incidents of a Travel in Egypt, Arabia Petraea and the Holy Land* [2 vols], New York [1st ed. 1837].

Anderson, M. S. (1966) *The Eastern Question 1774–1923*, New York.

Anderson, R. (1830) *Observations upon the Peleponnesus and Greek Islands, made in 1829*, Boston.

— (1863) *Memorial Volume of the First Fifty Years of the American Board of Commissioners for Foreign Missions*, Boston.

— (1872) *History of the Missions of the American Board of Commissioners for Foreign Missions to the Oriental Churches* [2 vols], Boston.

Anon (1832) 'Wie hoch ist der Berg Sinai?,' *Annalen der Erd-, Völker und Staatenkunde* 5, 176.

— (1863) 'Funeral of Professor Robinson,' *New York Evening Post*, January 30.

— (1863a) 'Robinson's Arch,' *New York Journal of Commerce*, January 31 [copied *Albany Evening Journal* Feb 7].

Antonius, G. (1965) *The Arab Awakening: The story of the Arab national movement*, New York [1st ed. 1946].

Appleton (1889) 'Wheaton, Henry,' in *Appleton's Cyclopædia of American Biography*, New York, VI, 450–51.

Armstrong, G. (1887) *Names and Places in the Old Testament and Apocrypha* (Rev. C. Wilson, C. R. Conder), London 1887.

Arundale, F. (1839) *Illustrations of Jerusalem and Mount Sinai. Including the Most Interesting Sites Between Grand Cairo and Beirut*, London.

Auji, H. (2016) *Printing Arab Modernity: Book Culture and the American Press in Nineteenth-Century Beirut*, Leiden.

Ayalon, A. (2016) *The Arabic Revolution: Cultural Production and Mass Readership*, Cambridge.

Bain, D. H. (2011) *Bitter Waters: America's Forgotten Naval Mission to the Dead Sea*, New York.

Barnes, J. (1984) and P. P. Barnes, 'Thomas Aspinwall: First Transatlantic Literary Agent,' *The Paper of the Bibliographic Society of America* 78.3, 321–31.

Bartal, I. (1983) and Y. Shavit, 'The Image of the Period: A Century of Changes,' in Y. Ben-Arieh and I. Bartal (eds), *The Last Phase of the Ottoman Rule (1799–1917)* [The History of Eretz Israel, 8], Jerusalem, 11–13 [Heb.].

Bartlett, J. R. (2008) *Mapping Jordan through two millennia* [Palestine Exploration Fund Annual, 10], Leeds.

— (2019) *Burchard of Mount Sion, OP: Descriptio Terrae Sanctae* [Oxford Medieval Texts], Oxford.

Bartlett, W. H. (1844) *Walks About the City and Environs of Jerusalem*, London.

Basalla, G. (1967) 'The Spread of Western Science,' *Science* 156, 611–22.

Beck, H. (1981) 'Carl Ritter — Christ und Geograph,' *Gesnerus* 38/1–2, 259–62.

Be'er, H. (2019), *Master Key* [ed. G. Ticotsky], Tel Aviv [Heb.].

Ben-Arieh, Y. (1972) 'Pioneer scientific exploration in the Holy Land at the beginning of the 19th century,' *Terrae Incognitae* 4, 95–110.

— (1972a) 'The Geographical Exploration of the Holy Land,' *Palestine Exploration Quarterly* 104, 81–92.

— (1973) 'The first surveyed maps of Jerusalem,' *Eretz-Israel* 11, 64–74 [Heb.].

— (1974) 'The Catherwood Map of Jerusalem,' *The Quarterly Journal of the Library of Congress* 31, 150–60.

— (1976) 'Patterns of Christian activity and dispersion in nineteenth-century Jerusalem,' *JHG* 2/1, 46–69.

— (1979) *The Rediscovery of the Holy Land in the 19th Century*, Jerusalem / Detroit.

— (1989) 'Perceptions and images of the Holy Land,' in R. Kark (ed.), *The land that became Israel. Studies in historical geography*, New York, London / Jerusalem, 37–53.
— (1991) 'Holy Land views in 19th century western travel literature,' in M. Davis and Y. Ben-Arieh (eds), *Western societies in the Holy Land* [With eyes toward Zion, 3], Westport, 10–29.
— (1997) *Painting the Holy Land in the Nineteenth Century*, Jerusalem.
— (2018) *The Making of Eretz Israel in the Modern Era 1799–1949: A Historical-Geographical Study*, Jerusalem [Heb.].
Ben-Dov, M. (1982) *The Dig at the Temple Mount*, Jerusalem.
Benjamin of Tudela (1840–41) *The itinerary of Rabbi Benjamin of Tudela* [ed. A. Ascher, 2 vols], London.
Bennett, J. A. (1987) *The Divided Circle: A History of Instruments for Astronomy, Navigation and Surveying*, Topsfield, MA.
Benzinger, I. (1903) 'Researchers in Palestine,' in H. V. Hilprecht (ed.), *Explorations in bible lands during the 19th century*, Philadelphia, 581–622.
Berggren, J. (1826) *Reisen in Europa und im Morgenlande* [trsl. F. H. Ungewitter], Leipzig.
— (1862) *Bibel und Josephus über Jerusalem und das Heilige Grab wider Robinson und neuere Sionspilger als Anhang zu Reisen im Morgenlande*, Lund.
Berghaus, H. (1835) Karte von Syrien, den Manen Jacotin's und Burckhardt's gewidmet, Gotha.
— (1835a) *Geographisches Memoir zur Erklärung und Erläuterung der Karte von Syrien*, Gotha.
— (1839) 'Extract from a letter of Professor Berghaus,' *JRGS* 9, 308–10.
— (1840) 'Über die Reisen von E. Robinson und E. Smith im Peträischen Arabien und im Gelobten Lande (Auszug eines Schreibens an Kapitain Washington, Sekretär der königl. geogr. Gesellschaft zu London),' *Almanach für das Jahr 1840. Der Belehrung und Unterhaltung auf dem Gebiete der Erd-, Länder-, Völker- und Staatenkunde* 4, 531–35.
Bertou, J. de (1838) 'Voyage de l'extrémité sud de la Mer Morte à la pointe nord du golfe Elanitique,' *Bulletin de la Société de Géographie de Paris* 10, 18–32.
— (1839) 'Notes on a journey from Jerusalem to Hebron, the Dead Sea, El Ghór, and Wádi 'Arabah to 'Akabah, and back by Petra, in April, 1838,' *JRGS* 9, 277–86.

Bewer, J. A. (1939) 'Edward Robinson as a Biblical Scholar,' *JBL* 58/4, 355–63.

Bindoff, S. T. (1934), E. F. M. Smith and C. K. Webster (eds), *British Diplomatic Representatives 1789–1852*, London.

Bird, I. (1972) Bible Work in Bible Lands: or, Events in the History of the Syria Mission, Philadelphia.

Blaser, K. (1981) 'Mission und Erweckungsbewegung,' *PN* 7, 128–46.

Bliss, F. J. (1907) *The Development of Palestine Exploration*, New York.

Bongars, J. (1611))ed.(*Gesta Dei per Francos, sive, Orientalium expeditionum et regni Francorum Hierosolomitani historia /*‛*a variis, sed illius aevi scriptoribus; nunc primum aut editis, aut ad libros veteres emendatis; auctores praefatio ad lectorem exhibet*, Hannover.

Bowring, J. (1840) *Report on the Commercial Statistics of Syria. Presented to both Houses of Parliament*, London.

[Brettell, J. C.] (1840?) Plan showing the bombardment and capture of Acre, 3 November 1840. Scale: 1 inch to about 600 feet. Compass indicator. Based on a rough sketch by Joseph C Brettell, Mining Engineer; lithographed by M Bellanti, Malta.

Breydenbach, B. von (1486) *Itinerarium Hierosolymitanum ac in Terram Sanctam*, Mogunt.

Brogiato, H. P. (2008) 'Gotha als Wissens-Raum,' in S. Lenz and F. Ormeling (eds), *Die Verräumlichung des Welt-Bildes: Petermanns Geographische Mitteilungen zwischen 'explorativer Geographie' und der 'Vermessenheit' europäischer Raumphantasien* [Friedenstein Forschungen, 2], Stuttgart, 15–39.

Brown, J. W. (1969) *The Rise of Biblical Criticism in America, 1800–1870: The New England Scholars*, Middletown, CT.

Brown, L. A. (1977) *The Story of Maps*, Toronto [1st ed. 1949].

Buckingham, J. S. (1821) *Travels in Palestine, through the Countries of Bashan and Gilead, East of the River Jordan: Including a visit to the cities of Geraza and Gamala, in the Decapolis* [2 vols], London.

— (1840) *Public Address, delivered by Mr Buckingham, in Defence of his Lectures on Palestine, against the criticism of the Reverend Eli Smith, published, anonymously, in The New York Observer, in 1839*, New York.

Burchardus de Monte Sion (1519) *Locorum Terrae Sanctae Descriptio*, Venice.

Burckhardt, J. L. (1822) *Travels in Syria and the Holy Land*, London.

Burckhardt-Sarasin, C. (1956) and H. Schwabe-Burckhardt, *Scheik Ibrahim (Johann Ludwig Burckhardt): Briefe an Eltern und Geschwister*, Basel.

Burford, R. (1835) *Description of the City of Jerusalem, and the Surrounding Country. Now exhibiting at the Panorama, Leicester Square. Painted by the Proprietor, Robert Burford, From Drawings taken in 1834, by Mr F. Catherwood, Architect*, London.

Carmel, A. (1981) *Christen als Pioniere im Heiligen Land: Ein Beitrag zur Geschichte der Pilgermission und des Wiederaufbaus Palästinas im 19. Jahrhundert*, Basel.

— (1985) 'The Activities of the European Powers in Palestine, 1799–1914,' *Asian and African Studies* 19/1, 43–91.

Carne, J. (1826) *Letters from the East. Written during a recent tour through Turkey, Egypt, Arabia, the Holy Land, Syria and Greece*, London.

Cary, [W.] (1811) 'Description of the Patent reflecting Semicircle, invented by Sir Howard Douglas, Bart,' *The Philosophical Magazine*, Series 1 38/161, 186–87.

Catalogue (2019) *Catalogue of the Library belonging to the Theological Institute in Andover*, Andover.

Catherwood, F. (1835) Plan of Jerusalem, London.

— (n.d.) Plan of Jerusalem by F. Catherwood, Architect, with the improved measurements of Professor Robinson, n.p.

Cheyne, T. K. (1971) *Founders of Old Testament Criticism: Biographical, Descriptive, and Critical Studies*, Jerusalem [1st ed. London 1893].

CMS (1896) *The Church Missionary Atlas, containing an account of the various countries in which the Church Missionary Society labours*, London.

Coakley, J. F. (2003) 'Homan Hallock, Punchcutter,' *Printing History* 45/1, 18–41.

— (2013) 'Edward Breath and the Typography of Syriac,' in G. Roper (ed.), *The History of the Book in the Middle East*, Farnham, Surrey, 545–68 [1st published: *Harvard Library Bulletin*, 6 (1995), 41–64].

Coleman, A. P. (1944) 'Talvj's Correspondence with James Gates Percival,' *The Slavonic and East European Review* 22, 83–90.

Collier, P. (2002) 'The RGS, exploration and empire and the contested nature of surveying,' *Area* 34/3, 273–83.

Conder, C. R. (1880) and H. H. Kitchener (eds), *The Survey of Western Palestine (1871–1877)*, London 1880–9 [Rep. Jerusalem 1998].

—(1881) and H. H. Kitchener, *The Survey of Western Palestine: Memoirs of the Topography, Orthography, Hydrography, and Archaeology*. I: *Galilee*, London.
— (1883) and H. H. Kitchener, *The Survey of Western Palestine: Memoirs of the Topography, Orthography, Hydrography, and Archaeology*. III: *Judaea*, London.
Conder, J. (1830) *The Modern Traveller. A Description, Geographical, Historical, and Topographical, of the Various Countries of the Globe ...*, I: *Palestine*, II: *Syria*, London.
Cooke, N. (1998) 'The Forgotten Egyptologist: James Burton,' in P. and J. Starkey (eds), *Travellers in Egypt*, London / New York, 85–94.
Cotoi, C. (2016) 'Cholera, Health for All, Nation-Building, and Racial Degeneration in Nineteenth-Century Romania,' *East Central Europe* 43, 161–87.
Crombie, K. (1991) *For the Love of Zion: Christian Witness and the Restoration of Israel*, London / Sydney / Auckland.
— (2006) *A Jewish Bishop in Jerusalem: The life story of Michael Solomon Alexander*, Jerusalem.
Ćurčin, M. (1905) *Das serbische Volkslied in der deutschen Literatur*, Leipzig.
Cvikel, D. (2012) Y. Kahanov, and H. Goren, 'The naval campaign for conquering Acre, December 1831,' *Cathedra* 145, 37–62 [Heb.].
Dabach, A. (2016) 'Checking the Accuracy of 19th Century Measuring Maps of Jerusalem, using Computerized Tools' [unpublished M.A. Thesis, Dept. of Geography, Hebrew University, Jerusalem [Heb.].
Davis, M. (1977) 'The Holy Land Idea in American Spiritual History,' in idem (ed.), *With Eyes Toward Zion* [Scholars Colloquium on American – Holy Land Studies], New York, 3–33.
Davis, J. (1996) *The Landscape of Belief: Encountering the Holy Land in Nineteenth-Century American Art and Culture*, Princeton.
Dawson, L. S. (1885) *Memoirs of Hydrography. Including Brief Biographies of the Principal Officers who have Served in H. M. Naval Surveying Service between the years 1750 and 1855*, I: *1750 to 1830*, II: *1830 to 1885*, Eastbourne.
Dawson, W. R. (1995) and E. P. Uphill, *Who Was Who in Egyptology* [3rd revised edition, ed. M. L. Bierbrier], London.
Dearman, J. A. (1979–80) 'Some Observations on Early Hebrew Works and Teachers in America: 1726–1823,' *Hebrew Studies* 20–1, 195–99.

— (1991) 'Edward Robinson: Scholar and Presbyterian Educator,' *American Presbyterians* 69/3, 163–74.

Dehrmann, M.-G. (2010) 'Talvjs *Auswanderer* – eine deutsche Philologin in Nordamerika,' in Talvj, *Die Auswanderer*, [Hannover], 349–73.

Delić, J. (2001) 'Therese von Jakob-Robinson (Talvj), gesehen in interkulturellen Prozessen der Epoche,' in G. Schubert and F. Krause (eds), *Talvj Therese Albertine Luise von Jakob-Robinson (1797–1870). Aus Liebe zu Goethe: Mittlerin der Balkanslawen*, Weimar, 49–67.

Deumlich, F. (1982) *Surveying Instruments* [trsl. W. Faig], Berlin / New York.

Dobrašinović, G. (2001) 'Therese von Jakob-Robinson (Talvj) und Vuk Stefanović Karadžić,' in G. Schubert and F. Krause (eds), *Talvj Therese Albertine Luise von Jakob-Robinson (1797–1870). Aus Liebe zu Goethe: Mittlerin der Balkanslawen*, Weimar, 69–93.

Driver, F. (2001) *Geography Militant. Cultures of exploration and empire*, Oxford.

Dumas, A. (1841) and A. Dauzats, *Quinze jours au Sinai*, Paris.

Durbin, J. P. (after 1838) *Explanation of the large Map of Jerusalem from actual survey by F. Catherwood, author of the Panorama of Jerusalem &c. With additions and corrections from late travellers*, [Philadelphia].

— (1845) *Observations in the East, Chiefly in Egypt, Palestine, Syria, and Asia Minor*, New York.

Edney, M. H. (1997) *Mapping an Empire. The geographical construction of British India, 1765–1843*, Chicago / London.

[Eds of the *NAR* (1843)] 'Art. IX: Travels in Egypt, Arabia Petraea, and the Holy Land. By the Rev. Stephen Olin…,' *NAR* 57/121, 491–95.

Elath, E. (1971) *Britain's Route to India: British Projects in 1834–1872 for linking the Mediterranean with the Persian Gulf by steam navigation on the Euphrates and by Euphrates Valley Railway*, Jerusalem [Heb.].

Eliav, M. (1996) 'The 1837 Earthquake in Safed: An Eyewitness Report,' *Cathedra* 79, 53–78 [Heb.].

— (1997) (ed.) *Britain and the Holy Land, 1838–1914: Selected documents from the British consulate in Jerusalem*, Jerusalem.

Ellenmeyer-Životić, O. (2001) 'Therese Albertine Luise von Jakobs Übersetzungen serbischer Volkspoesie und die Konzeptualizierung der Serbenbildes im 19. Jahrhundert,' in G. Schubert and F. Krause (eds), *Talvj Therese Albertine Luise von Jakob-Robinson (1797–1870). Aus Liebe zu Goethe: Mittlerin der Balkanslawen*, Weimar, 95–109.

Ellioston, J. (1839) *Lectures on the theory and practice of medicine, delivered in the University College, London* [eds J. C. Cooke and T. G. Thompson], London.

Elliott, C. B. (1839) *Travels in the Three Great Empires of Austria, Russia, and Turkey* [2 vols], Philadlephia.

Engelmann, G. (1960) 'Der Atlas Asien von Heinrich Berghaus,' *(Petermann's) Mittheilungen aus Justus Perthes' geographischer Anstalt* 104/4, 311–29.

— (1977) *Heinrich Berghaus. Der Kartograph von Potsdam* [Acta Historica Leopoldina, 10], Halle.

Ewald, H. (1837) 'Plan dieser Zeitschrift,' *ZKM* 1, 3–13.

Fabri, F. (1557) *Eigentliche Beschreibung der hin- und Widerfarth zu dem Heyligen Landt gen Jerusalem, (...) Anno M.D.LVII*, [Bautzen].

Faehndrich, J. (2017) 'The Land and the Map: C. W. M. van de Velse's *Map of the Holy Land*,' in H. Goren, J. Faehndrich and B. Schelhaas (with P. Weigel), *Mapping the Holy Land: The Foundation of a Scientific Cartography of Palestine* [Tauris Historical Geography Series], London / New York, 55–106.

Fagan, B. M. (1975) *The Rape of the Nile: Tomb Robbers, Tourists, and Archaeologists in Egypt*, New York.

Fahmy, K. (1997) *All the Pasha's Men: Mehmed Ali, His Army, and the Making of Modern Egypt*, Cambridge.

Falkenstein, K. K. (1840) *Geschichte der Buchdruckerkunst in ihrer Entstehung und Ausbildung*, Leipzig.

Farah, C. E. (2000) *The Politics of Interventionism in Ottoman Lebanon, 1830–1861*, Oxford / London / New York.

Feyrabend, S. (1609) (ed.) *Reyssbuch dess Heyligen Lands, das ist, Ein grundtliche Beschreibung aller und jeder Meer und Bilgerfahrten zum Heyligen Lande, so bisshero ... vorgenommen ... Beneben eingeführter auch eigentlicher Beschreibung dess gantzen Heyligen Lands Palaestinae ... Ferner auch der mächtigen Land Indien und Persien ... Endlich von gemeldter Ort und Landen jetziger Einwohner, Türcken und Araber ... Namen und Geschlecht der jenigen die solche Reyss in eigner Person vollbracht ...*, Frankfurt/Main.

Findling, J. E. (1989), *Dictionary of American Diplomatic History*, New York, etc. [2nd ed.].

Finnie, D. H. (1967), *Pioneers East: The Early American Experience in the Middle East*, Cambridge, MA.

Flavius, J. (1895) *Antiquities of the Jews*, in: *The Works of Flavius Josephus* [trsl. W. Whiston], Auburn / Buffalo.

Forsskål, P. (1775) *Flora Aegyptiaco-Arabica, sive, descriptiones plantarum, quas per Aegyptum Inferiorem et Arabiam Felicem Detexit, illustravit Petrus Forskal; accedit tabula Arabiae Felicis geographico-botanica* [ed. C. Niebuhr], Hauniae.

Freeman, T. W. (1980), 'The Royal Geographical Society and the development of geography,' in E. H. Brown (ed.), *Geography Yesterday and Tomorrow*, Oxford, 1–17.

Freytag, G. F. W. (1830–7) *Lexicon Arabico-Latinum: praesertium ex Djeuharri Firuzabadiique et aliorum Arabum operibus adhibitis Golii quoque et aliorum libris confectum: accedit index vocum Latinuarum locupletissimus* [5 vols], Halle.

Fück, J. (1955) *Die arabischen Studien in Europa bis in den Anfang des 20. Jahrhunderts*, Leipzig.

Furrer, K. (1878) 'Dr Titus Tobler,' *Zeitschrift des Deutschen Palästina-Vereins* 1, 49–60.

Gadsby, J. (1855) *My Wanderings: Being Travel in the East in 1846–7, 1850–1, 1852–3* [2 vols], London.

Gatch, M. McC. (1996) (ed.) *'so precious foundation': The Library of Leander van Ess at the Burke Library of Union Theological Seminary in the City of New York*, New York [bilingual, also German].

Gesenius, W. (1823) (ed.) *Johann Ludwig Burckhardt's Reisen in Syrien, Palästina und der Gegend des Berges Sinai* [2 vols], Weimar.

Gibson, S. (1996) and D. M. Jacobson, *Below the Temple Mount in Jerusalem: A sourcebook on the cisterns, subterranean chambers and conduits of the Ḥaram al-Sharīf* [BAR International Series 637], Oxford.

— (2020) 'Underground and Almost Invisible: The Archaeology of the Temple Mount in Jerusalem based on Archival Sources,' in J. R. Branham and B. St Laurent (eds), *Marking the Sacred: The Temple mount / Haram al-Sharif in Jerusalem*, Phildelphia [to be published].

Gichon, M. (2003) *Napoleon in the Holy Land* [ed. E. Melzer], Reut [Heb.].

Gidney, W. T. (1908) *The History of the London Society for Promoting Christianity Amongst the Jews, from 1809 to 1908*, London.

Gil Novales, A. (2019), 'Biography of José Hezeta Zenea (1788–1862),' *The Biography*, https://thebiography.us.

Giltner, J. H. (1988) *Moses Stuart: The Father of Biblical Science in America* [Biblical Scholarship in North America, 14], Atlanta.

Glass, D. (1997) *Malta, Beirut, Leipzig and Beirut Again: Eli Smith, the American Syria Mission and the Spread of Arabic Typography in 19th Century Lebanon*, Beirut.

Godlewska, A. (1988) *The Napoleonic survey of Egypt. A masterpiece of cartographic compilation and early nineteenth century fieldwork* [*Cartographica*, 25/1–2, Monograph 38/39], Toronto.

— (1994) 'Napoleon's geographers (1797–1815): imperialists and soldiers of modernity,' in idem and N. Smith (eds), *Geography and Empire*, Oxford / Cambridge MS, 34–53.

— (1999) *Geography unbound. French geographic science from Cassini to Humboldt*, Chicago / London.

Goldman, S. (1990–1) 'Isaac Nordheimer (1809–1842): "An Israelite Truly in Whom There Was No Guile",' *American Jewish History* 80/2, 1–14.

Goren, H. (1995) 'Titus Tobler's Legacy. Two Sources,' *Bulletin of the Anglo-Israel Archaeological Society* 14, 57–61.

— (1997) 'Johannes Rudolph Roth: Ein Leben für die Palästinaforschung,' *Journal des deutsches evangelische Institut für Altertumswissenschaft des heiligen Landes zu Jerusalem* 5, 22–44.

— (1997a) 'Nicolayson and Finn Describe the Expeditions and Death of Costigan and Molyneux,' *Cathedra* 85, 65–94 [Heb.].

— (1998) 'The chase after the Bible. Individuals and institutions and the study of the Holy Land,' in U. Wardenga and W. Wilczynski (eds), *Religion, ideology and geographical thought* [WPC Kielce Studies in geography, 3], Kielce, 103–15.

— (1999) 'Carl Ritter's contribution to Holy Land research,' in A. Buttimer, S. D. Brunn and U. Wardenga (eds), *Text and image. Social construction of regional knowledge* [Beiträge zur Regionalen Geographie, 49], Leipzig, 28–37.

— (1999a) 'Heinrich Kiepert in the Holy Land, spring 1870. Sketches from an exploration-tour of an historical cartographer,' in L. Zögner (ed.), *Antike Welten Neue Regionen. Heinrich Kiepert 1818–1899*, Berlin, 45–62.

— (2000) 'Pioneer Surveying and Mapping of the Galilee and the Hauran: The Map of the Sources of the Jordan River (1841) and Map of the *Ledja* (1838),' *Cathedra* 96, 111–30 [Heb.].

— (2001) 'Scientific organizations as agents of change. The Palestine Exploration Fund, the Deutsche Verein zur Erforschung Palästinas and nineteenth-century Palestine,' *JHG* 27\2, 153–65.

— (2001a) 'Franz Wilhelm Sieber,' in A. Tishby, R. Rubin, and N. Kadmon (eds), *Holy Land in Maps* [Israel Museum catalogue, 459], Jerusalem, 154–55.

— (2002) 'Sacred, But Not Surveyed: Nineteenth-Century Surveys of Palestine,' *IM* 54, 87–110.
— (2003) *'Zieht hin und erforscht das Land'. Die deutsche Palästinaforschung im 19. Jahrhundert* [Schriftenreihe des Instituts für Deutsche Geschichte der Universität Tel Aviv, 23, trsl. A. C. Naujoks], Göttingen.
— (2004) 'Charles Gaillardot, Renan's Man-in-the-Field: Physician and Scholar,' in M. Abitbol (ed.), *France and the Middle East: Past, Present and Future* [World Powers and the Middle East, The Harry S. Truman Institute], Jerusalem, 99–124.
— (2005) 'British surveyors in Palestine and Syria, 1840–1841' [International Cartographic Conference A Coruña 2005: Mapping approaches into a changing world. CD-ROM].
— (2008) 'August Petermann and Palestine cartography,' in S. Lenz and F. Ormeling (eds), *Die Verräumlichung des Welt-Bildes: Petermanns Geographische Mitteilungen zwischen 'explorativer Geographie' und der 'Vermessenheit' europäischer Raumphantasien* [Friedenstein Forschungen, 2], Stuttgart, 75–87.
— (2009) *'Echt katholisch und gut deutsch': Die deutschen Katholiken und Palästina 1838–1910* [trsl. M. Lemke], Göttingen.
— (2010) 'Nineteenth Century Jerusalem as a Test Case of European Involvement in the Near East: A Reappraisal,' in I. Bartal and H. Goren (eds), *The History of Jerusalem: The Late Ottoman Period (1800–1917)*, Jerusalem, 19–32 [Heb.].
— (2011) *Dead Sea Level: Science, Exploration and Imperial Interests in the Near East* [Tauris Historical Geography Series, 6], London.
— (2014) 'Edward Robinson – Father of the Scientific Study of Eretz Israel,' in E. Schiller and G. Barka'ai (eds), *Wonderful Land: Jerusalem and Eretz Israel, Papers and Studies* [*Ariel* 206–07], Jerusalem, 64–74 [Heb.].
— (2015) and B. Schelhaas 'Berghaus' map Part of Arabia Petraea and Palestine (1839) and the Royal Geographical Society: misuse or misunderstanding?,' *Terrae Incognitae* 47/2, 127–41.
— (2015a) and B. Schelhaas 'On an unknown measuring of Jerusalem in 1823,' *Die Erde* 146/1, 63–78.
— (2015b) and B. Schelhaas 'Colonial, Religious and Scientific Mapping. The Cartographic Practice of Charles William Meredith van de Velde,' in J. García Álvarez and J. Carlos Garcia (eds), *History of Geography and Colonialism*, Lisbon, 211–27.

— (2017) 'Edward Robinson, Eli Smith and the Cartography of Berghaus and Kiepert,' in H. Goren, J. Faehndrich and B. Schelhaas (with P. Weigel), *Mapping the Holy Land: The Foundation of a Scientific Cartography of Palestine* [Tauris Historical Geography Series], London / New York, 5–53.

— (2017a), J. Faehndrich and B. Schelhaas (with P. Weigel), *Mapping the Holy Land: The Foundation of a Scientific Cartography of Palestine* [Tauris Historical Geography Series], London / New York.

Grasshoff, H. (1966) 'Aus dem Briefwechsel Therese Jakob-Robinson an P. I. Köppen,' in W. Steinitz et al. (eds), *Ost und West in der Geschichte des Denkens und der kulturellen Beziehungen. Festschrift für Eduard Winter zum 70. Geburtstag*, Berlin, 477–88.

Grimm, J. L. (1830) Palaestina, 1:900.000, Berlin.

Gusrau, M. (2016) *Orientalen oder Christen? Orientalisches Christentum in Reiseberichten deutscher Theologen*, Göttingen.

Hall, C. (1841) 'Review of Robinson's biblical Researches,' *ABR* 6/2, 419–58.

Hallett, R. (1965) *The Penetration of Africa: European Enterprise and Exploration Principally in Northern and Western Africa up to 1830*, I: *To 1815*, London.

Hallock, H. (1929) *The New Arabic Type*, New York.

Hallote, R. S. (2006) *Bible, map, and spade. The American Palestine Exploration Society, Frederick Jones Bliss, and the forgotten story of early American Biblical archaeology*, Piscataway, NJ.

Hamilton, W. R. (1842) *Address of W. R. Hamilton, Esq., President of the Royal Geographical Society of London, to His Excellency Edward Everett, envoy extraordinary from the United States, on the occasion of presenting the gold medal awarded by the council to the Rev. Dr Edward Robinson of New York. With Mr Everett' Reply. May 23. 1842*, London.

Hamlin, C. (1893) *My Life and Times*, Boston [2nd ed.].

Handy, R. T. (1981) (ed.) *The Holy Land in American Protestant Life 1800–1949: A Documentary History*, New York.

— (1987) *A History of the Union Theological Seminary in New York*, New York.

Hansen, T. (1964) *Arabia Felix: The Danish Expedition of 1761–1767* [trsl. J. and K. McFarlane], London.

Harris, P. W. (1999) *Nothing but Christ: Rufus Anderson and the Ideology of Protestant Foreign Missions*, Oxford / New York.

Hasselquist, F. (1766) *Voyages and Travels in the Levant in the Years 1749, 50, 51, 52. Containing Observations in Natural History, Physick, Agriculture, and Commerce: Particularly on the Holy Land, and the Natural History of the Scriptures* [ed. C. Linnaeus], London.

Hecklinger, A. (1981) 'Der Rottweiler Stadtpfarrer und Orientalist D.Dr Philipp Wolff (1810–1894),' *Rottweiler Heimatblätter* 42/1.

Heim, H. J. (1879) *Dr Titus Tobler der Palästinafahrer. Ein appenzellisches Lebensbild*, Zürich.

Heller, G. P. (1836) '1) Palästina. Von Karl von Raumer, Professor in Erlangen (…). 2) Karte von Syrien, (…) von Heinrich Berghaus…,' *GABA* 2, Nr. 236–54, 25.11–21.12.

Henze, D. (2011) *Enzyklopädie der Entdecker und Erforscher der Erde* [6 vols], Darmstadt [1st ed. Graz 1978–2011].

Herbert, F. (1983) 'The Royal Geographical Society's Membership, The Map Trade, and Geographical Publishing in Britain 1830 to *ca* 1930: An Introductory Essay with Listing of some 250 Fellows in Related Professions,' *IM* 35, 67–95.

Herschel, J. F. W. (1851) (ed.) *A Manual of Scientific Enquiry, prepared for the use of Officers in Her Majesty's Navy, and Travellers in General*, London.

Hexelschneider, E. (2001) 'Talvj und das Verlagshaus F. A. Brockhaus in Leipzig,' in G. Schubert and F. Krause (eds), *Talvj Therese Albertine Luise von Jakob-Robinson (1797–1870). Aus Liebe zu Goethe: Mittlerin der Balkanslawen*, Weimar, 111–28.

Hirrel, L. P. (1998) *Children of Wrath. New School Calvinism and Antebellum Reform*, Lexington, Kentucky.

Hodge, A. A. (1880) *The Life of Charles Hodge D.D. LL.D. Professor in the Theological Seminary Princeton N. J.*, New York.

Homer (1822) *Homeri Iliadis Libri Novem Priores. Librique xvii, et xxh. Ex Recensione C. G. Heyna. Cum notis brevibus, quas ex Hene et sliis praecipue collegit, nonullas adjecit E. Robinson, A. M. in usum juventutis academicae,* Catskill and Andover.

Hooker, E. W. (1845) *Memoirs of Mrs Sarah L. Huntington Smith, Late of the American Mission in Syria*, New York [3rd ed.].

Howard, T. A. (2005) 'Deutsche Universitästheologie in den USA: Edward Robinson und Phillipp Schaff,' in P. Löser and C. Strupp (eds), *Universität der Gelehrten – Universität der Experten: Adaptionen deutscher Wissenschaft in den USA des neunzehnten Jahrhunderts* [Transatlantische historische Studien, 24], Stuttgart, 32–51.

Hunnisett, B. (1998) *Engraved on steel: The history of picture production using steel plates*, Aldershot / Brookfield, VT.

Ibbotson, J. D. (1933) *Edward Robinson and his Books: A Biographical and Descriptive Sketch*, Hamilton College.

ibn Shaddad, Baha' al-Din (1755) *Vita et res gestae sultani, almalichi al-nasiri, Saladini ... nec non excerpta ex historia universali Abulfedae* [trsl. A. Schultens], Leiden.

Ilgauds, H. J. (1995) and G. Münzel *Die Leipziger Universitätssternwarten auf der* Pleißenburg *und im Johannistal. Astronomische Schulen von Weltruf* [Leipziger Hefte], Beucha.

Irby, C. L. (1823) and J. Mangles *Travels in Egypt and Nubia, Syria, and Asia Minor: during the years 1817 and 1818*, London [new ed. 1868].

Jacotin, P. (1810) Carte topographique de l`Égypte et de plusieurs parties des pays limitrophes levee pendant l`expédition de l`Armee Français [47 sheets], Paris.

— (1818) 'Carte de l'Égypte,' in E.-F. Jomard (ed.), *Description de l'Égypte*, Paris.

Jagaillouz, S. (1989) *La médecine modern en Egypte. De Bonaparte au XXᵉ siecle*, Alexandrie.

Jampoler, A. C. A. (2005) *Sailors in the Holy Land: The 1848 American Expedition to the Dead Sea and the Search for Sodom and Gomorrah*, Annapolis, MD.

Jones, Y. (1973) 'British military surveys of Palestine and Syria 1840–1841,' *The Cartographical Journal* 10/1, 29–41.

Kaarsberg Wallach, M. (2001) 'Talvj und ihre kulturelle Mittlerrole in America,' in G. Schubert and F. Krause (eds), *Talvj Therese Albertine Luise von Jakob-Robinson (1797–1870). Aus Liebe zu Goethe: Mittlerin der Balkanslawen*, Weimar, 247–62.

Kalfatovic, M. A. (2011) 'Niles Notes of a *Howadji*: American Travellers in Egypt, 1837–1903,' in P. and J. Starkey (eds), *Unfolding the Orient: Travellers in Egypt and the Near East*, Reading, 239–59.

Kaplan, R. D. (1995) *The Arabists: The Romance of an American Elite*, New York, etc.

Kark, R. (1994), *American Consuls in the Holy Land, 1832–1914*, Detroit / Jerusalem.

— (2011), and H. Goren, 'Pioneering British exploration and scriptural geography: The Syrian Society/The Palestine Association,' *The Geographical Journal* 177/3, 264–74.

Karmon, Y. (1960) 'An analysis of Jacotin's map of Palestine,' *Israel Exploration Journal* 10/3, 155–73, 244–53.

Kawerau, P. (1958) *Amerika und die orientalischen Kirchen: Ursprung und Anfang der amerikanischen Mission unter den Nationalkirchen Westasiens* [Arbeiten zur Kirchengeschichte, 31], Berlin.

Keith Johnston, A. (1861) Palestine or the Holy Land, 1:712,649, in idem., *The Royal Atlas of Modern Geography ...*, Edinburgh / London.

Kellen, W. V. (1902) *Henry Wheaton, an Appreciation*, Boston.

Khatib, H. (2003) *Palestine and Egypt under the Ottomans: Paintings, Books, Photographs, Maps and Manuscripts*, London / New York.

Kieffer Memoir (1833) 'Memoir of M. Jean Daniel Kieffer,' *The Missionary Register*, 121–24.

Kiepert, H. (1841) 'Memoir on the Maps Accompanying this Work,' in E. Robinson and E. Smith, *Biblical Researches in Palestine, Mount Sinai and Arabia Petraea in 1838. A Journal of travels in the year 1838*, III, First Appendix, 29–55.

— (1841a) Plan of Jerusalem, sketched from Sieber and Catherwood, corrected by the Measurements of Robinson and Smith, London [also: Plan von Jerusalem. Entw. nach Sieber und Catherwood, berichtigt durch Beobeachtungen von Robinson und Smith. Entworfen u. gez. von H. Kiepert. In Stein gest. von H. Mahlmann; 1:10.000; 25 × 30 cm., in *Atlas zu Robinson's Palästina*].

Kitto, J. (1853) 'Moses Stuart,' *Journal of Sacred Literature and Biblical Record* 3/6, 308–28.

Klatzker, D. (1997) 'Sacred Journeys: Jerusalem in the Eyes of American Travelers before 1948,' in Y. Ben-Arieh and M. Davis (eds), *Jerusalem in the Mind of the Western World, 1800–1948* [With Eyes Toward Zion, V], Westport, Conn. / London, 47–58.

Klein, R. (1999) 'Aleppiner Handleshäuser zwischen Basar und Börse: Informationssuche und -verbreitung im Zeichen der Inkorporation (ca. 1780–1920),' in D. Rothermund (ed.), *Aneigung und Selbstbehauptung: Antworten auf die Europäische Expansion*, München, 345–72.

Körner, G. (1880) *Das deutsche Element in den Vereinigten Staaten von Nordamerika, 1818–1848*, Cincinnati.

Korte, J. (1743) *Reise nach dem weiland Gelobten, nun aber seit siebenzehn hundert Jahren unter dem Fluche liegenden Lande, wie auch nach Egypten, dem Berg Libanon, Syrien und Mesopotamien*, Halle.

Krafft, W. L. (1846) *Die Topographie Jerusalem's*, Bonn.

Krause, F. (1982) 'Das Rußlanderlebnis im Schaffen der Therese Albertine Luise von Jakob-Bobinson (Talyj),' *Zeitschrift für Slawistic* 27, 512–22.

— (1999) 'Das Amerikabild Therese Robinsons (Talvj) in ihrer Korrespondenz,' in G. Grimm and R. Pribić (eds), *Talvj, Tesla und die geistige Welt der Südslawen. Zum Gedächtnis von Nicola Pribić (1913–1992)* [Veröffentlichungen des Instituts für Geschichte Osteuropas und Südosteuropas der Universität München, 17], München, 49–58.

— (2001) 'Zur Forschung über die Talvj seit Jevto M. Milović: Ein Literaturbericht,' in G. Schubert and F. Krause (eds), *Talvj Therese Albertine Luise von Jakob-Robinson (1797–1870). Aus Liebe zu Goethe: Mittlerin der Balkanslawen*, Weimar, 167–79.

— (2001a) 'Die Korrespondenz der Therese von Jakob-Robinson (Talvj),' in G. Schubert and F. Krause (eds), *Talvj Therese Albertine Luise von Jakob-Robinson (1797–1870). Aus Liebe zu Goethe: Mittlerin der Balkanslawen*, Weimar, 266–319.

— (2001b) and G. Dobrašinović, 'Verzeichnis der Veröffentlichungen von und über Therese von Jakob-Robinson (Talvj) 1820–1999,' in G. Schubert and F. Krause (eds), *Talvj Therese Albertine Luise von Jakob-Robinson (1797–1870). Aus Liebe zu Goethe: Mittlerin der Balkanslawen*, Weimar, 321–37.

Kropej, M. (2013) 'The Cooperation of Grimm Brothers, Jernej Kopitar and Vuk Karadžić,' *Studia Mythologica Slavica* 16, 215–31.

Kruse F. (1854) et al. (eds) *Ulrich Jasper Seetzen's Reisen durch Syrien, Palästina, Phönicien, die Transjordan-Länder, Arabia Petraea und Unter-Aegypten* [3+1 vols], Berlin.

Kurz, M. (1998) and P. Linnant de Bellefonds, 'Linant de Bellefonds: Travels in Egypt, Sudan and Arabia Petraea (1818–1828),' in P. and J. Starkey (eds), *Travellers in Egypt*, London / New York, 61–69.

Kushner, D. (1986) (ed.) *Palestine in the late Ottoman period. Political, social and economic transformation*, Jerusalem.

Kutluoğlu, M. H. (1998) *The Egyptian Question (1831–1841): The Expansionist Policy of Mehmed Ali Paşa in Syria and Asia Minor and the Reaction of the Sublime Porte* [Middle Eastern and Balkan Studies Series, 2], Istanbul.

Laborde, L. de (1830) Carte de l'Arabie Petrée indiquant les Routes suivies par les Caravans, les Pélerins et les Voyageurs depuis l'époque reculée jusqu'a nos jours 1833, in idem, *Voyage de l'Arabie Pétrée par Leon de Laborde et Linant,* Paris [*sic*!].

— (1830a), *Voyage de l'Arabie Pétrée par Leon de Laborde et Linant,* Paris.

— (1836) *Journey Through Arabia Petraea, to Mount Sinai and the Excavated City of Petra, the Edom of the Prophecies*, London.

— (1841) *Commentaire géographique sur l'Exode et les Nombres*, Paris / Leipzig.

Lane, E. W. (1836) *An Account of the Manners and Customs of the Modern Egyptians*, London.

Laurens, H. (1989) *L'Expédition d'Égypte 1798–1801* [Bonaparte et l'Islam, le choc de cultures], Paris.

Lazarus, P. (1930), 'Neander, Johann August Wilhelm,' *Jüdisches Lexikon: Ein enzyklopädische Handbuch des jüdischen Wissens* ..., IV/1, Berlin, 438.

Leavy, M. R. (1992) 'Looking for the Armenians: Eli Smith's Missionary Adventure, 1830–1831,' *Transactions the Connecticut Academy of Arts and Sciences* 50/3, 189–275.

— (1993) *Eli Smith and the Arabic Bible* (Yale Divinity School Library, Occasional Publication, 4), New Haven CT.

Lee, C. E. (1840) 'On the Geology of Palestine, and the Destruction of Sodom and Gomorrah,' *ABR* 3/VI (April), 324–52.

Lenz, K. (2003) 'Die Bibliothek der Gesellschaft für Erdkunde zu Berlin,' *Die Erde*, Sonderheft 1, 29–33.

Licht, G. (1999) *Tochter und Vater: Therese Albertine Louise von Jakob-Robinson (1797–1870) und Ludwig Hainrich von Jakob (1759–1827)*, Halle.

— (2009) 'Einführung: Zwischen Auflehnung und Anpassung – Leben und Werk der Schriftstellerin Therese von Jakob Robinson (1797–1870),' in idem (ed.), *Therese Albertine Luise von Jakob Robinson Lesebuch* [Literatur aus Mitteldeutschland 2], Haale, 10–34.

Lieber, S. (1985) 'The purchase of the land for "Christ Church" by John Nicolayson,' *Cathedra* 38, 201–02 [Heb.].

Liebrenz, B. (2014) 'Früher Druck mit arabischen Typen in Leipzig, 17.-18. Jahrhundert,' in G. Roper (ed.), *Historical Aspects of Printing and Publishing in Languages of the Middle East: Papers from the Third Symposium on the History of Printing in the Languages and Countries of the Middle East, University of Leipzig, September 2008*, Leiden / Boston, 17–52.

Linan de Bellefonds, P. (2001) 'From Cairo to Petra: Léon de Laborde and L. M. A. Linant de Bellefondsd, 1828,' in P. and J. Starkey (eds), *Unfolding the Orient: Travellers in Egypt and the Near East*, Reading, 179–91.

Lindner, C. B. (2009) 'Negotiating the Field: American Protestant Missionaries in Ottoman Syria, 1823 to 1860' [unpublished PhD dissertation, University of Edinburgh].

Lindsay, A. W. C., Lord (1838) *Letters on Egypt, Edom, and the Holy Land* [2 vols], London.
— (1838a) 'Lord Lindsay's Letters from the East,' *DUM* 12, 568–92.
Lindt, A. (1981) 'Christian Friedrich Spittler aus Basel,' *PN* 7, 115–27.
Linke, M. (1984) 'Carl Ritter und die Royal Geographical Society I,' *Wissentschaftliche Zeitschrift der Martin-Luther-Universität Halle-Wittenberg, Mathematisch-naturwissenschaftliche Reihe* 33/4, 33–42.
Lipman, V. D. (1989), *Americans in the Holy Land through British Eyes: 1820–1917. A Documentary History*, London.
Liu, L. H. (2012) 'Henry Wheaton (1785–1848),' in B. Fassbender and A. Peters (eds), *The Oxford Handbook of the History of International Law*, Oxford 2012 [DOI: 10.1093/law/9780199599752.003.0057].
Livingstone, D. N. (1992) *The Geographical Tradition: Episodes in the History of a Contested Enterprise*, Oxford / Cambridge.
LJS (1839) 'An Abstract of the Thirty-First Report of the London Society for Promoting Christianity Amongst the Jews,' *JI* 5, 149–60.
Loew, J. (1838) *Be'er Hagolah*, Warsaw [Heb.].
Lord, W. F. (1901) *England and France in the Mediterranean 1660–1830*, London [new ed. Port Washington, NY 1970].
Luncz, A. M. (1980) 'The Values of the Coins,' in idem (ed.), *Luach Eretz Israel: a Selection of Papers* (Collected by E. Schiller), III, Jerusalem, 15–17 [Heb.].
Lynch, W. F. (1849) *Narrative of the Unites States' Expedition to the River Jordan and the Dead Sea*, Philadelphia.
Macalister, R. A. S. (1975) *A Century of Excavation in Palestine*, London.
Mackey, P. E. (1973) 'Reverend George Barrell Cheever: Yankee Reformer as Champion of the Gallows,' *Proceedings of the American Antiquarian Society*, 82/2 (1972), 323–42.
MacLean, J. (1841) 'Bacchus and Anti-Bacchus,' *Princeton Review*, 13, 267–306, 471–523.
Makdisi, U. (2008) *Artillery of Heaven: American missionaries and the failed conquest of the Middle East*, Ithaca / London.
Mangold, S. (2004) *Eine 'weltbürgerliche Wissenschaft' - Die deutsche Orientalistic im 19. Jahrhundert*, Stuttgart.
Manley, D. (2001) and P. Rée *Henry Salt: Artist, Traveller, Diplomat, Egyptologist*, London.
Marriott, J. A. R. (1940) *The Eastern Question: An Historical Study in European Diplomacy*, Oxford [1st ed. 1917].
Martens, R. (1949) *Eduard Rüppell. Leben und Werk eines Forschungsreisenden*, Frankfurt am Mein.

Maundrell, H. (1703) *A Journey from Aleppo to Jerusalem at Easter, A.D. 1697*, Oxford.

Maximilian (1978) Herzog in Bayern, *Wanderung nach dem Orient im Jahre 1838* [ed. W. Hansen], Pfaffenhofen [1st ed. München 1839].

Medvekova, O. (2014) 'Richard Pococke, or the Invention of Jerusalem for Tourists,' in B. Kühnel et al. (eds), *Visual Constructs of Jerusalem* [Cultural Encounters in Late Antiquity and the Middle Ages, 18], Turnhout, Belgium, 429–36.

Mehedinţi, M. (2012) and C. A. Sava 'Death as Statistics: Demographic Aspects of Plague in the Romanian Area (18th–19th Centuries),' *Romanian Journal of Population Studies*, 6/2, 33–51.

Meiner, A. (1930) *Geschichte des Verlages Johann Ambrosius Barth, 1780–1930*, Leipzig.

Merchiers, I. (2007) *Cultural Nationalism in the South Slav Habsburg Lands in the Early Nineteenth Century: The Scholarly Network of Jens Kopitar (1780–1844)* [Slavistische Beiträge, 455], München.

Mienert, M. (2017) and T. Keiderling 'Die Verlagsgründung,' in M. Mienert et al. (eds), *Baron der Englischer Bücher: Der Leipziger Verlag Bernhard Tauchnitz 1837–1973*, Beucha, 25–29.

Mill, H. R. (1930) *The Record of the Royal Geographical Society 1830–1930*, London.

Milović, J. M. (1941) *Talvjs erste Übertragungen für Goethe und ihre Briefe an Kopitar* [Veröffentlichungen des Slavischen Instituts an der Friedrich-Wilhelms-Universtät Berlin, ed. M. Vasmer, 33], Leipzig.

Mojašević, M. (1976) 'Eine Leistung Goethe zuliebe: Talvj, Goethe und das serbokroatische Volkslied,' *Goethe-Jahrbuch* 93, 164–89.

[Moore, G. H. (1837) and W. G. Beek] 'On the Dead Sea and Some Positions in Syria,' *JRGS* 7, 456.

Moresby, R. (1836) and T. G. Carless A Chart of the Red Sea, Comprising the part above Jiddah, Executed in the Years 1830, 31, 32, & 33 in the Hon[oura]ble Comp[an]y's Ship Palinarus, London.

Moscrop, J. J. (2000), *Measuring Jerusalem. The Palestine Exploration Fund and British interest in the Holy Land*, London.

Müller, H. H. (2011) and M. Nottscheid *Wissenschaft ohne Universität, Forschung ohne Staat. Die berliner Gesellschaft für deutsche Literatur (1888–1938)*, Berlin / Boston.

[Nicolayson, J. (1836)] 'Jerusalem. Journal of the Rev. J. Nicolayson,' *JI* 2, 1–21.

— (1837) 'Jerusalem. Journal of the Rev. J. Nicolayson at Beyrout,' *JI* 3, 25–29.

— (1837a) 'Journey from Beyrout to Jerusalem,' *JI* 3, 29–35.
— (1838) 'Jerusalem. Letter from the Rev. J. Nicolayson,' *JI* 4, 191–92.
Niebuhr, B. G. (1836) *The Life of Carsten Niebuhr, the Oriental Traveller* [trsl. E. Robinson], Edinburgh [1st published *ABR* October 1832, 585–656].
Niebuhr, C. (1969) *Beschreibung von Arabien*, Graz [1st ed. Kopenhagen 1772].
Noth, M. (1962) *Die Welt des Alten Testaments. Einführung in die Grenzgebiete der alttestamentlichen Wissenschaft*, Freiburg/Br.
Olin, S. (1844) 'Note to the Article IX of November 131,' *NAR* 58/122, 253–56.
— (1845) 'For the Commercial Advertiser,' *The Commercial Advertiser* 47, 2–5.
— (1845a) 'The Whole Matter Settled,' *Christian Advocate and Journal*, June 11, 19, 44.
— (1860) *Travels in Egypt, Arabia Petræa, and the Holy Land*, New York [1st ed. 1843].
Oliver, A. (2014) *American Travelers on the Nile: Early U.S. Visitors to Egypt, 1774–1839*, Cairo / New York.
Olshausen, J. (1842) 'Palästina und die südlich angränzenden Länder. Tagebuch einer Reise im J. 1838 ... Von E. Robinson und E. Smith ... Erster Band; zweyter Band; Dritten Band erste Abteilung. Halle 1841,' *Jahrbücher der Literatur* 98, 126–59.
Onn, A. (2010) and S. Weksler-Bdolach, 'Wilson's Arch in Light of Recent and Past Studies,' in D. Amit, O. Peleg-Barkat and G. D. Stiebel (eds), *New Studies in the Archaeology of Jerusalem and its Region: Collected Papers*, IV, Jerusalem, 84–100 [Heb.].
Oren, M. B. (2007) *Power, Faith, and Fantasy: America in the Middle East, 1776 to the Present*, New York / London.
Päßler, U. (2010) (ed.) *Alexander von Humboldt / Carl Ritter Briewechsel*, Berlin.
Palmer, H. P. (1935) *Joseph Wolff, his romantic life and travels*, London.
Partsch, J. (1901) 'Heinrich Kiepert. Ein Bild seines Lebens und seiner Arbeit,' *Geographische Zeitschrift* 7, 1–21, 77–94.
Paxton, J. D. (1839) *Letters from Palestine Written During a Residence There in the Years 1836, 7, and 8*, London.
— (1870) *Memoir of J. D. Paxton, D.D., Late of Princeton, Indiana*, Phildelphia.
Perrier, F. (1841) *La Syrie sous le gouvernement de Méhémet-Ali jusqu'en 1840*, Paris.

Perry, Y. (2003) *British mission to the Jews in nineteenth-century Palestine*, London.

— (2007) and E. Lev *Modern Medicine in the Holy Land: Pioneering British Medical Services in Late Ottoman Palestine*, London.

Petermann, J. H. (1865) *Reisen im Orient* [2 vols], Leipzig.

Pfullmann, U. (2001) *Durch Wüste und Steppe: Entdeckerlexikon arabische Halbinsel*, Berlin.

Phillips, C. J. (1969) *Protestant America and the Pagan World: The First Half Century of the American Board of Commissioners for Foreign Missions, 1810–1860*, Cambridge, Mass.

Pickering, J. (1820) *An essay on a uniform orthography for the Indian languages of North America*, Cambridge, N.E.

Plewe, E. (1978) (ed.) *Die Carl Ritter Bibliothek* [Erdkunde Wissen, 50], Wiesbaden.

Pococke, R. (1745) *A Description of the East, and some other countries*, II/1: *Observations on Palestine or the Holy Land, Syria, Mesopotamia, Cyprus, and Candia*, London.

Prentiss, G. L. (1889) *The Union Theological Seminary in the City of New York: Historical and Biographical Sketches of its First Fifty Years*, New York.

Pribić, N. R. (1973) 'TALVJ – Pioneer in Comparative Studies of Folklore in America,' in V. Terras (ed.), *American Contribution to the Seventh International Congress of Slavists Warsaw, August 21–27, 1972*, II: *Literature and Folklore*, The Hague / Paris, 495–504.

— (1979) 'Sima Milutinović Sarajlija und Therese Albertine Luise von Jacob (Talvj),' *Südost-Forschungen* 38, 139–51.

— (1985) 'TALVJ as Interpreter of South Slavic Folklore in America,' in *Papers for the V. Congress of Southeast European Studies, Belgrade, September 1984*, Columbus, Ohio, 293–99.

— (1999) 'Zur Entstehung von Talvj's Volksliederinterpretation,' in G. Grimm and R. Pribić (eds), *Talvj, Tesla und die geistige Welt der Südslawen. Zum Gedächtnis von Nicola Pribić (1913–1992)* [Veröffentlichungen des Instituts für Geschichte Osteuropas und Südosteuropas der Universität München, 17], München, 7–11.

— (1999a) 'TALVJs Interesse für die russische Sprache und Literatur,' in G. Grimm and R. Pribić (eds), *Talvj, Tesla und die geistige Welt der Südslawen. Zum Gedächtnis von Nicola Pribić (1913–1992)* [Veröffentlichungen des Instituts für Geschichte Osteuropas und Südosteuropas der Universität München, 17], München, 12–19.

Pringle, D. (2009) "Aqaba Castle in the Ottoman Period, 1517–1917,' in A. C. S. Peacock (ed.), *The Frontiers of the Ottoman World* [Proceedings of the British Academy, 156], New York, 95–112.
Prokesch-Osten, A. Graf von (1877) *Mehmed-Ali, Vize-König von Aegypten: Aus meinem Tagebuche 1826–1841*, Wien.
Promitzer, C. (2011) 'Stimulating the Hidden Dispositions of South-Eastern Europe. The Plague in the Russo-Turkish War of 1828–29 and the Introduction of Quarantine on the Lower Danube,' in D. T. Sechel (ed.), *Medicine Within and Between the Habsburg and Ottoman Empires, 18th–19th Centuries* [The Eighteenth Century and the Habsburg Monarchy, International Series, 2/2], Bochum, 79–107.
Quaresmio, F. (1639) *Historica, theologica et moralis Terrae Sanctae elvcidatio.⸢Cum triplice indice, et Elencho Concionum. Cum triplice indiceet Elencho Concionum* [2 vols], Antwerpen.
Raumer, K. von (1835) *Palästina*, Leipzig.
— (1843) *Beiträge zur biblischen Geographie* [Beilage zu des Verfassers 'Palästina'], Leipzig.
Recent Intelligence (1838) 'Recent Intelligence,' *MH* 34/6, 237.
— (1838a) 'Recent Intelligence,' *MH* 34/11, 443.
Rée, P. (2001) 'James Silk Buckingham (1786–1855): An Anecdotal Traveller,' in P. and J. Starkey (eds), *Unfolding the Orient: Travellers in Egypt and the Near East*, Reading, 169–78.
Reich, R. (2015) 'The Construction and Destruction of Robinson's Arch at the Temple Mount in Jerusalem,' *Eretz Israel* 31, 398–407 [Heb.].
Relandi, H. (1714) *Palaestina ex Monumentis veteribus Illustrata*, I–II, Trajecti Batavorum.
Rennell, J. (1831) *A Treatise on the Comparative Geography of Western Asia*, London.
Richter, H. (1983) (ed.) *Carl Ritter. Werk und Wirkungen*, Gotha.
Ritter, C. (1818) *Die Erdkunde im Verhältniß zur Natur und zur Geschichte des Menschen, oder allgemeine vergleichende Geographie…*, 2nd part, 3rd book: *Westasien*, 3rd unit: *Das Plateau von Soristan, oder das syrische Gebirgsland*, Berlin.
— (1842) 'Neue Entdeckungen über die Jordan-Quelle von Major Robe, und über einzelne Punkte von Palästina durch Wolcott, nach Mittheilungen von Herrn Robinson,' *MVGEB* 5, 124–27, pl. I.
— (1844) 'Herrn Rittm. v. Wildenbruch's Reiserouten in Syrien (Beirut, 24. December 1843),' *MVGEB* N.F. 1, 229–36, pl. V, VI.

— (1848–55) *Vergleichende Erdkunde der Sinai-Halbinsel, von Palästina und Syrien,* I–IV [Die Erdkunde im Verhältniß zur Natur und zur Geschichte des Menschen, oder allgemeine vergleichende Geographie, 4 vols], Berlin.
— (1850) *Der Jordan und die Beschiffung des Todten Meeres,* Berlin.
— (1866) *The Comparative Geography of Palestine and the Sinaitic Peninsula* [trsl. W. L. Gage, 4 vols], New York [Rep. 1968].
Robey, M. K. (2009) 'Sacred Geographies: Religion, Race, and the Holy Land in U.S. Literature, 1819–1920' [unpublished PhD. Diss., Rice Uni., https://scholarship.rice.edu/handle/1911/61800].
Robinson, E. (1831) 'Theological Education in Germany,' *ABR* 1/IV, 1–51, 201–26, 409–52, 613–37.
— (1831a), 'Genuineness of Isaiah, Chapters XL–LXVI,' *ABR* I/IV, 700–33.
— (1832) 'Exodus of the Israelites out of Egypt and their Wanderings in the Desert,' *ABR* 2\8, 743–97.
— (1832a) 'Additions by the American Editor,' in C. Taylor (ed.), *Calmet's Dictionary of the Holy Bible* [Rev. Robinson, E.], Boston, 400–20.
— (1837) 'Art, IV. – 1. *Voyage de l'Arabie Pertrée*, par MM. Léon de Laborde *et* Linant … Paris. 1830–1834. 2. *Journey through Arabia Petræa, to Mount Sinai and the excavated City of Petra, the Edom of the Prophecies*. By M. Léon de Laborde. London; Murray. 1836,' *NAR* 44/95, 382–426.
— (1839) 'A Brief Report of Travels in Palestine and the Adjacent Regions in 1838; undertaken for the illustration of Biblical Geography, by the Rev. Prof. E. Robinson and Rev. E. Smith,' *ABR* 1/II, 400–30.
— (1839a) 'Extracts from a Journal of Travels in Palestine &c., in 1838; undertaken for the Illustration of Biblical Geography,' *JRGS* 9, 295–310.
— (1839b) 'Kurzer Bericht über eine Reise nach Palästina und der Umgegend im Jahr 1838, in Bezug auf die biblische Geographie unternommen von E. Robinson und E. Smith,' *ZKM* 2, 325–83.
— (1840) 'On the Dead Sea, and the Destruction of Sodom and Gomorrah,' ABR New Series III\5, 324–52.
— (1840a) and E. Smith 'Plan of Jerusalem: Description,' *New York Observer and Chronicle*, Dec. 12, 200.
— (1841) and E. Smith *Biblical Researches in Palestine, Mount Sinai and Arabia Petraea in 1838. A Journal of travels in the year 1838* [3 vols], Boston.

— (1841a) and E. Smith *Palästina und die südlich angrenzenden Länder: Tagebuch einer Reise im Jahre 1838 in Bezug auf die biblische Geographie unternommen* [3 vols], Halle.
— (1841b) and E. Smith *Biblical Researches in Palestine, Mount Sinai and Arabia Petraea in 1838* [3 vols], London.
— (1843) 'Researches in Palestine. Compiled by the Editor from various communications received at different times from the Rev. Eli Smith and the Rev. S. Wolcott. With a map of the country around the Sources of the Jordan,' *BS* 1, 9–88.
— (1843a) 'Notes on Biblical Geography,' *BS* 1, 563–66.
— (1844) 'Letter from Dr Robinson,' *NAR* 59/124, 253–58.
— (1844a) 'Remains of the Ancient Bridge Between the Jewish Temple and Mount Zion,' *BS* 1/4, 794–800.
— (1856) *Later Biblical Researches in Palestine and the adjacent Regions*, London.
— (1857) et al. *Neuere biblische Forschungen in Palästina und in den angrenzenden Ländern: Tagebuch einer Reise im Jahre 1852*, Berlin.
— (1859) 'Preface,' in Talvj, *Historical View of the Language and Literatures of the Slavic Nations*, New York, v–ix.
— (1863) 'Death of Dr Robinson,' *The Evangelist* 33/6 (February 5).
— (1865) *Physical Geography of the Holy Land* [ed. T. Robinson], Boston.
Robinson, G. (1837) *Three Years in the East: Being the Substance of a Journal Written during a Tour and Residence in Greece, Egypt, Palestine, Syria, and Turkey, in 1829–1830, 1831, and 1832* [2 vols], London / Paris.
— (1837a) *Travels in Palestine and Syria* [2 vols], London.
Robinson, M. A. (1862–64) 'Edward Robinson,' in *Memorial Biographies of the New England Historic Geneological Society*, V, 146–61 [copied as: *Memoir of the Rev. Edward Robinson, D.D., L.L.D.*, Cambridge 1895].
— (1874) 'Einleitung,' in *Gesammelte Novellen von Talvj. Nebst einer Auswahl bisher ungedruckter Gedichte und einer biographischen Einleitung*, Leipzig, I, VII–XXIX.
Rochfort Scott, C. (1837) *Rambles in Egypt and Candia, with Details of the Military Power and Resources of Those Countries, and Observations on the Government, Policy, and Commercial System of Mohammed Ali*, London.
Rödiger, E. (1839) 'Ueber die aramäische Vulgärsprache der heutigen syrischen Christen,' *ZKM* 2, 77–93.

— (1839a) 'Ueber die neusyrische Sprache,' *ZKM* 2, 314–16.
— (1840) 'Syrische Briefe,' *ZKM* 3, 218–25.
— (1840–2) and A. F. Pott, 'Kurdische Studien. I. Allgemeine Ansicht der kurdischen Sprache, Statistik und Litteratur. II. Lautlehre,' *ZKM*, 3 (1840), 1–63; 'Kurdische Studien. III. Naturgeschichtliches aus der Kurdischen und andern Sprachen Westasiens,' *ZKM* 4, 1–42, 259–80.
Röhricht, R. (1890) *Bibliotheca geographica Palaestina. Chronologisches Verzeichnis der von 333 bis 1878 verfassten Literatur über das Heilige Land. Mit dem Versuch einer Kartographie*, Berlin [rep. Jerusalem 1963].
Roper, G. (1998) 'The Beginning of Arabic Printing by the ABCFM, 1822–1841,' *Harvard Library Bulletin*, 9/1, 50–68.
— (1998a) 'Texts from Nineteenth-Century Egypt: The Role of E. W. Lane,' in P. and J. Starkey (eds), *Travellers in Egypt*, London / New York 1998, 244–54.
Rosenfeld, M. (1929), 'Lebrecht, Fürchtegott,' *Jüdisches Lexikon: Ein enzyklopädisches Handbuch der jüdischen Wissens*, III, Berlin, 1003.
Rosenmüller, E. F. K. (1841) *The biblical Geography of Asia Minor, Phoenicia, and Arabia ... containing an abstract of the more important geographical illustrations of Messrs. Robinson and Smith*, Edinburgh.
Rubinstein, S. (1997) *Coinage, Measures and Weights in Eretz-Israel from the Beginning of the 19th Century to the Period of transition from Ottoman to British Rule*, Jerusalem [Heb.].
Rüppell, E. (1838) *Reise in Abyssinien*, I, Frankfurt/Main.
Ruffle, John (1998) 'The Journeys of Lord Prudhoe and Major Orlando Felix in Egypt, Nubia and the Levant 1826–1829,' in P. and J. Starkey (eds), *Travellers in Egypt*, London / New York 1998, 75–84.
Russegger, J. (1841–3) *Reisen in Europa, Asien und Afrika*, 1: *Reise in Griechenland, Unteregypten, im nördlichen Syrien und südöstlichen Kleinasien* [2 vols], Stuttgart.
Sanudo, M. (1972) *Liber Secretorum fidelium Crucis super Terrae Sanctae recuperatione et conservatione...*, Jerusalem [printed in Bongars, Gesta dei, Hannover 1611].
Schäbler, B. (1995) 'Ulrich Jasper Seetzen (1767–1811), Jeveraner Patriot, aufgeklärter Kosmopolit und Orientreisender,' in *Ulrich Jasper Seetzen (1767–1811) Leben und Werk: Die arabischen Länder und die Nahostforschung im napoleonischen Zeitalter* [Vorträge des Kolloquiums vom 23. und 24. September 1994, Veröffentlichungen der Forschungs- und Landesbibliothek Gotha, 33], Gotha, 113–34.

Schaff, P. (1891) 'Robinson, Edward,' in J. McClintock and J. Strong (eds), *Cyclopedia of Biblical, Theological and Ecclesiastical Literature*, 9, New York, 50–53.

Schelhaas, B. (2017) 'August Petermann's Palestine Maps,' in H. Goren, J. Faehndrich and B. Schelhaas (with P. Weigel), *Mapping the Holy Land: The Foundation of a Scientific Cartography of Palestine* [Tauris Historical Geography Series], London / New York, 107–39.

Schienerl, J. (2000) *Der Weg in den Orient: Der Forscher Ulrich Jasper Seetzen: Von Jever in den Jemen (1802–1811)*, Oldenburg.

Schlatter, W. (1916) *Geschichte der Basler Mission 1815–1915. Mit besonderer Berücksichtigung der ungedruckten Quellen* [3 vols], Basel.

Scholz, J. M. A. (1823) *Biblisch-kritische Reise in Frankreich, der Schweiz, Italien, Palästine und im Archipel in den Jahren 1818, 1819, 1820, 1821: Nebst einer Geschichte des Textes des N[euen] T[estaments]*, Leipzig / Sorau.

Schubert, G. H. von (1839) *Reise in das Morgenland in den Jahren 1836 und 1837* [3 vols], Erlangen.

— (1841) 'Palästina und die südlich angränzenden Länder. Tagebuch einer Reise im Jahre 1838 von E. Robinson und E. Smith,' *GABA* 13, 681–86.

Schubert, G. (2001) and F. Krause (eds), *Talvj Therese Albertine Luise von Jakob-Robinson (1797–1870). Aus Liebe zu Goethe: Mittlerin der Balkanslawen*, Weimar.

— (2001a) 'Einleitung,' in G. Schubert and F. Krause (eds), *Talvj Therese Albertine Luise von Jakob-Robinson (1797–1870). Aus Liebe zu Goethe: Mittlerin der Balkanslawen*, Weimar, 13–18.

Schütz, C. (1988) *Preussen in Jerusalem (1800–1861): Karl Friedrich Schinkels Entwurf der Grabeskirche und die Jerusalempläne Friedrich Wilhelms IV* [Die Bauwerke und Kunstdenkmäler von Berlin], Berlin.

Schwake, N. (1983) *Die Entwicklung des Krankenhauswesens der Stadt Jerusalem vom beginn des 19. Jahrhunderts bis zum beginn des 20. Jahrhunderts* [2 vols], Herzogenrath.

Schwarz, H. (1924) *Karl Christoph Traugott Tauchnitz: Buchhändler, Buchdrucker und Schriftgiesser zu Leipzig. Geboren 29. October 1761 Gestorben 14. Januar 1836: Ein Lebensbild*, Leipzig.

Scientific Proceedings (1847) 'Scientific Proceedings: Second Meeting of the Ethnological Society,' *The Literary World* 2, 42 (Nov 20), 382.

Searight, S. (1979) *The British in the Middle East*, London / The Hague.

— (2003) 'The charting of the Red Sea,' *History Today* 53/3, 40–47.

Sechel, D. T. (2011) 'Contagion Theories in the Habsburg Monarchy (1770–1830),' in idem (ed.), *Medicine Within and Between the Habsburg and Ottoman Empires, 18th–19th Centuries* [The Eighteenth Century and The Habsburg Monarchy, International Series, 2/2], Bochum, 55–77.

Seetzen, U. J. (1805–6) 'Fortgesetzte Reise-Nachrichten,' *MCEHK* 12, 57–73; 13, 502–13; 14, 159–69.

— (1810) Charte von Palästina reducirt aus den von dem Herrn D. Seetzen an Ort und Stelle entworfenen Handzeichnungen, Gotha.

— (1810a) *Verzeichniss der für die orientalische Sammlung in Gotha zu Damask, Jerusalem u. s. w. angekauften orientalischen Manuscripte und gedruckten Werke, Kunst- und Naturprodukte u. s. w.* [ed. E. F. K. Rosenmüller], Leipzig.

Shamir, S. (1984) 'Egyptian Rule (1832–1840) and the beginning of the modern period in the history of Palestine,' in A. Cohen and G. Baer (eds), *Egypt and Palestine: A Millennium of Association (1868–1948)*, Jerusalem, 214–31.

Shepherd, N. (1987) *The Zealous Intruders. The Western rediscovery of Palestine*, London.

Sieber, F. W. (1818) Karte von Jerusalem, und seiner naechsten Umgebungen, geometrisch aufgenomen, Prague.

— (1823) *Reise von Cairo nach Jerusalem und wieder Zurück, nebst Beleuchtung einiger heiligen Orte*, Prague.

Silberman, N. A. (1982) *Digging for God and country. Exploration, archeology, and the secret struggle for the Holy Land, 1799–1917*, New York.

Sim, K. (1969) *Desert Traveller: The life of Jean Louis Burckhardt*, London.

Simon, A. (1998) *Sigmund Feyerabend's Das Reyssbuch Dess Heyligen Lands: A Study in Printing and Literary History* [Wissensliteratur im Mittelalter, 32], Wiesbaden.

Simpson, W. (1879) 'The Visit of Bonomi, Catherwood, and Arundale, in the Haram Es Sheriff at Jerusalem in 1833,' *Palestine Exploration Fund, Quarterly Statement* 11, 51–53.

[Smith, C. F.] (1841), 'Fall of St. Jean D'Acre', *The Nautical Magazine and Naval Chronicle* 10, 3–4.

Smith, E. (1833) and H. G. O. Dwight, *Researches of the Rev. E. Smith and Rev. H. G. O. Dwight in Armenia: including a journey through*

Asia Minor, and into Georgia and Persia, with a visit to the Nestorian and Chaldean Christians of Oormiah and Salmas, Boston.
— (1839) 'Notices of the Bedaween Tribes of Arabs,' *MH* 35/3, 81–91.
— (1840) *Address on the Missionary Character*, Boston.
— (1841) 'Essay on the Pronounciation of the Arabic,' in E. Robinson and E. Smith, *Biblical Researches in Palestine, Mount Sinai and Arabia Petraea in 1838. A Journal of travels in the year 1838*, III, 2nd Appendix, Boston, 89–111 [also ESP, ABC 60 (119)].
— (1841a) 'Lists of Arabic Names of Places in Palestine and the Adjacent Regions,' in E. Robinson and E. Smith, *Biblical Researches in Palestine, Mount Sinai and Arabia Petraea in 1838. A Journal of travels in the year 1838*, III, 2nd Appendix, Boston, 112–96 [also ESP, ABC 60 (119)].
— (1842) *Verzeichnisse arabischer Namen von Ortschaften in Palästina und den angrenzenden Gegenden*, [Halle].
— (1843) 'Ancient Temple on Mount Lebanon,' *BS* 1, 557–63.
— (1845) 'Concluding Remarks,' in E. W. Hooker, *Memoirs of Mrs Sarah L. Huntington Smith, Late of the American Mission in Syria*, New York [3rd ed.], 360–93.
— (1846) *Toleration of the Turkish Empire*, Boston ['Turkish Toleration,' *BSTR* 3/X, 390–97].
— (1846a) 'The Wines of Mount Lebanon,' *BSTR* 3/X, 385–89.
— (1900) and C. V. A. Van Dyck, *Brief Documentary History of the Translation of the Scriptures into the Arabic Language*, Beirut.
Smith, G. A. (1974) *The Historical Geography of the Holy Land*, London [1st ed. 1894].
Smith, H. B. (1863) and R. D. Hitchcock, *The life, writings and character of Edward Robinson*, New York [Reprint New York 1977].
Smith, J. R. (1997) *Introduction to Geodesy: The History and Concepts of Modern Geodesy*, New York, etc.
Smith, J. (1993) 'The Schmalcalders of London and the Priddis Dial,' *The Journal of the Royal Astronomical Society of Canada* 87/1, 4–13.
Sobel, D. (1995) *Longitudes: the true story of a lone genius who solved the greatest scientific problem of his time*, New York.
Soler, R. (2014) *Edward Robinson (1794–1863) et l'émergence de l'archéologie biblique*, Paris.
Stanley, A. P. (1879) 'An American Scholar. An Address to the Students of the Union Theological Seminary of the Presbyterian Church of America, at the Seminary in New York, Oct. 29, 1878,' in idem, *Ad-*

dresses and Sermons, delivered during a visit to the United States and Canada in 1878, New York, 23–33.

Starkey, P. (2019) 'Rabbi Benjamin of Tudela: a 12th-Century Traveller to the Middle East,' in N. Cooke (ed.), *Journeys Erased by Time: The rediscovered footprints of travelers in Egypt and the Near East*, Oxford 2019, 1–15.

Steig, R. (1891) 'Breiefwechsel zwischen Goethe und Therese von Jacob,' *Goether Jahrbücher* F/M 12, 33–77.

— (1894) 'Breiefwechsel zwischen Jacob Grimm und Therese von Jacob,' *Preußische Jahrbüche* 76, 345–66.

Stern, S. (1988) 'Titus Tobler, 19th Century Researcher of Eretz Israel,' *Cathedra* 48, 30–45 [Heb.].

— (1988a) and H. Goren, 'A Biliography,' *Cathedra* 48, 46–48.

Stevens, C. J. (1956) 'John Pickering's "Uniform Orthography",' *Quarterly Journal of Speech* 42/2, 139–43.

Stiehler, J. (2013) 'Naturrecht und liberale Staatswissenschaft. Ludwig Heinrich von Jakob – Philosoph, Ökonom und Staatswissenschaftler in Halle und Russland,' in D. Recknagel and S. Wöller (eds), *'Vernunft, du weißt allein, was meine Pflichten sind': Naturrechtslehre in Halle*, Halle (Saale), 76–91.

Stinespring, W. F. (1939) 'The critical faculty of Edward Robinson,' *JBL* 58/4, 379–87.

Stock, E. (1899) *A History of the Church Missionary Society, its environs, its men and its work* [3 vols], London.

Stoddard, R. (2009) 'The Rev. Eli Smith, 1801–1857: Evangelical Orientalist in the Levant,' *Theological Review* 30/2, 202–22.

Stoneman, R. (1987) *Land of lost gods: the search for classical Greece*, Norman, OK.

Strathern, P. (2009) *Napoleon in Egypt*, New York.

Struthers, G. (1834) *Memoirs of American Missionaries*, Glasgow / Edinburgh / London.

Stuart M. (1823) *Hebrew Grammar with a Praxis on select portions of Genesis and the Psalms*, Andover.

Stuecher, D. D. (1990) *Twice Removed: The Experience of German-American Woman Writers in the 19th Century* [New German-American Studies, I], New York, etc.

Sykes, W. H. (1838) 'On the Use of Common Thermometers to determine heights,' *JRGS* 8, 435–42.

Talvj (1825–26) *Volkslieder der Sereben, Metrisch übersetzt und historisch eingeleitet* [2 vols], Leipzig.

— (1834) 'Historical View of the Slavic Languages in its Various Dialects, with Special Reference to Theological Literature,' *ABR* 4, 328–412, 417–532.

— (1834a) *Ueber die Indianischen Sprachen Amerikas. Aus dem Englischen des Amerikaners John Pickering übersetzt und mit Anmerkungen begleitet von Talvj*, Leipzig.

— (1836) 'Popular Poetry of the Teutonic Nations,' *NAR* 42/91, 265–339.

— (1836a) 'Slavic Popular Poetry,' *NAR* 43/92, 85–120.

— (1840) *Die Unächtheit der Lieder Ossian's und des Macpherson'schen Ossian's inbesondere*, Leipzig.

— (1840a) *Versuch einer geschichtlichen Charakteristik der Volkslieder germanischer Nationen mit einer Uebersicht der Lieder aussereuröpaischer Völkerschaften*, Leipzig.

— (1850) *Historical View of the Languages and Literature of the Slavic Nations with a Sketch of the Popular Poetry* [2 vols], New York / London.

Taylor, A. R. (1957) *The American Protestant mission and the awakening of modern Syria, 1820–1870*, Washington, DC.

Taylor, C. (1832) (ed.) *Calmet's Dictionary of the Holy Bible* [Rev. E. Robinson], Boston.

Taymanova, M. (1998) 'Alexandre Dumas in Egypt: Mystification or Truth?,' in P. and J. Starkey (eds), *Travellers in Egypt*, London / New York, 181–88.

Tejirian, E. H. (2012) and R. S. Simon *Conflict, Conquest and Conversion: Two thousand years of Christian missions in the Middle East*, New York.

Temple, D. H. (1855) *Life and Letters of Rev. Daniel Temple, for twenty-three years a missionary of the A.B.C.F.M. in Western Asia*, Boston.

Theological Faculty (1834) Leipzig University, 'Outlines of a Course of Theological Study for the use of Students' [trsl. E. Robinson], *ABR* 4, 127–37.

Thompson, J. (1996) 'Edward William Lane's "Description of Egypt",' *International Journal of Middle East Studies* 28/4, 565–83.

— (1998) and A. T. Thompson, 'Between Two Lost Worlds: Frederick Catherwood,' in P. and J. Starkey (eds), *Travellers in Egypt*, London / New York, 131–39.

— (2010) *Edward William Lane, 1801–1876. The Life of the Pioneering Egyptologist and Orientalist*, Cairo / New York.

Thomson, W. McC. (1880) *The Land and the Book; or, Biblical illustrations drawn from the manners and customs, the scenes and scenery, of the Holy Land*, New York [1st ed. 1859].

Thornton, T. P. (2016) *Nathaniel Bowditch and the Power of Numbers: How a Nineteenth-Century Man of Business, Science and the Sea Changed American Life*, Chapel Hill, NC.

Tibawi, A. L. (1966) *American interests In Syria, 1800–1901: a study of educational, literary and religious work*, Oxford.

Tillett, S. (1984) *Egypt Itself: The Career of Robert Hay, Esquire of Linplum and Nunraw, 1799–1863*, London.

Tobler, T. (1839) *Lustreise ins Morgenland* [2 vols], Zürich.

— (1847) 'Ausflug von Bethlehem nach den Labyrinth in Chareitun,' *Das Ausland* 20, Nr. 57–63.

— (1849) *Bethlehem in Palästina: Topographisch und historisch nach Anschau und Quellen geschildert*, St Gallen / Bern.

— (1851) *Golgatha: seine Kirchen und Klöster*, St Gallen / Bern.

— (1852) *Die Siloahquelle und der Oelberg*, St Gallen.

— (1853–4) *Zwei Bücher Topographie von Jerusalem und seinen Umgebungen* [2 vols], Berlin.

— (1867) *Bibliographia geographica Palaestina. Zunächst kritische Übersicht gedruckter und ungedruckter Beschreibungen der Reisen ins Heilige Land*, Leipzig.

Travis, A. S. (2012) *Dreamland: American Travelers to the Holy Land in the 19th Century* (Exhibition catalog), Jerusalem.

Tresse, R. (1936) 'L'installation du premier consul d'Angleterre a Damas (1830–1834),' *Revue d'histoire des colonies* 24, 359–80.

Tucher, H. (1561) *Reyssbeschreibung*, Frankfurt 1561 [1st ed. Augsburg 1482].

Turner, G. L'E. (1983) *Nineteenth-Century Scientific Instruments*, London.

— (1998) *Scientific Instruments, 1500–1900: an introduction* [ed. J. Gilbert and T. Ayers], Berkeley, CA / London.

Union Theological Seminary (1850), *Triennial Catalogue of the Officers and Alumni of the Union Theological Seminary in the City of New York. June, 1850*, New York.

van Adrichom, C. (1590) *Theatrum Terrae Sanctae et biblicarum historiarum:ſ cum tabulis geographicis aere expressis*, Cologne.

van Cootwijk, J. (1619) *Itinerarium Hierosolymitanum et Syriacum:ſ in quo variarum gentium mores et instituta: insularum, regionum, vrbium situs ... vnà cum euentis, quæ auctori ... acciderunt, dilucidè recensentur...*, Antwerpen.

VanDeMark, B. (2012) *American Sheikhs: Two Families, Four Generations, and the Story of America's Influence in the Middle East*, Amherst NY.

Varnhagen von Ense, K. A. (1840) 'Versuch einer geschichtlichen Characteristic (...), von Talvj. Leipzig 1840. Bei F. A. Brockhaus,' *Jahrbücher für wissenschaftliche Kritik*, 683–92.

Vasmer, M. (1939) *Bausteine zur Geschichte der deutsch-slavischen geistigen Beziehungen*, I, Berlin.

Vivian, C. (2012) *Americans in Egypt, 1770–1915: Explorers, Consuls, Travellers, Soldiers, Missionaries, Writers and Scientists*, Jefferson, NC / London.

Vogel, L. I. (1993) *To see a promised land. Americans and the Holy Land in the nineteenth century*, University Park, PA.

Voigt, I. E. (1914) *The Life and Works of Mrs Therese Robinson*, Chicago.

von Hagen, V. W. (1948) *Maya Explorer: John Lloyd Stephens and the Lost Cities of Central America and Yucatan*, Norman, OK.

— (1950) *Frederick Catherwood Archt.*, New York.

— (1968) *F. Catherwood Architect-Explorer of Two Worlds*, Barre, MS.

von Lindenau, B. (1810) 'Über bey diesem Heft befindliche Karte von Palästina,' *MCEHK* 12, 542–51.

von Löher, F. (1886) 'Talvj, ein Deutsches Frauen-Leben,' in idem, *Beiträge zur Geschichte und Völkerkunde*, 2, Frankfurt, 451–66.

von Suchem, L. (1477) *Von dem gelobten Land und Weg gegen Jherusalem*, [Augsburg].

Wagner, L. (1897) *TALVJ 1797–1870. Biographische Skizze zur Erinnerung an ihren hundertsten Geburgstag*, Oressburg.

Wahl, A. (1825) *Greek and English Lexicon of the New Testament, from 'Clavis Philologia' of Christ* (trsl. E. Robinson), Andover.

Watson, C. M. (1915) *Fifty years' work in the Holy Land. A record and a summary, 1865–1915*, London.

Weigel, P. (2014) H. Goren, B. Schelhaas, and J. Faehndrich (eds) *Das Heilige Land in Gotha: Der verlag Justus Perthes und die Palästinakartographie im 19. Jahrhundert* [Veröffentlichungen der Forschungsbibliothek Gotha, 50], Gotha.

Weigel Williams, H. (1994?) *The Edward Robinson Collection, Catalogue* [2 vols], Hamilton College.

— (1997) *TALVJ (Therese Albertine Luise von Jacob Robinson). An Extraordinary Nineteenth-Century Faculty Wife* [UTS, The Burke Library Occasional Publication, 5], New York.

— (2003) (ed.) *Letters from Edward Robinson to Moses Stuart 1826–1830*, Clinton, NY.

— (2007) *Therese Von Jakob: A Biographical Portrait*, New York / London / Shanghai.

— (2009) (ed.) *Robinson's Letter – Journal (1826–1829). Written from Europe by Edward Robinson to His Sister Elisabeth*, New York / Bloomington.

von Wildenbruch, L. (1846) 'Profilzeichnungen nach barometrischen Nivellements in Syrien,' *MVGEB* N.F. 3, 270–72, pl. III.

Wilken, F. (1807–32) *Geschichte der Kreuzzüge nach morgenländischen und abendländischen Berichten* [7 vols], Leipzig.

Willermus Tyrensis (1560) *Historia Rerum in partibus etc. seu Historia Belli Sacri*, Basel.

Williams, G. (1849) *The Holy City: Historical, Topographical, and Antiquarian Notices of Jerusalem* [2 vols], London.

Williams, J. B. (1972) *British Commercial Policy and Trade Expansion 1750–1850*, Oxford.

Williams, J. G. (1996) 'Edward Robinson (1794–1863): Pioneer of Biblical Archaeology,' *Hamilton Alumni Review* LXI\I, 19–21.

— (1997) *The Education of Edward Robinson* [Burke Library Occasional Publication, 4], New York.

— (1999) *The Times and Life of Edward Robinson: Connecticut Yankee in King Solomon's Court* [Biblical Scholarship in North America, 19], Atlanta.

Willis, R. (1849) *The Architectural History of the Church of the Holy Sepulchre*, London.

Winter, E. (1962) 'Eine Propagandistin der vergleichenden slawischen Literatur: Therese Jacob (Talvj),' in L. Bóka (ed.), *Acta Litteraria. Academiae scientiarum Hungaricae*, V: *Conférence de Littérature Comparée, Budapest, 26–29 octobre 1962*, Budapest, 313–19.

Withers, C. W. J. (2013) 'Science, scientific instruments and questions of method in nineteenth-century British geography,' *Transactions of the Institute of British Geographers* 38/1, 167–79.

Wokoeck, U. (2009) *German Orientalism: The Study of the Middle East and Islam from 1800 to 1945*, London / New York.

Wolcott, S. (1846) 'Geographical Notes on Palestine,' *BSTR* 3, 398–403.

Wolff, J. (1835) *Researches and Missionary Labours Among the Jews, Mohammedans, and other sects, during his travels between the years 1831 and 1834 from Malta to Egypt...*, London.

Zachariae (von Lingenthal), C. E. (1840) *Dr E. Zachariä's Reise in den Orient in den Jahren 1837 und 1838, ueber Wien, Venedig, Florenz, Rom, Neapel, Malta, Sicilien und Griechenland nach Saloniki, dem Berg Athos, Konstantinopel und Trapezunt*, Heidelberg.

Zaun, J. (2000) 'Pistor & Martins, the builders of the Berlin meridian circles,' in W. R. Dick and C. Fritze (eds), *300 Year of Astronomy in Berlin and Potsdam. A Collection of Essays on the Occasion of the Founding Anniversary of the Berlin Observatory* [Acta Historica Astronomiae, 8], Thun / Frankfurt a. M., 91–106.
Zögner, L. (1979) (ed.) *Carl Ritter in Seiner Zeit 1779–1859* [Ausstellung der Staatsbibliothek Preußischer Kulturbesitz, Berlin], Berlin.
— (1999) (ed.) *Antike Welten Neue Regionen. Heinrich Kiepert 1818–1899*, Berlin.

D. Biographies in bibliographical dictionaries

ADB Historische Kommission bei der Königlichen Akademie der Wissenschaften (ed.) (1875–1912) *Allgemeine Deutsche Biographie* [56 vols], Leipzig.
DAB Johnson, A. (ed.) (1957), *Dictionary of American Biography* [10 vols], New York [1st ed. 1927].
BLKO Wurzbach, C. von (1856–91), *Biographisches Lexicon des Kaiserthums Oesterreich* [40 vols], Vienna.
NDB Historischen Kommission bei der Bayerischen Akademie der Wissenschaften (ed.) (1953), *Neue Deutsche Biographie* [26 vols], Berlin.
ODNB H. G. C. Matthew et al. (eds) (2004), *The Oxford Dictionary of National Biography* [60 vols], Oxford [also digital edition].

ADB
ADB Asher (1875) Mühlbrecht, O. 'Asher, Adolf,' ADB 1, 619 https://www.deutsche-biographie.de/pnd124478271.html#adbcontent.
ADB Barth Mühlbrecht, O. (1875) 'Barth, Johann Ambrosius,' ADB 2, 99–100 https://www.deutsche-biographie.de/pnd132235080.html#adbcontent.
ADB Beer Siegfried, C. (1875) 'Beer, Eduard Friedrich Ferdinand,' ADB 2, 247–48 https://www.deutsche-biographie.de/pnd100030750.html#adbcontent.
ADB Blumhardt Hartmann, J. (1875) 'Blumhardt, Christian Gottlieb,' ADB 2, 755 https://www.deutsche-biographie.de/pnd12241697X.html#adbcontent.

ADB Bohlen Leskien, A. (1876) 'Bohlen, Peter von,' ADB 3, 61 https://www.deutsche-biographie.de/pnd116231742.html#adbcontent.
ADB Breitkopf und Härtel Hase, O. (1876) 'Breitkopf und Härtel,' ADB 3, 296–303 https://www.deutsche-biographie.de/sfz98538.html#adbcontent.
ADB Buch Gümbel, W. von (1876) 'Buch, Leopold von,' ADB 3, 464–75 https://www.deutsche-biographie.de/pnd116817569.html#adbcontent.
ADB De Wette Holtzmann, H. (1877) 'De Wette, Martin Leberecht,' ADB 4, 101–05 https://www.deutsche-biographie.de/pnd118631977.html#adbcontent.
ADB Delitzsch Dalman, G. (1903) 'Delitzsch, Franz,' ADB 47, 651–52 https://www.deutsche-biographie.de/pnd118524550.html#adbcontent.
ADB Eß Reusch, H. (1877) 'Eß, Leander van,' ADB 6, 378–79 https://www.deutsche-biographie.de/pnd11943881X.html#adbcontent.
ADB Falkenstein Schnorr von Carolsfeld, F. (1877) 'Falkenstein, Constantin Karl,' ADB 6, 556 https://www.deutsche-biographie.de/pnd100346340.html#adbcontent.
ADB Fleischer Goldziher, I. (1904) 'Fleischer, Heinrich Leberecht,' ADB 48, 584–94 https://www.deutsche-biographie.de/pnd11660042X.html#adbcontent.
ADB Francke Kramer (1878) 'Francke, August Hermann,' ADB 7, 219–31 https://www.deutsche-biographie.de/pnd118534688.html#adbcontent.
ADB Freitag Wolff, P. (1878) 'Freytag, Georg Friedrich Wilhelm,' ADB 7, 373–74 https://www.deutsche-biographie.de/pnd100126707.html#adbcontent.
ADB Hammer-Purgstall Schlechta von Wssehrd, O. Freiherr von (1897) 'Hammer-Purgstall, Joseph Freiherr von,' ADB 10, 482–87 https://www.deutsche-biographie.de/pnd118545426.html#adbcontent.
ADB Haupt Scherer, W. (1880) 'Haupt, Moriz,' ADB 11, 72–80 https://www.deutsche-biographie.de/pnd119034743.html#adbcontent.
ADB Hengstenberg Ranke, O. von (1880) 'Hengstenberg, Wilhelm,' ADB 11, 737–47 https://www.deutsche-biographie.de/pnd118773917.html#adbcontent.
ADB Jakob Prantl, C. von (1881) 'Jakob, Ludwig Heinrich von,' ADB 13, 689–90 https://www.deutsche-biographie.de/pnd11871158X.html#adbcontent.
ADB Lebrecht Ritter, I. (1883) 'Lebrecht, Fürchtegott,' ADB 18, 97–98 https://www.deutsche-biographie.de/pnd116850159.html#adbcontent.

ADB Perthes Pallmann, Heinrich (1887) 'Perthes, Wilhelm,' ADB 25, 401–02 https://www.deutsche-biographie.de/pnd132553155.html#adbcontent.

ADB Robinson, T. Beneke, O. (1889) 'Robinson, Therese,' ADB 28, 724–25 https://www.deutsche-biographie.de/pnd118620614.html#adbconten.

ADB Rödiger Siegfried, C. (1889) 'Roediger, Emil,' ADB 29, 26–30 https://www.deutsche-biographie.de/pnd116581050.html#adbcontent.

ADB Rosenmüller Siegfried, C. (1889) 'Rosenmüller, Ernst Friedrich Karl,' ADB 215–17 https://www.deutsche-biographie.de/pnd116624981.html#adbcontent.

ADB Rüppell Stricker, W. (1889) 'Rüppell, Eduard,' ADB 29, 707–14 https://www.deutsche-biographie.de/pnd11669811X.html#adbcontent.

ADB Russegger Gümbel, W. von (1890) 'Russeger, Joseph Ritter von,' ADB 30, 14–15 https://www.deutsche-biographie.de/pnd116710713.html#adbcontent.

ADB Tholuck Frank, G. (1894) 'Tholuck, August,' ADB 38, 55–59 https://www.deutsche-biographie.de/pnd118757199.html#adbcontent.

ADB Tittmann Tschackert, P. (1894) 'Tittmann, Johann August Heinrich,' ADB 38, 384–85 https://www.deutsche-biographie.de/pnd1174015,01.html#adbcontent.

ADB Wahl Siegfried, C. (1896) 'Wahl, Christian Abraham,' ADB 40, 591–92 https://www.deutsche-biographie.de/pnd117114073.html#adbcontent.

ADB Wilken Stoll, A. (1898) 'Wilcken, Friedrich,' ADB 43, 236–41 https://www.deutsche-biographie.de/pnd117384259.html#adbcontent [both Wilken and Wilcken used].

ADB Zachariae Fischer, W. (1898) 'Zachariae, Karl Eduard Z. von Lingenthal,' ADB 44, 653–57 https://www.deutsche-biographie.de/pnd118772120.html#adbcontent.

BLKO*:

*No authors mentioned

BLKO Hammer-Purgstall (1861) 'Hammer-Purgstall, Joseph Freiherr von,' BLKO 7, 267–89.

BLKO Kopitar (1869) 'Kopitar, Bartholomäus,' BLKO 12, 437–42.

BLKO Russegger (1874) 'Russegger, Joseph Ritter von,' BLKO 27, 292–97.

DAB:
DAB Appleton, D. Peterson, A. E. 'Appleton, Daniel,' DAB I, 326–27.
DAB Appleton, W.H. Peterson, A. E. 'Appleton, William Henry,' DAB I, 334.
DAB Bliss Archer, J. C. 'Bliss, Daniel,' DAB I, 369–70.
DAB Cheever Persons, F. T. 'Cheever, George Barrell,' DAB II/2, 48–49.
DAB Dwight Archer, J. C. 'Dwight, Harrison Gray Otis,' DAB III/1, 565–66.
DAB Everett Pearson, H. G. 'Everett, Edwaed,' DAB III/2, 223–26.
DAB Hamlin Wright, W. L. 'Hamlin, Cyrus,' DAB IV/2, 195–96.
DAB Hitchcock William, A. B. 'Hitchcock, Roswell Dwight,' DAB V/1, 79–780.
DAB Hodge Nichols, R. H., 'Hodge, Charles,' DAB V/1, 98–99.
DAB Homes Peterson, A. E. 'Homes, Henry Augustus,' DAB V/1, 191–92.
DAB Howe Gregorie, A. K. 'Howe, George,' DAB V/1, 286–87.
DAB Jessup Archer, J. C. 'Jessup, Henry Harris,' DAB V/2, 61.
DAB Kirkland, J. T. Morison, S. E. 'Kirkland, John Thornton,' DAB V/2, 431.
DAB Kirkland, S. Ibbotson, J. D. 'Kirkland, Samuel,' DAB V/2, 432–34.
DAB Muhlenberg Genzmer, G. H. 'Muhlenber, Henry Augustus,' DAB VII/1, 309–10.
DAB Nordheimer Pool, D. de S. 'Nordheimer, Isaac,' DAB VII/1, 547–48.
DAB Olin Garber, P. N., 'Olin, Stephen,' DAB VII/2, 13–14.
DAB Patton Violette, E. M. 'Patton, William,' DAB VII/2, 317–18.
DAB Peters Persons, F. T. 'Peters, Absalom,' DAB VII/2, 502.
DAB Pickering Genzmer, P. N. 'Pickering, John,' DAB VII/2, 564–65.
DAB Prentiss Rockwell, W. W. 'Prentiss, George Lewis,' DAB VIII/1, 189–90.
DAB Robinson Albright, W. F. 'Robinson, Edward,' DAB VIII/2, 40–41.
DAB Robinson, T. Kubler, E. A. 'Robinson, Therese Albertine Louise von Jakob,' DAB VIII/2, 55–56.
DAB Silliman Warren, C. H. 'Silliman, Benjamin,' DAB IX/1, 160–63.
DAB Skinner Faris, P. P. 'Skinner, Thomas Harvey,' DAB IX/1, 201–02.
DAB Smith Albright, W. F., 'Smith, Eli,' DAB IX/1, 257–58.
DAB Stuart Albright, W. F. 'Stuart, Moses,' DAB IX/2, 174–75.

DAB Thomson Nichols, R. H. 'Thomson, William McClure,' DAB IX/2, 490.

DAB Wheaton Hicks, F. C., 'Wheaton, Henry,' DAB X/2, 39–42.

NDB:

NDB Barth Meiner, A. (1953) 'Barth, Wilhelm Ambrosius,' NDB 1, 600–01 https://www.deutsche-biographie.de/pnd116119578.html#ndbcontent.

NDB Brockhaus, F. A. Meiner, A. (1955) 'Brockhaus, Friedrich Arnold,' NDB 2, 623–24 https://www.deutsche-biographie.de/pnd119501201.html#ndbcontent.

NDB Brockhaus, H. Meiner, A. (1955) 'Brockhaus, Heinrich,' NDB 2, 624–25 https://www.deutsche-biographie.de/pnd116552433.html#ndbcontent.

NDB Gesenius Zobel, H.-J. (1964) 'Gesenius, Wilhelm,' NDB 6, 340–41 https://www.deutsche-biographie.de/pnd119096552.html#ndbcontent.

NDB Kopitar Vintr, J. (1979), 'Kopitar, Bartholomäus,' NDB 12, 566 https://www.deutsche-biographie.de/pnd118565 29X.html#ndbcontent.

NDB Neander Selge, K.-V. (1999), 'Neander, August,' NDB 19, 10–11 https://www.deutsche-biographie.de/pnd118738240.html#ndbcontent.

ODNB:

ODNB Carne Boase, G. C. rev. E. Baignet, 'Carne, John,' ODNB 4714.

ODNB Hamilton Anderson, R. E., 'Hamilton, William Richard,' ODNB 12147.

ODNB Herschel Crowe, M. J., 'Herschel, Sir John Frederick William,' ODNB 13101.

ODNB Irby Laughton, J. K. rev. A. Lambert, 'Irby, Charles Leonard,' ODNB 14443.

ODNB Lindsay Brigstocke, H., 'Lindsay, Alexander William Crawford,' ODNB 16686.

ODNB Mill Bendall, C. rev. B. Lolo, 'Mill, William Hodge,' ODNB 18712.

ODNB Williams Courtney, P. rev. T. Adams, 'Williams, George,' ODNB 29505.

ODNB Washington Laughton, J. K. rev. R. O. Morris, 'Washington, John,' ODNB 28807.

INDEX

Index of Names

Index of Places, Organizations, and Events*

* Not included are obvious countries (UK, Britain, US, USA); continents; Palestine – Holy Land – Eretz Israel; East – Near East – Middle East; *Biblical Researches* (book); Bible, Biblical; Scriptures; Mission, missionaries.